Physical Oceanography

A Mathematical Introduction with MATLAB®

Physical Oceanography

A Mathematical Introduction with MATLAB®

Reza Malek-Madani

CRC Press is an imprint of the
Taylor & Francis Group, an informa business
A CHAPMAN & HALL BOOK

CRC Press
Taylor & Francis Group
6000 Broken Sound Parkway NW, Suite 300
Boca Raton, FL 33487-2742

Printed in the United States of America on acid-free paper
Version Date: 20120229

International Standard Book Number: 978-1-58488-830-7 (Hardback)

Visit the Taylor & Francis Web site at
http://www.taylorandfrancis.com

and the CRC Press Web site at
http://www.crcpress.com

to Jo, Behzad, and Darob

and

to all my students

Contents

Preface

This book is about the interplay between applied mathematics and the field of geophysical fluid dynamics. Its primary goals are to demonstrate how one uses the basic tenets of multivariate calculus to derive the governing equations of fluid dynamics in a rotating frame, and how one uses methods from linear algebra and partial differential equations to solve some of the basic initial-boundary value problems that have become the hallmark of physical oceanography. MATLAB® is the key tool used throughout the book. Special care has been taken to take advantage of this software's matrix algebraic functions, its differential equation solvers, and its visualization capabilities in almost every section of every chapter. In fact, it is the use of MATLAB that allows us to consider this highly interdisciplinary material at a level that I hope is accessible to an undergraduate student.

The book is intended for advanced undergraduates, those who have already completed courses in calculus, differential equations and linear algebra. Despite requiring these prerequisites, several of the early chapters are dedicated to reviewing materials from these three topics, with varying degrees of depth and completion. While the material in these chapters may be familiar to the reader, basic use of MATLAB as well as simple examples that introduce features of fluid flows populate the illustrations and exercises.

Physical oceanography, at least the part of it that we are concerned with in this book, is characterized by the fact that the fluid flows of interest are occurring on a planet that rotates, and that this rotation can balance the forces acting on the fluid particles in such a delicate fashion to produce exquisite phenomena such as the Gulf Stream, the Jet Stream, internal waves, and the Madden–Julian Oscillation, to name a few. Much of the development in this book is motivated by the desire to explain how the existence of the Gulf Stream can be explained by the proper balance between the Coriolis force, wind stress, and molecular frictional forces. It is precisely because of the role that rotation plays in oceanography that this field is fundamentally different from rectilinear fluid flows, flows that we typically observe and measure in laboratories.

Although the Coriolis effect is part of our daily experience, it is difficult for most of us to develop an intuitive sense for its impact on the behavior of motion of particles, fluid or solid. After all, the measurements we make are most often carried out on the planet itself and are therefore relative to a rotating frame. By contrast, laboratory observations are made in an inertial frame, at least at time scales that are much faster than the time scale associated with the rotation of our planet. It is because of this lack of familiarity that it is difficult to appreciate the "apparent" forces that the planet is exerting on us, unless we float untethered for several days (as icebergs do), an experience that most of us have not had. It was therefore a particularly noteworthy moment when the early practitioners of physical oceanography finally sorted out how a current such as the Gulf Stream comes about and remains relatively stable for centuries. The ten years between 1945 and 1955 form a period when some of the most exciting applications of mathematics appeared in physical oceanography; the seminal papers of H. Stommel in 1948 and W. Munk in 1950, on the "western intensification" of ocean currents, ushered in a new era of applications of mathematics, which is the focus of this book.

The material in this book is not exhaustive, neither in mathematical methods nor in oceanographic topics. Our goal has been instead to concentrate on introducing a set of applications that are motivated by some of the questions we would like to investigate about our environment, and the type of questions where mathematics could play a critical role in their investigation. The choice of topics in many of the chapters was motivated by the desire to direct students to topics that have appeared in research manuscripts, most as journal articles published in the past few decades, and, by providing the basic mathematical tools, to invite students to begin to consult and read some of these arcticles. The new twist for us is the availability of MATLAB, which enables us to take a fresh look at many of the fundamental problems that define physical oceanography today.

Most chapters in the book contain a few projects. All projects have a significant component of MATLAB programming in them. Our hope is that these projects may suggest templates for capstone projects or honors theses for those students who are inclined to pursue a special project in applied mathematics. Most of these projects, in one way or another, are influenced by some aspect of research presented by the founders of mathematical modeling in physical oceanography, starting with the aforementioned Stommel amd Munk, but also works by G. Veronis, E. Lorenz, J. G. Charney, J. Pedlosky, A. Robinson, and A. Gill, to name a few. I believe their writing styles are accessible and inviting. Students of mathematics can benefit enormously from spending valuable time with the papers cited in this book and developing their intuition by consulting and understanding the work of pioneering experts.

In addition to introducing the basic mathematical and computational concepts, an attempt has been made in this book to introduce and to point to some of the work of current applied mathematicians who are making significant impact in developing tools for modern physical oceanography. Works by A. Majda, M. Ghil, H. Dijsktra, S. Wang, S. Wiggins, and C. K. R. T. Jones, among others, have motivated several of the projects throughout the book. The final chapter of the book is dedicated to describing several areas of current mathematical research.

Chapter 1

An Introduction to MATLAB®

MATLAB is a powerful computer language that provides an environment for numerical computation as well as graphical display of outputs. This chapter contains a brief introduction to this software, primarily concentrating on plotting graphs of functions, data management and communication of results. More detailed aspects of this software will be brought up throughout the text as needed.

There is a large library of excellent introductory texts on MATLAB. In addition, many authors, researchers and teachers have made their favorite approach to introducing MATLAB available on the World Wide Web. The reference section of this chapter contains a listing of books and web sites that the reader may find useful.

1.1 A Session in MATLAB

After invoking a session of MATLAB, we receive the prompt line

```
>>
```

at which point we are ready to call on the variety of resources that MATLAB makes available.

MATLAB has a very large number of built-in functions or programs that are available to us once we have learned a few basic notions regarding its syntax. An important feature of MATLAB's syntax is that commands and built-in functions are *case sensitive*, so that `plot` and `Plot` are treated as two distinctly different entities; the former is a central built-in utility that we will use routinely to plot graphs of functions, while the latter has no meaning at this point in our session and its invoking will result in error messages. More on this point later when we actually introduce the `plot` command.

One of the central utilities in MATLAB is the `help` command. When invoked by itself

```
help
```

it provides a long listing of available MATLAB commands and toolboxes on your computer. A typical listing looks like this

```
HELP topics:

My Documents\MATLAB    - (No table of contents file)
matlab\general         - General purpose commands.
matlab\ops             - Operators and special characters.
matlab\lang            - Programming language
                                  constructs.
matlab\elmat           - Elementary matrices and
                                  matrix manipulation.
matlab\randfun         - Random matrices and
                                   random streams.
matlab\elfun           - Elementary math functions.
matlab\specfun         - Specialized math functions.
....
....
```

Note that the listing you will see on the screen could be different, depending on the version of MATLAB, and the extra packages you may have purchased with the stand-alone version of MATLAB. The output will, however, look somewhat similar. In particular, the output has several live links attached to it, each of which is linked to another document with more information. For example, by clicking on

```
matlab\elfun           - Elementary math functions.
```

we arrive at another listing of linked pages

```
Basic operations.
    max          - Largest component.
    min          - Smallest component.
    mean         - Average or mean value.
    median       - Median value.
    std          - Standard deviation.
    ....
    ....
```

hinting that more detailed information is available. Clicking on `max` (notice, by the way, that this command is all in lowercase), we receive the following information:

```
MAX    Largest component.
    For vectors, MAX(X) is the largest element in X. For
matrices, MAX(X) is a row vector containing the maximum
```

```
element from each column. For N-D arrays, MAX(X) operates
along the first non-singleton dimension.

    [Y,I] = MAX(X) returns the indices of the maximum
values in vector I. If the values along the first non-
singleton dimension contain more than one maximal element,
the index of the first one is returned.

    .....
    .....
```

As its name suggests, `max` is capable of determining the maximum value among a set of numbers. Finally, note the link

```
doc max
```

which takes us to a summary of usage of `max`, illustrated by several examples.

Caution: Despite the fact that "`max`" is referred to as "`MAX`" in MATLAB's `help` documentation, its proper syntax is the all lowercase version. For example, MATLAB gives the correct answer to

```
max([1 2 3 4])
```

but

```
MAX([1 2 3 4])
```

leads to the error message

```
??? Undefined function or method 'MAX' for input arguments
of type 'double'.
```

If the reader is a beginner MATLAB user, this is the time to browse MATLAB's `help` documentation extensively to appreciate what a powerful resource is available to you.

The next section introduces the structure of the basic algebraic operations of multiplication, division and exponentiation in MATLAB before proceeding to defining and plotting simple functions.

Problems 1.1

Use MATLAB's `help` command and review the structure of each of the following commands. After browsing the basic narrative of each document, proceed to the examples, if there are any, and execute them in MATLAB. Finally, bring up the full documentation of each command through the `doc ...` link at the end of each document.

1. `plot`. Referring to the example at the end of the document, copy and paste the entire set of lines into a MATLAB Command Window to generate the output. Next

   (a) copy and paste the following altered three lines of the example

   ```
   x = -pi:pi/10:pi;
   y = tan(sin(x)) - sin(tan(x));
   plot(x,y)
   ```

   (b) replace the first line of the program in the above problem by

   ```
   x=-pi:pi/100:pi;
   ```

   and rerun the entire program. Report on how the output changes relative to the one obtained in 1a).

   (c) Change the domain to $(-10\pi, 10\pi)$ by replacing the first line of the program with

   ```
   x=-0*pi:pi/100:10*pi;
   ```

2. `rand` and `random`. What is the difference between two commands?

   (a) What does

   ```
   random('Normal',1,2)
   ```

   return? What does this value mean?

   (b) Enter the following line into MATLAB:

   ```
   max(sin(rand(0,1)), sin(rand(0,1)), sin(rand(0,1)))
   ```

   What is the output and what does this value mean? Next, enter the line again. Is the value obtained different from the first one this line has executed? Why?

1.2 Operations .*, ./ , and .^

MATLAB is a high-level computer language capable of manipulating arrays of numbers. An expression such as

```
A=[1 2 3 -1]
B=[0 3 -2 7]
A+B
```

defines a row of numbers for `A`, another row of numbers for `B`, and sums the two arrays. MATLAB's output is

```
A =

     1     2     3    -1


B =

     0     3    -2     7


ans =

     1     5     1     6
```

displaying both `A` and `B` and concluding with their sum. If instead we enter the following three lines

```
A=[1 2 3 -1];
B=[0 3 -2 7];
A+B
```

the output becomes

```
ans =

     1     5     1     6
```

that is, without `A` and `B` being displayed. This is due to the addition of the punctuation mark ";" at the end of the lines that define `A` and `B`. This output-suppression function of ";" will be quite important in future applications when we expect that the output of certain commands could result in very lengthy arrays.

A different usage of ";" is shown below:

```
A=[1 2; 3 -1]
B=[0 3; -2 7]
A+B
```

leads to displaying `A` and `B` as arrays with 2 rows and 2 columns, i.e., 2×2 matrices. Their sum is now displayed as an array having 2 rows and 2 columns:

```
A =

     1     2
     3    -1


B =

     0     3
    -2     7


ans =

     1     5
     1     6
```

Pointwise or entry-wise multiplication, division and exponentiation of arrays are carried out in MATLAB in much the same way that these operations are handled on single numbers but with one important caveat; MATLAB is capable of applying these operations to the entire arrays simultaneously. For instance, to multiply the following two arrays pointwise

```
A=[1 2 3 -1]; B=[0 3 -2 7];
```

we apply MATLAB's `.*` command to them:

```
A.*B
```

resulting in the array

```
ans =

     0     6    -6    -7
```

Note that each entry of the resulting array `A.*B` is the pointwise multiplication of the corresponding entries of `A` and `B`. Similarly, pointwise division of the two arrays

```
A=[1 3 4 6]; B=[4 6 -7  12];
```

is accomplished by applying "`./`" as follows:

```
A./B
```

resulting in the array

```
ans =

    0.2500    0.5000   -0.5714    0.5000
```

The pointwise exponentiation of the two arrays

```
A=[1 2 3 -1]; B = [2 3 3 2];
```

is accomplished by applying ".^" as follows:

```
A.^B
```

whose result is

```
ans =

     1     8    27     1
```

One of the remarkable features of MATLAB is how easy it is to extend pointwise algebraic operations between two arrays to the evaluation of functions of arrays. For example, given the array `A = [1 2 3 7]`, one computes the polynomial expression $A^2 - 3A$ as follows:

```
A=[1 2 3 7];
A.^2 - 3*A
```

to get

```
ans =

    -2    -2     0    28
```

Similarly, one can compute $\sin(A), \cos(A), \exp(A), \ln(A)$, just to mention a few elementary functions we routinely encounter:

```
A=[1 2 3 7];
sin(A)
cos(A)
exp(A)
log(A)
```

with the results being

```
ans =

    0.8415    0.9093    0.1411    0.6570
```

```
ans =

    0.5403   -0.4161   -0.9900    0.7539


ans =

  1.0e+003 *

    0.0027    0.0074    0.0201    1.0966


ans =

         0    0.6931    1.0986    1.9459
```

Problems 1.2

1. Consider the array `a=[1; -1; 2]`. Enter this array to MATLAB and complete the following operations and report on MATLAB's response:

```
a*a
a.*a
a^2
a.^2
1/a
1./a
sin(a)
exp(a)
```

2. Consider the array

$$a = \begin{bmatrix} 1 & -1 \\ 2 & 3 \end{bmatrix}.$$

Carry out the following operations in MATLAB and report your results.

```
a^2
a.^2
a*a
a.*a
```

```
1/a
1./a
sin(cos(a))
sin(sin(sin(cos(a))))
exp(a)
log(a)
```

1.3 Defining and Plotting Functions in MATLAB

There are several ways of defining a function in MATLAB. If a relatively simple definition of f is available, the **inline** command is the easiest way to proceed. For instance, to define the function

$$f(x) = ax^2 + bx + c$$

to MATLAB enter

```
f = inline('a*x^2 + b*x + c', 'a', 'b', 'c', 'x')
```

MATLAB's response is

```
f =

     Inline function:
     f(a,b,c,x) = a*x^2+b*x+c
```

This statement indicates that we have succeeded in defining f as a function of the four variables a, b, c and x; although we typically think of f as a function of x with a, b and c as its parameters, as far as MATLAB is concerned all four variables make equal contribution to f. To plot the graph of f when $a = 1$, $b = -2$ and $c = 3$, say, we apply the **ezplot** function of MATLAB:

```
ezplot(@(x) f(1,-2,3,x))
```

After receiving the warning

```
Warning: Function failed to evaluate on array inputs;
vectorizing the function may speed up its evaluation
and avoid the need to loop over array elements.
```

MATLAB plots the graph of $f(1, -2, 3, x)$ by selecting its own default domain of $(-2\pi, 2\pi)$, as shown in Figure 1.1. The above warning is about using "^" in place of '.^" in the definition of f. Had we used

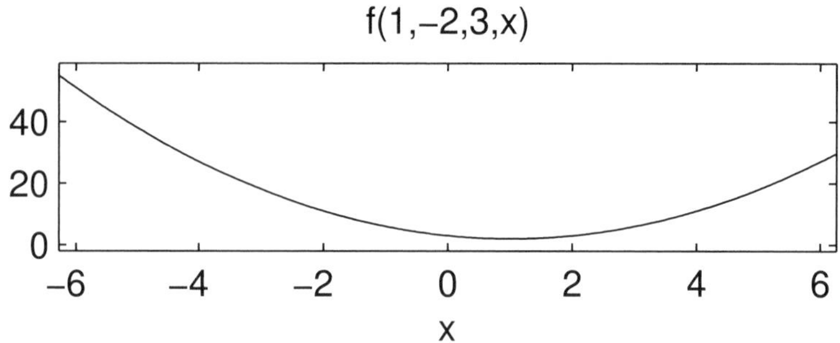

FIGURE 1.1: The output of `ezplot`.

```
f = inline('a*x.^2 + b*x + c', 'a', 'b', 'c', 'x')
```

we would not have received the warning; the use of "`.^`" allows MATLAB to apply its array capabilities cell by cell in `ezplot` to evaluate x^2 optimally.

A few comments about the usage of `ezplot`:

i) The expression `@` in `@(x)`, referred to as the *function handle constructor* in MATLAB's documentation, signifies that x is the variable of interest in the function expression. If instead we enter

```
ezplot(@(a) f(a, -2, 3, 4))
```

we receive the graph of f with a varying in the interval $(-2\pi, 2\pi)$ with other variables as specified.

ii) If we want to graph f in a domain different from the default domain, we may designate the new domain as an option in `ezplot`:

```
ezplot(@(x) f(1,-2, 3, x),[-6*pi 6*pi])
```

An alternative way to plot the graph of a function is to define the domain directly and then invoke the `plot` command. For example, to plot the graph of the function $f(x) = x\sin(2x)$ on the interval $(0, 4\pi)$, we first define the domain to MATLAB as follows:

```
h=0.1;
x=0: h: 4*pi;
```

Once these two lines are executed in MATLAB, the expressions `h` and `x` will remain available for our use during the remainder of the MATLAB

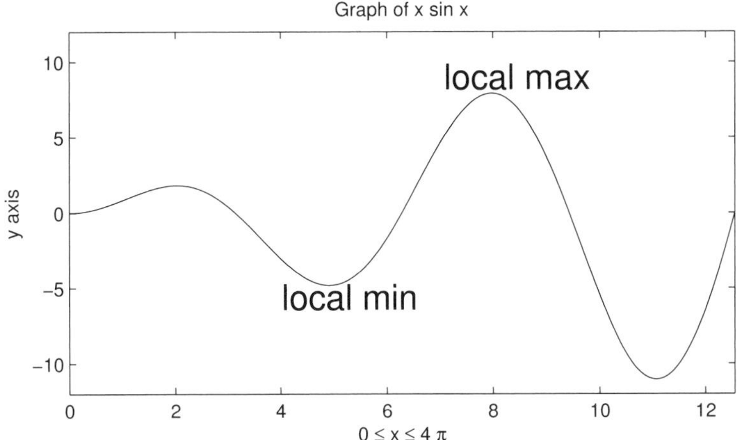

FIGURE 1.2: Graph of $x \sin x$.

session. Note, by the way, that no output actually gets printed to the screen after these lines are executed. That is because each line ended with a ";", the punctuation mark that in MATLAB signifies suppression of the output. As discussed earlier, this feature is an important tool in cases where viewing the output is not particularly interesting, as is the case in this example where 1200 entries of `x` would otherwise be displayed on the screen, had we not ended the line that defines `x` with ";".

To plot f we apply the `plot` command:

```
plot(x, x.*sin(2*x))
```

Note the "." before "*" in the above expression. MATLAB plots the graph as in Figure 1.2. The additional features in the figure are obtained by the self-explanatory lines in the following program:

```
h=0.01;
x=0:h:4*pi;
plot(x,x.*sin(x))
axis([0 4*pi -12 12])
xlabel('0 \leq x \leq 4 \pi'); ylabel('y axis')
title('Graph of x sin x')
text(4.01318,-5.6,'local min', 'FontSize', 18)
text(7.07867,9,'local max','FontSize', 18)
```

We often need to plot graphs of several functions on the same screen. The function `hold on` does the job in MATLAB. To plot the graphs of $\sin x$, $\sinh x$, and x^3 on the same screen, we enter

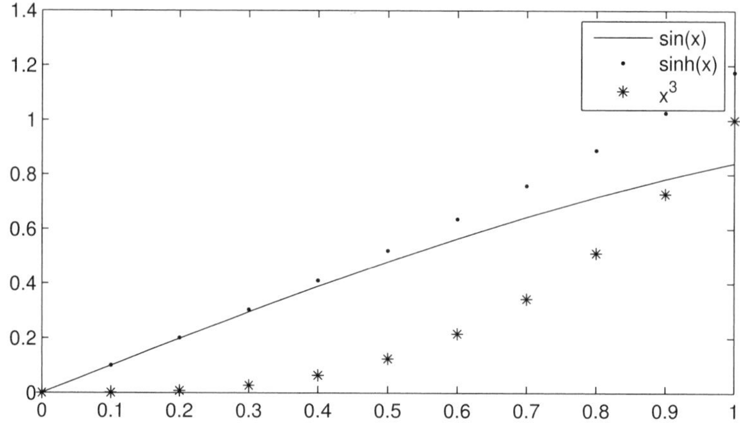

FIGURE 1.3: Graphs of $\sin x$, $\sinh x$ and x^3 using `hold on`.

```
clf
x=0:0.1:1;
plot(x,sin(x),'-')
hold on
plot(x,sinh(x),'.')
hold on
plot(x,x.^3,'*')
legend('sin(x)','sinh(x)','x^3');
```

The output is shown in Figure 1.3. The `hold on` command is the right tool for plotting several graphs on the same screen. It is often desirable, however, to plot several graphs on the same screen but on different axes. This is particularly important when the functions, such as velocity, salinity, and temperature, have different dimensions. MATLAB's `subplot` command is ideal for such a setting. `subplot`'s syntax is `subplot(mnp)`, which results in subdividing the screen into an $m \times n$ partition, with the p-th graph plotted in the mn subdivision. For example, to plot the graphs $f(x) = xe^{\sin(x)}$ and its first and second derivatives, we proceed as follows:

```
f=inline('x.*exp(sin(x))','x')
h=0.01;
x=0:h:4*pi;
subplot(311);
plot(x,f(x));
title('graph of x.*exp(sin(x))');
subplot(312)
```

```
plot(1/h*diff(f(x)));
title('graph of the first derivative');
subplot(313)
plot(1/h^2*diff(diff(f(x))))
title('graph of the second derivative');
```

Figure 1.4 shows the output of this code. In addition to `subplot`, this code uses two other MATLAB commands we have not encountered before: `clf` simply clears the graphics screen to make it ready for the next graph to be displayed, and `diff` is the internal function in MATLAB that computes the finite difference of the array it acts on. For example

```
diff([1 3.1 5.4 7])
```

results in the new array

```
2.1000    2.3000    1.6000
```

the differences between the consecutive entries of the original array. Notice that the original array has 4 entries, while the resulting array after the action of `diff` has one fewer entry. This fact is the main reason why we chose `subplot` to plot the three graphs in Figure 1.4: The two arrays `x` and `f(x)` have the same size, so `plot(x,f(x))` makes good sense and results in the first graph in Figure 1.4. The second graph in Figure 1.4, however, involves `1/h*diff(f(x)))`, whose length is one fewer than `f(x)` and `x`, hence the command

```
plot(x,1/h*diff(f(x)))
```

results in the error message

```
??? Error using ==> plot
Vectors must be the same lengths.
```

On the other hand, the command

```
plot(1/h*diff(f(x)))
```

results in the second graph in Figure 1.4 by plotting the array `1/h*diff(f(x))` against the array `[1 2 3 ... n]`, where n is the length of `1/h*diff*f(x))`.

`plot` is also capable of plotting the graphs of parametrized curves. For instance, to draw the graph of the curve whose parametrization is given by

$$\mathbf{r}(t) = \langle \sin^2 t, \cos t \rangle. \qquad t \in (0, 2\pi),$$

we proceed as follows:

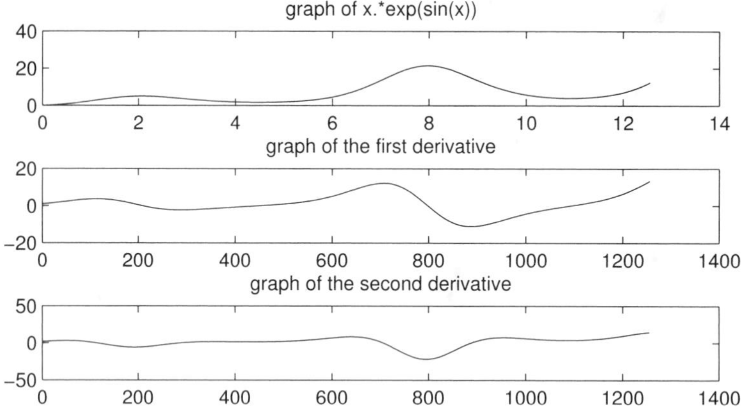

FIGURE 1.4: Graphs of $xe^{\sin x}$ and its derivatives using `subplot`.

```
t=(0:0.01:2*pi);
plot(sin(t).^2, cos(t))
```

Problems 1.3

1. Plot the graph of the following functions. In each case label the axes and provide a title.

   (a) xe^{-x}

   (b) $f(x) = \sin(5x)$. What is the period of this function, i.e., what is the smallest value of $T > 0$ for which $f(x+T) = f(x)$?

   (c) $g(x) = \sin(2x)+3\sin(3x)$. What is the period of this function?

   (d) $h(x) = \sin x + \sin\sqrt{2}x$. Draw the graph of this function on the intervals $(0, 5)$, $(0, 10)$, $(0, 50)$. Do these graphs give any indication as to whether h is periodic or not? Why?

   (e) $\sin(2\tan x) - 3\tan(\sin x)$

   (f) $J_0(x)$, where J_0 is the Bessel function of the first kind of order zero. (Hint: Use `help` to learn about `besselj`.)

   (g) $\frac{x^2-2x+3}{x^3+3x^2-x+4}$

   (h) $x \ln x$

2. Let $f(x) = x^n$.

(a) Draw the graph of f for $n = 0, 1, 2, ..., 5$ for $x \in (0, 2)$ on the same screen.

(b) Explore the `for` command in MATLAB and use this function to plot the graphs of f when n ranges from 0 to 10.

3. Consider the function $f(x) = e^{-x^2}$ with $x \in (-3, 3)$.

   (a) Plot the graph of this function.

   (b) Compare this graph to the graphs of the functions $g(x) = f(x+2)$ and $h(x) = f(\frac{x}{2})$ over the same domain and describe the scale change among these three graphs.

4. Draw on the same screen the graph of

$$\frac{1}{2}(f(x - 2t) + f(x + 2t)),$$

   for t ranging between 0 and 4 at increments of 0.5, where $f(x) = \sin x$.

5. Consider the function

$$f(x) = e^{-x} \int_0^x \sin(y^2)\, dy$$

   (a) Define this function to MATLAB. (Hint: Use `help` and read about `quad`, which is one of several internal functions that compute integrals functions. Combine `quad`, `quadl` or `quadv` with `inline` to define f.)

   (b) Evaluate f at $x = 0$, $x = 1$ and $x = 2$.

   (c) Plot the graph of f over the interval $(0, 10)$.

6. Plot the graph of the following curves.

   (a) $\mathbf{r}(t) = \langle \sin t, \cos t \rangle$; $t \in (0, 2\pi)$
   (b) $\mathbf{r}(t) = \langle 5\sin t, \cos t \rangle$; $t \in (0, 2\pi)$
   (c) $\mathbf{r}(t) = \langle \sin 5t, \cos t \rangle$; $t \in (0, 2\pi)$
   (d) Plot the above three curves on the same screen.
   (e) $\mathbf{r}(t) = \langle t + \sin t, t + \cos t \rangle$; $t \in (0, 2\pi)$
   (f) $\mathbf{r}(t) = \langle t + \sin 5t, \cos t \rangle$; $t \in (0, 2\pi)$
   (g) $\mathbf{r}(t) = \langle t + \sin 5t, t - \cos t \rangle$; $t \in (0, 2\pi)$
   (h) $\mathbf{r}(t) = \langle 1 + 3\sin^2 t, 1 - 4\cos^2 t \rangle$; $t \in (0, 2\pi)$
   (i) $\mathbf{r}(t) = \langle \sin^5 t, \cos^5 3t \rangle$; $t \in (0, 2\pi)$
   (j) $\mathbf{r}(t) = \langle \sin^3 t, \cos^3 t \rangle$; $t \in (0, 2\pi)$
   (k) $\mathbf{r}(t) = \langle \sin^3 t, \cos(10t + 1) \rangle$; $t \in (0, 2\pi)$
   (l) $\mathbf{r}(t) = \langle \sin^5 t, \cos(1 + 2t) \rangle$; $t \in (0, 2\pi)$

1.4 3-Dimensional Plotting

Plotting graphs of curves and surfaces in three dimensions is quite similar to two-dimensional plots. The command `plot3` replaces `plot` for curves in R^3. For example, the graph of the curve $\mathbf{x} = \langle t, \cos t, \sin 2t \rangle$ is obtained as follows:

```
t=0:0.01:10;
```

and then invoke `plot3`:

```
plot3(t, cos(t), sin(2 t)
```

To plot graphs of surfaces $z = f(x, y)$ on a domain B, we must first create a mesh for the domain. The command `meshgrid(x,y)` creates a two-dimensional grid for B using the arrays `x` and `y`. For example, consider $f(x, y) = J_3(\sqrt{x^2 + y^2})$, where J_3 is the third Bessel function of the first kind. To draw the graph of f in the rectangular domain $(x, y) \in (-5, 5) \times (-3, 3)$, first we create two arrays X and Y, each appropriately related to the above domain

```
[X, Y] = meshgrid(-5:0.5,-3:0.1:3);
```

The command `mesh`, when combined with the above arrays `X` and `Y` through

```
Z=besselj(3,sqrt(X.*X+Y.*Y));
mesh(X,Y,Z)
```

renders Figure 1.5. The command

```
[contours, h]=contour(X,Y,Z)
```

computes the level curves of this surface and places the level values in `h`. When we next apply the option (see MATLAB's documentation on `contour` to read about this particular option)

```
set(h,'ShowText','on','TextStep',get(h,'LevelStep')*2)
```

we get the graph in Figure 1.6.

There are many other internal functions in MATLAB that aid in visualizing curves and surfaces, among them `contour3`,

```
pcolor, surf, surfl, shading, colormap
```

Use `help` on these commands for more information.

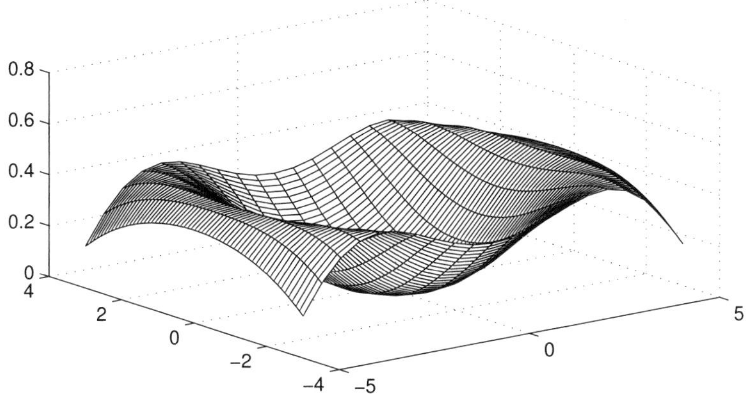

FIGURE 1.5: An output generated by `mesh`.

Problems 1.4

1. Draw the graphs of the following curves. In each case choose a range for the parameter t.

   (a) $\mathbf{r}(t) = \langle 1+t, 1-2t, t\rangle$

   (b) $\mathbf{r}(t) = \langle t, \sin t, \cos^2 t\rangle$

   (c) $\mathbf{r}(t) = \langle t\sin(t^2), t^2\cos(t), t\rangle$

   (d) $\mathbf{r}(t) = \langle \sinh^2 t, t, \cosh t\rangle$

   (e) $\mathbf{r}(t) = \langle t-t^2, \frac{1}{t+1}, 0\rangle$

2. Draw the graphs of each of the following 3D curves.

   (a) A circle of radius one centered at the origin and located in the xy-plane.

   (b) A circle of radius two centered at the origin and located in the $z=5$ plane.

   (c) Graphs of three circles of radius one on the same screen, one located in the xy-plane, another in the plane $z=1$, and the third in the plane $z=4$.

   (d) Graphs of three circles of radius one on the same screen, one located in the xy-plane and centered at $(0,3,0)$, another in the plane $z=1$ and centered at $(-2,1,1)$, and the third in the plane $z=4$ and centered at $(3,5,4)$.

   (e) The ellipse located in the xy plane centered at the origin with major and minor axes of 4 and 5, respectively.

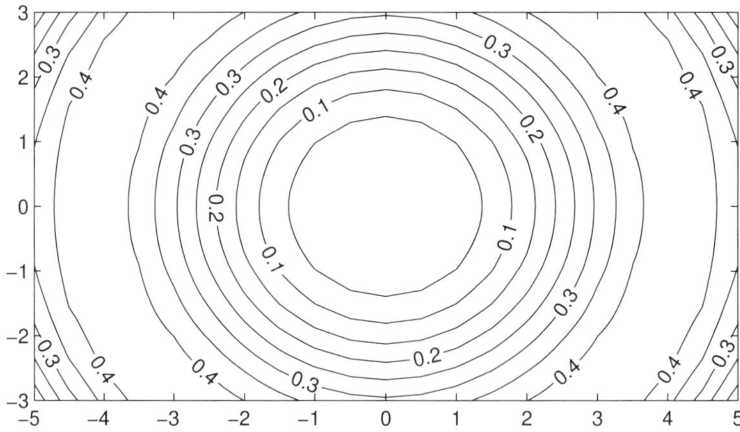

FIGURE 1.6: An output generated by `contour` and the option `set`.

(f) The curve of intersection of $x^2 + y^2 = 1$ and $z = x$

3. Draw the graphs of each of the following surfaces.

(a) $z = 2x^2 - 3y^2$, on the domain $(-2, 2) \times (-3, 3)$

(b) $z = \sqrt{(x-1)^2 + y^2}$, on the domain $(-2, 3) \times (-3, 2)$

(c) $z = \cos(x^2 + y^2)$, on the domain $(-2\pi, 2\pi) \times (-\pi, \pi)$

(d) $z = \sqrt{1 + \sin(x^2 + y^2)}$, on the domain $(-2\pi, 2\pi) \times (-\pi, \pi)$ on the same screen with the graph of the function defined in the previous problem.

(e) $z = \frac{\sin(x^2+y^2)}{2+\cos(x)}$, on the domain $(-2\pi, 2\pi) \times (-2\pi, 2\pi)$

(f) $z = \sin(x^2+y^2)\cos(2y)$, on the domain $(-2\pi, 2\pi) \times (-2\pi, 2\pi)$

(g) $z = \frac{\sin\sqrt{x^2+y^2}}{\sqrt{x^2+y^2}}$, on the domain $(-\pi, \pi) \times (-\pi, \pi)$.

1.5 M-Files

There are two ways to define functions in MATLAB. One way is to use the `inline` command, which works well for relatively simple functions. A second way is to place the definition of a function in a separate file, which we call an M-file, and call up this file from within MATLAB's Command Window. The latter approach will be the preferred one throughout this text.

To see an example of an M-file at work, consider the function

$$f(x) = ax^2 + bx + c$$

which you recall we defined in MATLAB using the `inline` command:

```
f = inline('a*x^2 + b*x + c', 'a', 'b', 'c', 'x')
```

Alternatively, we now define this function in an M-file as follows: From within a MATLAB Command Window first we select **File**, followed by **New**, followed by **Blank M-File**, to get a blank page in which we will enter the following lines:

```
function y=poly(x, a, b, c)

y=a*x.^2 + b*x + c;
```

Save the M-file by accessing **File**, followed by **Save As**, in the M-file window. MATLAB will select the name `poly` by default, the name we pre-selected by its use in the first line of the M-file, and saves the file as `poly.m` in the MATLAB directory selected by MATLAB's Command Window. Both the name of the M-file as well as its location can be altered by the user. Returning to the Command Window, if we now enter

```
poly(2,1,2,1)
```

MATLAB returns the expected result.

Because we defined the function `poly` using the ".^" operation, plotting the graph of `poly` is as simple as

```
x=1:0.01:5;
plot(x,poly(x,1,2,1))
```

Problems 1.5

1. Define the function $f(x) = \cosh(\sin x^2)$ in an M-file. Plot the graph of this function in the interval $(0, 3\pi)$.

2. Define $f(x) = x^2$ in one M-file and $g(x) = \sin 2x$ in another M-file. Use these two M-files to compute $g(f(2))$.

1.6 Loops and Iterations in MATLAB

We are often faced with repeating a computation until a desired result is reached. The repetition could come from summing a series, or simply displaying the frames of a graph to generate the sensation of an animation, or a recursive process when we are seeking a fixed point of a function, say. MATLAB's

```
for ... then
```

command is suited well for these types of operations. The sequence of `statements` in the following code

```
for i=1:n
        statement
        statement
        ....
        ....
        statement
end
```

will get executed as many times as it takes for the index `i` to run through its range, starting with $i = 1$, and ending with $i = n$. Here is an example to illustrate this point. Consider the sum

$$S = \sum_{n=1}^{100} \frac{1}{n^2}$$

whose value will be close to $\frac{\pi^2}{6}$, the exact value of the infinite series $\sum_{n=1}^{\infty} \frac{1}{n^2}$. The following code computes the sum:

```
format long            % to display numbers with 15 digits
S=0;                   % initializes S
n=100;
%
for i=1:n
    S=S+1/i^2;
end
%
S
```

resulting in 1.634983900184892. Replacing n by 1000 and rerunning the code results in 1.643934566681562. Noting that $\frac{\pi^2}{6} = 1.64493406684823$, we try the code with $n = 1,000,000$ to get better accuracy. The result is 1.644933066848770, providing us with 6 significant digits of accuracy.

Alternatively, we could use the `for .... end` structure to generate an array X that begins with 1 and ends with $\frac{1}{n^2}$, and then apply MATLAB's internal function `sum` to sum the entries of the array:

```
n=100;
X=[]; % Initialize X as the empty array
%
for i=1:n
    X=[X 1/i^2];
end
%
sum(X)
```

The result is 1.634983900184892, which is the same as the output of the first code.

Remark: Note the use of `X=[X 1/i^2]`. This line appends the entry `1/i^2` to the array `X` every time the loop is executed, so the array `X` starts as an empty array, has the entry 1 appended to it after the first go around, then the entry $\frac{1}{4}$ is appended, and so on and so forth. It is one of the remarkable features of MATLAB that the size of an array can be changed in the middle of a code, a feature that we will make use of often throughout this text.

A second example of usage of `for ... end` arises in plotting and displaying several graphs. To illustrate, consider the functions $f_n(x) = J_n(x)$, where J_n is the n-th Bessel function of the first kind, which is labeled `besselj(n,x)` in MATLAB. To plot the graphs of the first ten J_n on the interval $(0, 10)$, we proceed as follows:

```
clf
n=10;
x=0:0.01:10;
for i=1:n
    plot(x,besselj(i,x))
    hold on
end
```

The output is shown in Figure 1.7. The graphs displayed in 1.7 can also be shown as an array of plots using the `subplot` command, as well as may be animated using the `for ...end` command by combining it with the `getframe` command. To see an example of this application, consider the following code, which plots the graphs of $\sin kx$ and then animates them (look up the use of `pause` and `getframe` before proceeding to run this code):

```
x=0:0.01:2*pi;
for j=1:10
    for k = 1:16
        plot(x,sin(k*x))
        axis([0 2*pi -1 1])
        M(k) = getframe;
        pause(0.1)
    end
    pause(5)
end
```

As a final example, let's consider how `for ... end` could be used to find a fixed point of function g, a point a that remains fixed under the action of g, that is,

$$g(a) = a.$$

One approach to finding a fixed point of a function is to come up with a sequence of points $\{x_0, x_1, x_2, ...\}$ such that x_n converges to a. We can generate one such sequence by the following iterative idea: assuming that we have a general idea where a may be located, for example by plotting the graph of g, we start the iteration scheme somewhere relatively close to a. We call this initial guess x_0. The second term of the sequence, denoted by x_1, is simply the image of x_0 under g, that is,

$$x_1 = g(x_0).$$

The third element of the sequence, x_2, is obtained as the image of x_2 under g, i.e., $x_2 = g(x_1)$. In general, the i-th element x_i is the image of the previous entry x_{i-1}:

$$x_i = f(x_{i-1}).$$

As you can see, if the sequence $\{x_n\}$ converges to a point a, $x_n \to a$ and if g is a continuous function, then $g(a) = a$. This algorithm is quite simple and when it works, it works quite well. Here is a code that finds the smallest positive zero of the function $f(x) = x^2 - 3x + 2$, which is $x = 1$ (note that a is a zero of f if an only if a is a fixed point of $g(x) = f(x) + x$):

```
g=inline('x^2 - 3*x +2+x','x')
xold=1.5; % initial guess
result=[];
for i=1:10
    xnew=g(xold);
    result=[result; xold xnew];
    xold=xnew;
end
result
```

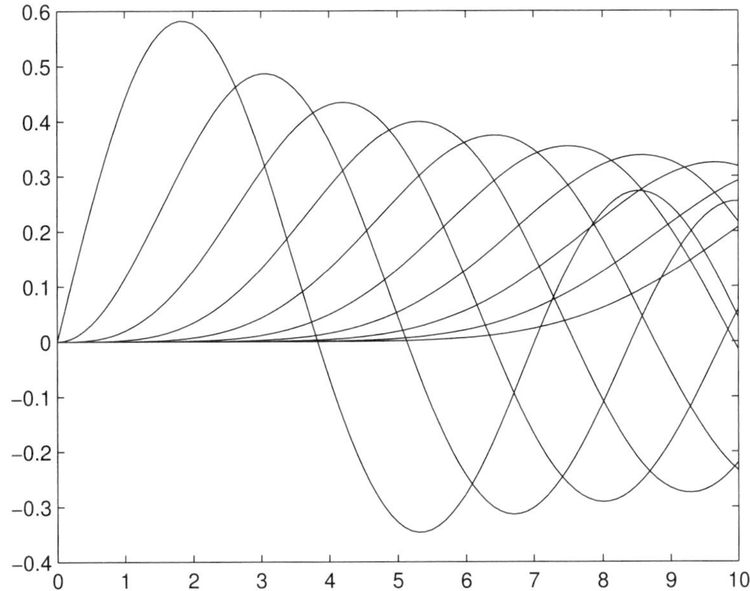

FIGURE 1.7: The graphs of Bessel functions J_n, $n = 1, ..., 10$, generated by combining MATLAB's for .. end command with `hold on`.

MATLAB returns

```
result =

   1.500000000000000   1.250000000000000
   1.250000000000000   1.062500000000000
   1.062500000000000   1.003906250000000
   1.003906250000000   1.000015258789063
   1.000015258789063   1.000000000232831
   1.000000000232831   1.000000000000000
   1.000000000000000   1.000000000000000
   1.000000000000000   1.000000000000000
   1.000000000000000   1.000000000000000
   1.000000000000000   1.000000000000000
```

clearly showing a fast convergence of the sequence to $x = 1$. We are not going into the details to describe the conditions on f under which one is guaranteed a fixed point. An interested reader can consult the text by Gilbert Strang, listed in the reference section, for information on the mathematical analysis of the fixed-point approach discussed here.

Problems 1.6

1. Generate Figure 1.7.

2. Plot the graphs of the functions $f_n(x) = x^n$ on the interval $(0, 1)$ for $n = 1, 2, ..., 10$ on the same screen.

3. Referring to the code in the fixed-point algorithm and the function f used in the example, alter the initial guess x_0 and report on whether the sequence converges, if it converges fast, or if it diverges:

   (a) $x_0 = 1.6$, $x_0 = 1.7$, $x_0 = 1.8$, $x_0 = 1.9$, $x_0 = 1.99$, $x_0 = 2$, $x_0 = 2.1$.

   (b) $x_0 = 0.9$, $x_0 = 0.5$, $x_0 = 0.1$, $x_0 = 0$ (what happens in this case?), $x_0 = -0.1$.

   (c) Is there a range of initial guesses x_0 for which the sequences converges to the other fixed point of g, or equivalently the second zero of f, namely $x = 2$?

4. Consider the sequence of Fibonacci numbers, 1, 1, 2, 3, 5, 8, ..., where the nth term of the sequence x_n is the sum of the two previous terms of the sequence, that is

$$x_n = x_{n-1} + x_{n-2}, \quad n = 3, 4, ...$$

   and $x_1 = x_2 = 1$. Write a MATLAB code to compute the 100th Fibonacci number.

5. Consider the function f defined by

$$f(x) = \int_0^x e^{-t} \tan t \, dt.$$

   Define this function in MATLAB and plot its graph on the interval $(0, 1)$. (Hint: Read about the `quad` command in MATLAB and then use it to define f, and follow that with `for ... end` to plot its graph.)

1.7 Conditional Statements in MATLAB

In Problem 4 of Section 1.6 we encountered the Fibonacci sequence and explored how to use MATLAB to generate the first n entries of this sequence. In addition to and in combination with the `for .... end` command there are two *conditional* commands in MATLAB that are helpful in controlling how information is passed through various statements we wish to execute. The first conditional command, `while ... end`, whose syntax is

```
while Condition
     Statement
     Statement
     ....
     ....
     Statement
end
```

gives the user the freedom to have MATLAB execute the `Statements` in the body of the command until the `Condition` constraint is met. A typical usage of this statement is as follows

```
S=0;
while S < 100
     S=S^2 + 1
end
```

The output shows that `S` takes on values `1, 2, 5, 26, 677`: note that the final value of `S` is larger than 100. Why doesn't that contradict the `S < 100` condition set in `while`?

The second conditional command in MATLAB has the structure `if ... else ... end` or `if ... elseif ... elseif ... end` structure. Generally these commands look like

```
if  Condition
      Statement
      Statement
      ....
      ....
      Statement
elseif Condition
         Statement
         Statement
         ....
         ....
         Statement
end
```

A typical use for the `if ... else ... end` structure is in defining functions that need to satisfy certain conditions in different parts of their domain. The absolute value function

$$f(x) = \begin{cases} -x & \text{if} \quad x < 0, \\ x & \text{otherwise.} \end{cases} \tag{1.1}$$

is an example of such a function. Here is one way of defining this function:

```
function y=absvalue(x)
%
if x < 0
    y = -x;
else
    y = x;
end
```

If we save these lines in an M-file labeled **absvalue.m**, we can then call it up in MATLAB's Command Window for evaluation. For example, here is how one proceeds to plot a graph of this function (look up the use of `length` in MATLAB):

```
x=-2:0.1:2;
for i=1:length(x)
    z(i) = absvalue(x(i));
end
plot(x,z)
```

which leads to the familiar graph of $|x|$.

Note that we did **not** use the array approach to plotting the graph of f, that is, we did not use

```
plot(x,absvalue(x))
```

because if we had, we would get the wrong graph! In fact, MATLAB will return the graph of $f(x) = x$ rather than $f(x) = |x|$. To see where the issue lies, try

```
absvalue([-2 2])
```

MATLAB returns the unexpected array

```
[-2 2]
```

rather than `[2 2]`. One reason for this is that MATLAB's `if` command is not "vectorized" in the way we have used it. To come up with a better definition of the absolute value function in MATLAB we take a different approach based on the concept of *relational operators*. A relational operator provides an output of "0" if the statement is false, and a "1" if the statement is true. For instance

```
-3 > 0
```

results in 0, while

```
-3 < 0
```

results in 1. The significance of relational operators is that they apply equally well to arrays. Hence,

```
[-2 2] > 0
```

returns `[0 1]`. We can then use these relations to define functions that have different definitions on different parts of their domain. For example, the unit step function defined by

$$\text{UnitStep}(x) = \begin{cases} 0 & \text{if} \quad x < 0, \\ 1 & \text{otherwise.} \end{cases} \tag{1.2}$$

can now be defined in MATLAB's Command Window simply by

```
UnitStep = @(x) (x>=0)
```

(recall the use of `@`, the *function handle constructor*, which is quite powerful in defining functions.) Or, equivalently, we can define an M-file, called `UnitStep.m`, as follows:

```
function y=UnitStep(x)
%
y=(x>=0);
```

In either case we have succeeded in defining a function that is "off" as long as $x < 0$ and is turned "on" when $x \geq 0$. To test that this function is array-enabled, we try the expression

```
UnitStep([-3 -2 0 1 3])
```

which returns the expected array

```
ans =

     0     0     1     1     1
```

With `UnitStep.m` in hand, we can define the absolute value function as follows:

$$f(x) = -x(1 - \text{UnitStep}(x)) + x\text{UnitStep}(x),$$

which can defined in MATLAB's Command Window by

```
VecAbsAal = @(x) -x.*(1-UnitStep(x))+x.*UnitStep(x)
```

or as an M-file, named `VecAbsVal.m`, as

```
function y = VecAbsVal(x)
%
y = -x.*(1-UnitStep(x))+x.*UnitStep(x);
```

To test that we have the right definition, try

```
x=-3:0.1:3;
plot(x,VecAbsVal(x))
```

to get the familiar graph of the absolute value function.

Problems 1.7

1. Plot the graphs of the following functions first using the `if ... then` command and next using the `UnitStep` function:

$$\text{a) } f(x) = \begin{cases} 2x - 1 & \text{if } x > 0, \\ x & \text{otherwise} \end{cases},$$

$$\text{b) } g(x) = \begin{cases} 1 & \text{if } x < -1, \\ -1 & \text{if } -1 \le x < 2 \\ 2 & \text{otherwise.} \end{cases}$$

2. Let $f(x) = x$ be defined on the interval $(0, 1)$. Extend this function as an even function to the interval $(-1, 0)$. Use the conditional capabilities of MATLAB to define the extended function and to plot its graph.

3. Let $f(x) = x^2$ be defined on the interval $(0, 1)$. Extend this function as an odd function to the interval $(-1, 0)$, and then periodically with period 2 to the rest of the real line. Use the conditional capabilities of MATLAB to define the extended function and to plot its graph on the interval $(-8, 8)$.

1.8 Fourier Series in MATLAB

We are often faced with computing a series of the form

$$S_N(x) = \sum_{n=1}^{N} a_n \sin \frac{n\pi x}{L} \tag{1.3}$$

where N could be a relatively large integer, say $N = 100$, L is a given real number defining a domain of interest, and a_n's are coefficients that may be computed beforehand; a typical example is when a_n's are the Fourier sine coefficients of a given function f defined on the interval $(0, L)$, that is,

$$a_n = \frac{2}{L} \int_0^L f(x) \sin \frac{n\pi x}{L} \, dx. \tag{1.4}$$

The following process and code shows one way we may go about computing and displaying results of the partial sum of the type shown in (1.3); start with a general M-file that generates the Fourier coefficients a_n's in (1.4):

```
function a=FourierCoefficients(N,L,f)
%
for n=1:N
    a(n)=2/L*quad(@(x) f(x).*sin(n*pi*x/L),0,L);
end
```

We note that in place of `quad` we could use a host of other integrating functions within MATLAB, including `quadv`, `quadl` and `quadgk`. The latter is in fact the integrator of choice when dealing with highly oscillatory integrands, which is often the case when we compute high frequency Fourier coefficients.

For concreteness, let's consider the function f defined by $f(x) = x^2 + 1$ whose Fourier sine series we seek in the interval $(0, 5)$. First let $N = 8$ and compute S_8 (see (1.3)) as follows:

```
L=5;
f=inline('1 + x.^2','x');
a=FourierCoefficients(8,L,f)
```

These lines gives us the first eight Fourier sine coefficients:

```
a =

   10.7384   -7.9577    5.4907   -3.9789    3.3861   -2.6526
    2.4367   -1.9894
```

To see how S_8 compares with f itself, we plot their graphs:

```
x=0:0.01:L;
plot(x,f(x))
hold on
plot(x,sum(a*sin((1:8)'*pi*x/L),1))
```

Figure 1.8 shows the graph of S_8 superimposed on the graph of f, showing the expected behavior of the graph of S_8, weaving around the graph of f. Since f does not satisfy the boundary conditions of S_8, that is since $f(0)$ and $f(5)$ do not vanish, the well-known Gibbs phenomenon appears at both ends of the interval $(0, 5)$. This effect is observed better in Figure 1.9 when we display the graphs of S_{16} and S_{32}. The latter figure is obtained by executing the following lines in MATLAB:

```
clf;
f=inline('1+x.^2','x');
L=5;
x=0:0.01:L;
plot(x,f(x));
hold on
for i=1:3
    N=2^(2+i);
    a=FourierCoefficients(N,L,f);
    plot(x,sum(a*sin((1:N)'*pi*x/L),1))
end
```

The extension of the previous MATLAB programs to two or three space dimensions is routine. For instance, the N-th Fourier sine partial sum of a function $f(x, y)$ in the domain $(0, a) \times (0, b)$ is given by

$$S_N(x, y) = \sum_{m=1}^{N} \sum_{n=1}^{N} a_{mn} \sin \frac{m\pi x}{a} \sin \frac{n\pi y}{b}, \tag{1.5}$$

where the coefficients a_{mn} are obtained by computing the following double integrals:

$$a_{mn} = \frac{4}{ab} \int_0^b \int_0^a \left[f(x, y) \sin \frac{m\pi x}{a} \sin \frac{n\pi y}{b} \right] dx\, dy \tag{1.6}$$

Consider, for example, the function f given by

$$f(x, y) = e^{-(1-x)^2(1-y)^2} \sin \pi x \sin 2\pi y + $$
$$+ e^{-(1.5-x)^2(2.3-y)^2} \sin 2\pi x \sin \pi y, \tag{1.7}$$

in the domain $(0, 2) \times (0, 3)$. Figure 1.10 shows the contours of this function. The Fourier sine approximation of this function with $N = 10$, say, is obtained by numerically computing the integrals in the formula (1.6) for a_{mn}:

$$a_{mn} = \frac{1}{3} \int_0^3 \int_0^2 \left(e^{-(1-x)^2(1-y)^2} \sin \pi x \sin 2\pi y + \right.$$

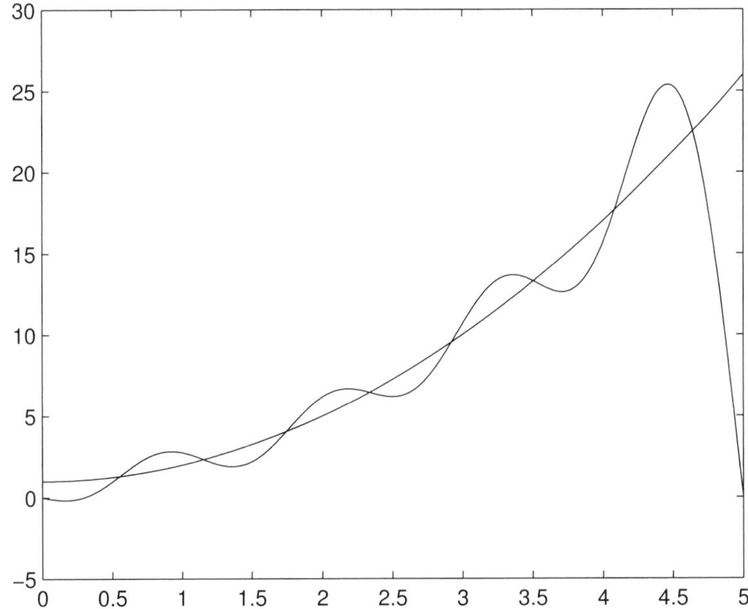

FIGURE 1.8: The graphs of $f(x) = 1 + x^2$ and its 8-th Fourier sine partial sum S_8 in the interval $(0, 5)$.

$$e^{-(1.5-x)^2(2.3-y)^2} \sin 2\pi x \sin \pi y\Big) \sin \frac{m\pi x}{2} \sin \frac{n\pi y}{3}\, dx\, dy. \tag{1.8}$$

the following MATLAB code computes the coefficients in (1.8) and plots the contours of $S_{10}(x, y)$ defined in (1.5), see Figure (1.11), as well as the contours of the error between f and S_{10}, which is shown in Figure 1.12:

```
clear all
clf
f=inline('exp(-(1-x).^2.*(1-y).^2).*sin(pi*x).*sin(2*pi*y)+
  exp(-(1.5-x).^2.*(2.3-y).^2).*sin(2*pi*x).*
     sin(pi*y)','x','y');
% the above lines need to be entered as a single line
for n=1:10
    for m=1:10
        a(m,n) = 4/(2*3)*dblquad(@(x,y) f(x,y).*
               sin(m*pi*x/2).*sin(n*pi*y/3),0,2,0,3);
% the above 2 lines need to be entered as a single line
    end
```

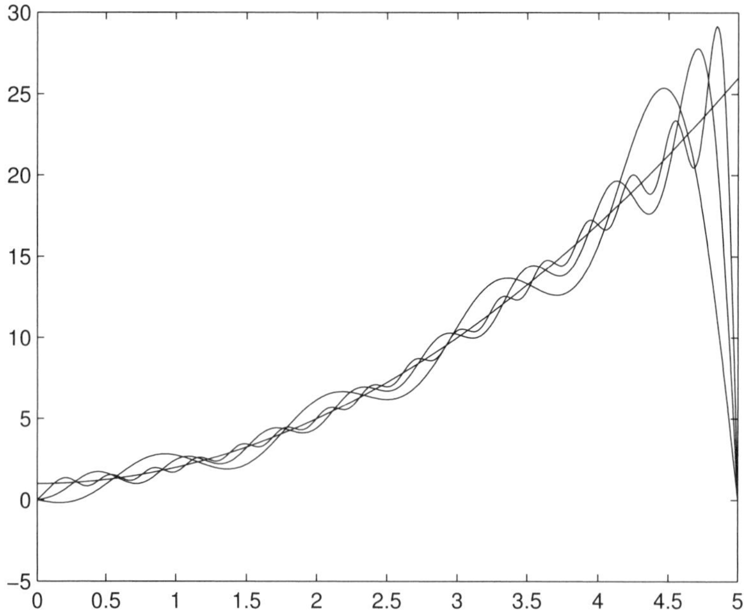

FIGURE 1.9: The graphs of $f(x) = 1 + x^2$ and its eighth, sixteenth and thirty second Fourier sine partial sums in the interval $(0, 5)$.

```
end
S=0;
[x,y]=meshgrid(0:0.01:2,0:0.01:3);
for n=1:10
    for m=1:10
        S=S+a(m,n)*sin(m*pi*x/2).*sin(n*pi*y/3);
    end
end
contour(x,y,S)
exact=f(x,y);
contour(x,y,exact)
error=max(max(abs(exact-S)))
contour(x,y,abs(exact-S))
```

Problems 1.8

1. Generate the graphs in Figures 1.8 and 1.9.

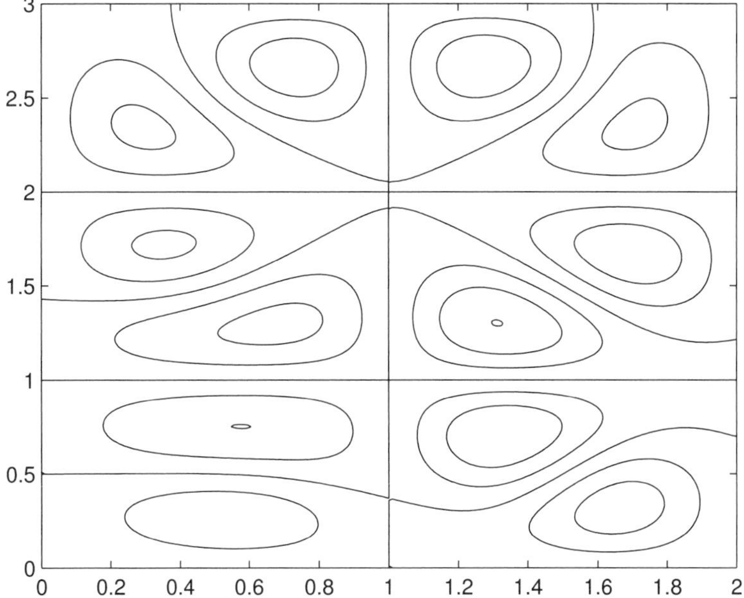

FIGURE 1.10: Sample contours of the function f defined in (1.7).

2. Define the absolute error E between two $M \times N$ arrays A and B by
$$E = \max_{1 \leq i \leq M} \max_{1 \leq j \leq N} |a_{ij} - b_{ij}| \tag{1.9}$$
Referring to f, S_8, S_{16} and S_{32} in Figures 1.8 and 1.9, compute the error E between f and each of the three partial sums. In each case use 100 points to sample f and the partial sums.

3. Use MATLAB and find the eighth, the sixteenth, and the thirty second partial sums of Fourier sine series of each function f defined below. In each case plot the graphs of all four functions on the same screen. Compute the error E_8, E_{16} and E_{32} (see the definition in (1.9)), where the domain is sampled at a hundred points.

   (a)
$$f(x) = \begin{cases} 1 & \text{if } 0 < x < 2, \\ 3 & \text{if } 2 < x < 5. \end{cases}$$

   (b)
$$f(x) = \begin{cases} x & \text{if } 0 < x < 1, \\ 2 - x & \text{if } 1 \leq x < 2. \end{cases}$$

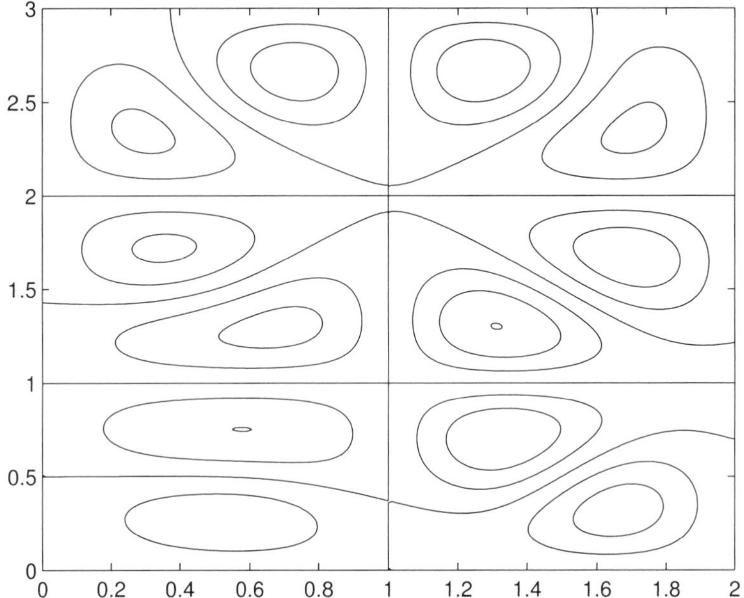

FIGURE 1.11: Sample contours of the function S_{10} as defined in (1.5).

(c) $f(x) = \cos x$, $x \in (0, 1)$.

4. The Fourier cosine series of a function f in the interval $(0, L)$ is defined in much the same way as its Fourier sine series; the main difference being that $\cos \frac{n\pi x}{L}$ replaces the equivalent sine functions;

$$C_N = \sum_{n=0}^{N} b_n \cos \frac{n\pi x}{L} \tag{1.10}$$

where (note that sum in (1.10) begins with $n = 0$)

$$b_0 = \frac{1}{L} \int_0^L f(x)\, dx,$$

and

$$b_n = \frac{2}{L} \int_0^L f(x) \cos \frac{n\pi x}{L}\, dx, \quad n = 1, 2, ... \tag{1.11}$$

Find the Fourier cosine partial sum of the following functions. In each case compute the associated Fourier sine series and plot the graphs of the f, S_N and C_N on the same screen.

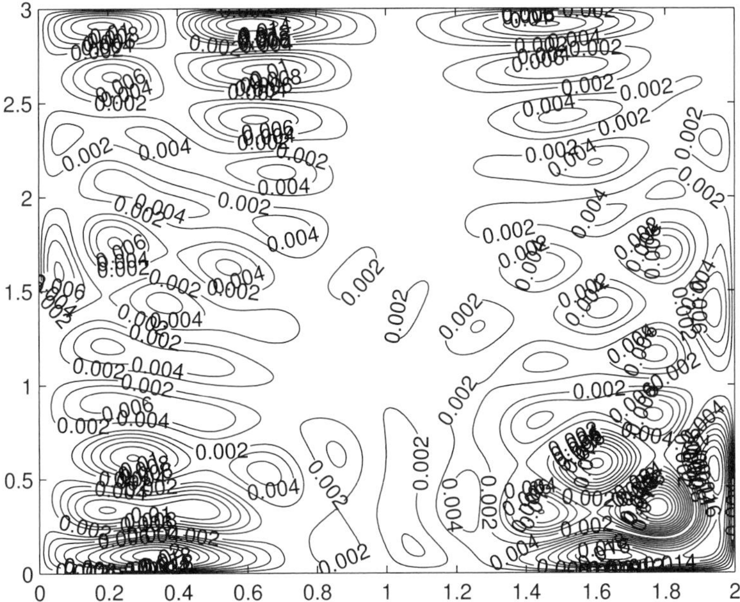

FIGURE 1.12: Contours of the absolute error between f and S_{10}.

(a) $f(x) = 1$, defined in the interval $(0, 2)$, and $N = 8$.

(b)

$$f(x) = \begin{cases} 1 & \text{if } 0 < x < 1, \\ 2 & \text{if } 1 \leq x < 3, \end{cases}$$

and $N = 16$.

(c) $f(x) = x^2$ defined in the interval $(0, 2)$, and $N = 32$.

5. Compute the Fourier sine series of the following functions. In each case plot the contours of the function f and its specified partial sum.

(a) $f(x, y) = xy(1 - x)(2 - y)$, in the domain $(0, 1) \times (0, 2)$, and $N = 8$.

(b) $f(x, y) = \sin(x^2 + y^2)$, in the domain $(0, 1) \times (0, 1)$, and $N = 32$.

(c) $f(x, y) = \begin{cases} 1 & \text{if } (x-2)^2 + (y-2)^2 < 1, \\ 0 & \text{otherwise}, \end{cases}$ in the domain $(0, 4) \times (0, 4)$, and $N = 32$.

1.9 Solving Differential Equations

We will be solving ordinary differential equations throughout this text and will often apply MATLAB's ODE solvers extensively. `ode45` is the main internal function we will employ, which we now introduce in the context of simple examples.

`ode45` uses a numerical algorithm (based on the standard Runge-Kutta scheme) and is capable of computing very accurate approximate solutions to linear as well as nonlinear initial value problems for systems of differential equations. Here is an example. Let $y(t)$ be a solution to the forced nonlinear pendulum equation

$$y'' + \alpha y' + \sin y = A\cos\omega t, \quad y(0) = 0, \quad y'(0) = 1, \tag{1.12}$$

where α, ω and A are physical constants. To prepare this initial value problem for `ode45` we must first convert the second order differential equation to a first order system: Define x by $x(t) = y'(t)$. Then $x'(t) = y''(t)$, which from the original differential equation yields $x' = -\alpha x - \sin y + A\cos\omega t$. Thus the initial value problem is equivalent to the following first order system:

$$y' = x, \quad x' = -\alpha x - \sin y + A\cos\omega t, \qquad y(0) = 0, \quad x(0) = 1. \tag{1.13}$$

The following M-file, labeled `pendulum.m`, contains the information in (1.13):

```
function yprime=pendulum(t,z);
global alpha omega A
yprime=[z(2); -alpha*z(2)-sin(z(1))-A*cos(omega*t)];
```

In the above M-file we have used the array `z`, with its two components `[z(1) z(2)]`, to stand for the two unknowns (y, x) in (1.13). We have also introduced `global`, which is quite a useful function in MATLAB, allowing for sharing values for the announced parameters, in this case `alpha`, `omega`, and `A`, to have the same values in the M-file as they have in the Command Window. Next, in the Command Window we enter the lines

```
global alpha omega A
%
alpha = 0.1;
omega=3.2;
A=14;
```

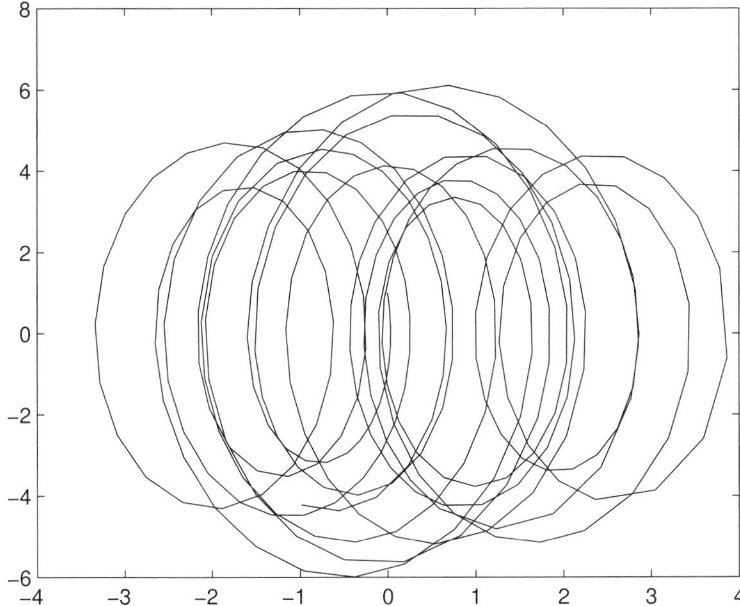

FIGURE 1.13: The phase plane diagram of the solution to $y'' + \alpha y' + \sin y = A \cos \omega t$, with $\alpha = 0.1$, $\omega = 3.2$ and $A = 14$.

just to have concrete values for these parameters. Finally we enter the line

```
[t,z] = ode45(@pendulum,[0 30],[0;1]);
```

which solves the initial value problem in (1.13) for 30 units of time for the stated initial parameter values. To get the phase portrait of this particular solution, i.e., the graph of y versus x, we plot the first column in `z` versus its second column:

```
plot(z(:,1), z(:,2))
```

Here ":" is the wild card character in MATLAB; thus `z(:,1)` accesses all entries in the first column of `z`. Similarly, `z(:,2)` stands for the entire second column of `z`. Figure 1.13 shows the output.

Problems 1.9

1. Find the solution of the following equation analytically and by using `ode45`. Compare the solutions by graphing both solutions on the same screen. In each case supply a final value of t.

   (a) $y'' + y = 0$, $y(0) = 0$, $y'(0) = 1$.

   (b) $x'' + x' + x = 0$, $x(0) = 1$, $x'(0) = 0$.

   (c) $x'' + x = \sin 2t$, $x(0) = x'(0) = 0$.

   (d) $y'' + y = \sin t$, $x(0) = x'(0) = 0$.

2. Use ode45 and draw the trajectories, i.e., $x(t), x'(t)$ of the following differential systems. In each case supply a final value of t.

   (a) $y'' + 0.1y' + \sin y = \cos t, y(0) = 0, y'(0) = 0.$

   (b) $y'' + 0.1y' + \sin y = \cos 20t, y(0) = 0, y'(0) = o.$

1.10 Concluding Remarks

We have barely touched on what a powerful resource MATLAB is. We will continue introducing the capabilities of this language throughout the text, but cannot emphasize enough that so many teachers, authors and researchers have written excellent online MATLAB tutorials, which are often available on the World Wide Web free of cost. Here we mention just a handful that deserve special attention for a beginner user, realizing that the URL for these web sites may have changed and the reader is strongly encouraged to consult his or her favorite search engine to start a new search for MATLAB tutorials:

1. http://www.uib.no/People/ngftf/CV/Misc/mat_eng.pdf

2. http://mechanical.poly.edu/faculty/vkapila/matlabtutor.htm

3. http://www.maths.dundee.ac.uk/~ftp/na-reports/MatlabNotes.pdf

4. http://www.math.ufl.edu/help/matlab-tutorial/

In addition to these resources, two excellent books, *Numerical Computing with MATLAB* and *Experiments with MATLAB*, by Cleve Moler, the founder of MATLAB, are available for download.

1.11 References

1. Pratap, R., *Getting Started with MATLAB 7: A Quick Introduction for Scientists and Engineers*, The Oxford Series in Electrical and Computer Engineering, Oxford, 2005.

2. Higham, D., Higham, N., *MATLAB Guide*, SIAM, Philadelphia, 2005

3. Strang, G., *Introduction to Applied Mathematics*, Wellesley-Cambridge Press, 1987.

4. Moler, C., *Numerical Computing with MATLAB*, may be downloaded from www.mathworks.com/moler.

5. Moler, C., *Experiments with MATLAB*, may be downloaded from www.mathworks.com/moler.

Chapter 2

Matrix Algebra

In this chapter we develop the basic concepts and tools in matrix algebra, including vector spaces and subspaces, systems of algebraic equations, determinants and inverse of matrices, Gaussian elimination, and eigenvalues and eigenvectors. Each mathematical topic is supplemented with the elementary MATLAB functions that relate to it.

2.1 Vectors and Matrices

A **vector** is a quantity that has magnitude and direction, while a scalar is a quantity with magnitude only. As is standard in science, scalars are often denoted by Latin or Greek letters, such as a, b, α and β, and vectors are displayed in boldface – $\mathbf{x}$, $\mathbf{y}$, and $\mathbf{e_2}$. Physical concepts such as force, velocity, and acceleration are represented by vectors, while quantities such as mass, pressure, temperature, and salinity are examples of scalars.

We adopt the natural geometric interpretation of a vector $\mathbf{v}$ in the plane or the three-dimensional space, as an arrow that begins at the origin of the coordinate axes, is parallel to the direction of the vector, and has its length equal to the magnitude of the vector. In this setting, we use the coordinates of the endpoint of the arrow to identify the vector. For example

$$\mathbf{v} = \langle 1, -2, 2 \rangle$$

is the vector that originates at $(0, 0, 0)$ and ends up at $(1, -2, 2)$. Note the use of $\langle$ and $\rangle$ to denote a vector, while (and) are used to denote coordinates of positions. With this interpretation in mind, the *length* or the *magnitude* of the vector $\mathbf{v} = \langle a_1, a_2, a_3 \rangle$, denoted by $||\mathbf{v}||$, is the distance from $(0, 0, 0)$ to (a_1, a_2, a_3):

$$||\mathbf{v}|| = \sqrt{a_1^2 + a_2^2 + a_3^2}. \tag{2.1}$$

Although vectors in physical settings typically have two or three components, we often encounter vectors that have n components, where n can be any positive integer, so a vector $\mathbf{v}$ may appear as

$$\mathbf{v} = \langle a_1, a_2, ..., a_n \rangle.$$

The magnitude of $\mathbf{v}$ is determined the same way as in (2.1)

$$||\mathbf{v}|| = \sqrt{a_1^2 + a_2^2 + ... + a_n^2} = \sqrt{\sum_{i=1}^{n} a_i^2}.$$

Matrices are simply rectangular arrays of numbers delimited by square brackets. A few examples are

$$A = \begin{bmatrix} \sqrt{2} & 2 \\ -3.1 & 4 \end{bmatrix}, \quad B = \begin{bmatrix} a_1 & a_2 & a_3 \\ b_1 & b_2 & b_3 \end{bmatrix}, \quad C = \begin{bmatrix} \sin x & \sin 2x \\ \sin 2x & \sin 4x \\ \sin 3x & \sin 6x \end{bmatrix},$$

or

$$D = \begin{bmatrix} 4 & -1 & 0 & 0 & 1+i \\ -1 & 4 & -1 & 0 & 0 \\ 0 & -1 & 4 & -1 & 0 \\ 0 & 0 & -1 & 4 & -1 \\ 1-i & 0 & 0 & -1 & 4 \end{bmatrix}, \quad E = \begin{bmatrix} 1/2 & 1/3 & 1/4 \\ 1/3 & 1/4 & 1/5 \\ 1/4 & 1/5 & 1/6 \end{bmatrix}.$$

Each of these matrices has a certain number of rows and columns; Matrix A is a 2 by 2 matrix, consisting of two rows and two columns. Similarly, matrices B, C, D and E are 2 by 3, 3 by 2, 5 by 5 and 3 by 3, respectively. From now on we use the notation $m \times n$ to denote an m by n matrix.

Each entry of a matrix is identified by the row and column positions it occupies, so the $(1,1)$ (pronounced "one one") entry of A is $\sqrt{2}$, its $(1,2)$ entry is 2, its $(2,1)$ entry -3.1, and so on. It is common to denote the (i,j)-th entry of a matrix A by a_{ij}, so, referring back to matrix A defined above, we write

$$a_{11} = \sqrt{2},\ a_{12} = 2,\ a_{21} = -3.1,\ a_{22} = 4.$$

Vectors could also be viewed as matrices having only one row or one column. For instance the vector $\mathbf{v} = \langle 1, -2, 2 \rangle$ can also be identified by the matrices V_1 or V_2 defined as

$$V_1 = \begin{bmatrix} 1 & -2 & 2 \end{bmatrix}, \quad V_2 = \begin{bmatrix} 1 \\ -2 \\ 2 \end{bmatrix}.$$

The appropriate identification is often determined by the problem context.

2.2 Vector Operations

Based on experience with modeling physical problems, we have an understanding and appreciation of where vectors come from and why we should represent and manipulate them mathematically. The typical operations of vector addition and scalar multiplication, as well as the various vector multiplications, have their origin in well-known physical settings. To a large extent, similar interpretations exist in matrix algebra, which we now address.

Given a vector $\mathbf{v}$ and a scalar α, an element in the set of real numbers R or complex numbers C, we define $\alpha\mathbf{v}$, the *scalar product* of α and $\mathbf{v}$, as the vector that has magnitude $|\alpha|\,||\mathbf{v}||$ and is otherwise parallel to $\mathbf{v}$. If $\mathbf{v} = \langle a_1, a_2, a_3\rangle$, we find that

$$\alpha\mathbf{v} = \langle \alpha a_1, \alpha a_2, \alpha a_3\rangle.$$

The *sum* or *vector addition* $\mathbf{v}_1 + \mathbf{v}_2$ of the two vectors $\mathbf{v}_1$ and $\mathbf{v}_2$ is the vector we obtain as the main diagonal of the parallelogram constructed based on the two vectors $\mathbf{v}_1$ and $\mathbf{v}_2$. This geometric construct is equivalent to the following algebraic operation: Let $\mathbf{v}_1 = \langle a_1, a_2, a_3\rangle$ and $\mathbf{v}_2 = \langle b_1, b_2, b_3\rangle$. Then

$$\mathbf{v}_1 + \mathbf{v}_2 = \langle a_1 + b_1,\ a_2 + b_2,\ a_3 + b_3\rangle.$$

Similarly, the sum of two vectors $\mathbf{v}_1 = \langle a_1, a_2, ..., a_n\rangle$ and $\mathbf{v}_2 = \langle b_1, b_2, ..., b_n\rangle$ is the vector

$$\mathbf{v}_1 + \mathbf{v}_2 = \langle a_1 + b_1, a_2 + b_2, ..., a_n + b_n\rangle.$$

There are two *vector multiplications* that we need to consider. First, the *dot* or the *inner* product of the two vectors $\mathbf{v}_1$ and $\mathbf{v}_2$, denoted by $\mathbf{v}_1 \cdot \mathbf{v}_2$, is defined by

$$\mathbf{v}_1 \cdot \mathbf{v}_2 = ||\mathbf{v}_1||\,||\mathbf{v}_2|| \cos\theta,$$

where θ is the angle between the two vectors. This operation has two significant geometric interpretations: a) when $||\mathbf{v}_1|| = 1$, $\mathbf{v}_1 \cdot \mathbf{v}_2$ equals the length of the *projection* of $\mathbf{v}_2$ on $\mathbf{v}_1$, and b) $\mathbf{v}_1 \cdot \mathbf{v}_2$ vanishes if and only if the vectors are *orthogonal*. In component form the dot product takes the form

$$\mathbf{v}_1 \cdot \mathbf{v}_2 = \sum_{i=1}^{3} a_i b_i,$$

which has the straightforward extension to n-dimensional vectors:

$$\mathbf{v}_1 \cdot \mathbf{v}_2 = \sum_{i=1}^{n} a_i b_i.$$

The second way to consider the product of two vectors $\mathbf{v}_1 = \langle a_1, a_2, a_3 \rangle$ and $\mathbf{v}_2 = \langle b_1, b_2, b_3 \rangle$ is by forming their *vector* or *cross* product. Denoted by $\mathbf{v}_1 \times \mathbf{v}_2$, the cross product of $\mathbf{v}_1$ and $\mathbf{v}_2$ is a vector whose magnitude is

$$||\mathbf{v}_1||\,||\mathbf{v}_2|| \sin\theta$$

and whose direction is perpendicular to both $\mathbf{v}_1$ and $\mathbf{v}_2$, and uniquely determined by the right-hand rule. In component form this vector takes the form

$$\mathbf{v}_1 \times \mathbf{v}_2 = \langle a_2 b_3 - a_3 b_2,\ a_3 b_1 - a_1 b_3,\ a_1 b_2 - a_2 b_1 \rangle. \tag{2.2}$$

Geometrically, this operation provides information about how close two vectors are to being parallel. In particular, $\mathbf{v}_1 \times \mathbf{v}_2 = \mathbf{0}$ if and only if $\mathbf{v}_1$ and $\mathbf{v}_2$ are parallel.

Problems 2.2

1. Let $\mathbf{a} = \langle 1, -2 \rangle$, $\mathbf{b} = \langle 3, -1 \rangle$.

   (a) Find $||\mathbf{a}||$, $||\mathbf{b}||$, $||3\mathbf{a} + 2\mathbf{b}||$,

   (b) $\mathbf{a} \cdot \mathbf{a}$, $\mathbf{a} \cdot \mathbf{b}$, $(2\mathbf{a}) \cdot (-3\mathbf{b})$,

   (c) Extend $\mathbf{a}$ and $\mathbf{b}$ to be three-dimensional vectors by setting their third components to zero. Compute $\mathbf{a} \times \mathbf{b}$, $\mathbf{b} \times \mathbf{a}$, $\mathbf{a} \times \mathbf{a}$, $(\mathbf{a} + \mathbf{b}) \times \mathbf{a}$.

2. Let $\mathbf{a} = \langle 2.1, -2, 3 \rangle$, $\mathbf{b} = \langle 3.2, -1.1, 4.3 \rangle$.

   (a) Find $||\mathbf{a}||$, $||\mathbf{b}||$, $||\alpha\mathbf{a} + \mathbf{b}||$, where α is a real number,

   (b) $\mathbf{a} \cdot \mathbf{a}$, $||\mathbf{a}||$, $\mathbf{a} \cdot \mathbf{b}$,

   (c) Compute $\mathbf{a} \times \mathbf{b}$, $\mathbf{a} \times \mathbf{a}$, $\mathbf{b} \times \mathbf{a}$,

   (d) Compute $\mathbf{a} \cdot (\mathbf{b} \times \mathbf{c})$ where $\mathbf{c} = \langle 1, 1, 1 \rangle$.

3. Let $\mathbf{a} = \langle a_1, a_2, a_3 \rangle$, $\mathbf{b} = \langle b_1, b_2, b_3 \rangle$, and $\mathbf{c} = \langle c_1, c_2, c_3 \rangle$.

   (a) Show that $\mathbf{a} \times \mathbf{b} = -\mathbf{b} \times \mathbf{a}$.

   (b) Is $\mathbf{a} \times (\mathbf{b} \times \mathbf{c}) = (\mathbf{a} \times \mathbf{b}) \times \mathbf{c}$? Either prove the result or give a counterexample.

4. Let $\mathbf{a} = \langle a_1, a_2, a_3 \rangle$, $\mathbf{b} = \langle b_1, b_2, b_3 \rangle$, and $\mathbf{c} = \langle c_1, c_2, c_3 \rangle$. Show that

(a) $\mathbf{a} \cdot \mathbf{a} = ||\mathbf{a}||^2$

(b) $\mathbf{a} \cdot (\mathbf{b} \times \mathbf{c}) = (\mathbf{a} \times \mathbf{b}) \cdot \mathbf{c}$.

(c) $\mathbf{a} \cdot (\mathbf{b} \times \mathbf{c})$ is the volume of the parallelepiped constructed from $\mathbf{a}$, $\mathbf{b}$ and $\mathbf{c}$.

(d) $\mathbf{a} \times (\mathbf{b} \times \mathbf{c}) = (\mathbf{a} \cdot \mathbf{c})\mathbf{b} - (\mathbf{a} \cdot \mathbf{b})\mathbf{c}$.

5. Let $\mathbf{e}_1 = \langle 1, 0, 0 \rangle$, $\mathbf{e}_2 = \langle 0, 1, 0 \rangle$, and $\mathbf{e}_3 = \langle 0, 0, 1 \rangle$.

   (a) Show that $\mathbf{e}_i \cdot \mathbf{e}_j = \delta_{ij}$, where δ is the *Kronecker delta function*, and is defined by $\delta_{ij} = 1$ if $i = j$ and zero otherwise.

   (b) Show that $\mathbf{e}_1 \times \mathbf{e}_2 = \mathbf{e}_3$, $\mathbf{e}_2 \times \mathbf{e}_3 = \mathbf{e}_1$ and $\mathbf{e}_3 \times \mathbf{e}_1 = \mathbf{e}_2$.

   (c) Let $\mathbf{a} = \langle a_1, a_2, a_3 \rangle$. Note that $\mathbf{a} = a_1\mathbf{e}_1 + a_2\mathbf{e}_2 + a_3\mathbf{e}_3$. Let $\mathbf{b} = \langle b_1, b_2, b_3 \rangle$. Using the information in parts (a) and (b), show that $\mathbf{a} \times \mathbf{b} = (a_2b_3 - a_3b_2)\mathbf{e}_1 + (a_3b_1 - a_1b_3)\mathbf{e}_2 + (a_1b_2 - a_2b_1)\mathbf{e}_3$. Compare this result with the formula in (2.2).

6. Let $\mathbf{u} = \langle u_1, u_2, u_3 \rangle$ and $\mathbf{e}_3$ as defined in Problem (5a). Compute $\mathbf{w} = \mathbf{e}_3 \times (\mathbf{e}_3 \times \mathbf{u})$. How is $\mathbf{w}$ related to $\mathbf{u}$ geometrically?

2.3 Matrix Operations

The concepts of *matrix addition* and *scalar multiplication* are borrowed directly from their counterparts in vectors: Given two $m \times n$ matrices A and B, their sum, $A + B$, is another $m \times n$ matrix whose ij-th entry is the sum of the ij-th entries of A and B:

$$A = [a_{ij}], \quad B = [b_{ij}], \text{ then } \quad A + B = [a_{ij} + b_{ij}].$$

Similarly, the scalar product of α and A, denoted by αA, is defined as the matrix whose ij-th entry is αa_{ij}:

$$A = [a_{ij}], \quad \alpha \in R, \quad \text{then} \quad \alpha A = [\alpha a_{ij}].$$

The operation of dot product of vectors is the basis for the definition of matrix multiplication. Consider the two matrices A and B, with A an $m \times p$ matrix and B a $p \times n$ matrix. We define C, the product of A and B, as the $m \times n$ matrix whose (i, j)-th entry is the dot product of the i-th row of A and the j-th column of B, i. e.,

$$c_{ij} = \sum_{k=1}^{p} a_{ik}b_{kj}.$$

For example, consider the matrices

$$A = \begin{bmatrix} 1 & -2 & 2 \\ 0 & 1 & -1 \end{bmatrix}, \qquad B = \begin{bmatrix} 0 & 1 \\ -1 & 1 \\ 1 & 3 \end{bmatrix}.$$

Then

$$C = AB = \begin{bmatrix} 1 & -2 & 2 \\ 0 & 1 & -1 \end{bmatrix} \begin{bmatrix} 0 & 1 \\ -1 & 1 \\ 1 & 3 \end{bmatrix} = \begin{bmatrix} 4 & 5 \\ -2 & -2 \end{bmatrix},$$

a fact we verify in MATLAB by entering the following lines:

```
A=[1 -2 2;0 1 -1]
B=[0 1;-1 1;1 3]
A*B
```

One of the important features of matrix multiplication is that this operation is not *commutative*, that is, AB does not necessarily equal BA. To see an example, simply compute BA in the previous example:

$$BA = \begin{bmatrix} 0 & 1 \\ -1 & 1 \\ 1 & 3 \end{bmatrix} \begin{bmatrix} 1 & -2 & 2 \\ 0 & 1 & -1 \end{bmatrix} = \begin{bmatrix} 0 & 1 & -1 \\ -1 & 3 & -3 \\ 1 & 1 & -1 \end{bmatrix}.$$

Note that BA looks quite different from AB, including having a different size and shape.

The definition of matrix multiplication is intimately related to how one represents systems of linear algebraic equations. Consider, for instance, the system of equations

$$\begin{cases} 2x + 3y - z &= 1, \\ -3x + 2y + 4z &= -2, \\ x + y + z &= 0. \end{cases} \tag{2.3}$$

The left side of each equation in (2.3) is the dot product of two vectors, one vector consisting of the variables x, y and z, and the other the vector of the coefficients. In this way (2.3) is rewritten as

$$\begin{cases} \langle 2, 3, -1\rangle \cdot \langle x, y, z\rangle &= 1, \\ \langle -3, 2, 4\rangle \cdot \langle x, y, z\rangle &= -2, \\ \langle 1, 1, 1\rangle \cdot \langle x, y, z\rangle &= 0. \end{cases} \tag{2.4}$$

Recalling that the dot product of two vectors is at the essence of matrix multiplication, we now rewrite (2.4) in matrix notation. First we construct a 3×3 matrix A, each row of which consists of the coefficients of x, y and z in a corresponding equation:

$$A = \begin{bmatrix} 2 & 3 & -1 \\ -3 & 2 & 4 \\ 1 & 1 & 1 \end{bmatrix}.$$

Next write the variables x, y and z as a 3×1 column vector and denote it by $\mathbf{x}$:

$$\mathbf{x} = \begin{bmatrix} x \\ y \\ z \end{bmatrix},$$

and finally we construct a second column vector to include the input variables and denote it by $\mathbf{b}$:

$$\mathbf{b} = \begin{bmatrix} 1 \\ -2 \\ 0 \end{bmatrix}.$$

The system of linear equations (2.3) is now equivalent to the matrix equation

$$A\mathbf{x} = \mathbf{b}, \tag{2.5}$$

as it can easily be verified.

This strategy generalizes to any system of linear equations. Consider the system of linear equations consisting of m equations in n unknowns:

$$\left\{ \begin{array}{rcl} a_{11}x_1 + a_{12}x_2 + \ldots + a_{1n}x_n & = & b_1, \\ a_{21}x_1 + a_{22}x_2 + \ldots + a_{2n}x_n & = & b_2, \\ \ldots & = & \ldots \\ \ldots & = & \ldots \\ a_{m1}x_1 + a_{m2}x_2 + \ldots + a_{mn}x_n & = & b_m. \end{array} \right. \tag{2.6}$$

This system is written in the form $A\mathbf{x} = \mathbf{b}$ with

$$A = \begin{bmatrix} a_{11} & a_{12} & \ldots & a_{m1} \\ a_{21} & a_{22} & \ldots & a_{m2} \\ \ldots & \ldots & \ldots & \ldots \\ a_{m1} & a_{m2} & \ldots & a_{mn} \end{bmatrix}, \quad \mathbf{x} = \begin{bmatrix} x_1 \\ x_2 \\ \ldots \\ x_m \end{bmatrix} \quad \mathbf{b} = \begin{bmatrix} b_1 \\ b_2 \\ \ldots \\ b_m \end{bmatrix} \tag{2.7}$$

In many examples of physical significance the matrix A in (2.7) will be square (i.e., $m = n$). One such example is

$$\left\{\begin{array}{ccc} 2x_1 - x_2 & = & h^2 f(t_1), \\ -x_1 + 2x_2 - x_3 & = & h^2 f(t_2), \\ -x_2 + 2x_3 - x_4 & = & h^2 f(t_3), \\ \dots & = & \dots \\ \dots & = & \dots \\ -x_{n-1} + 2x_n & = & h^2 f(t_n), \end{array}\right. \tag{2.8}$$

where the typical i-th equation is

$$-x_{i-1} + 2x_i - x_{i+1} = h^2 f(t_i). \tag{2.9}$$

System (2.8), as we see later, results from the numerical discretization of the boundary value problem $-x''(t) = f(t)$, with $x(a) = x(b) = 0$. This system is of the form (2.5) with

$$A = \begin{bmatrix} 2 & -1 & 0 & 0 & \dots & \dots & 0 \\ -1 & 2 & -1 & 0 & 0 & \dots & 0 \\ 0 & -1 & 2 & -1 & 0 & \dots & 0 \\ \dots & \dots & \dots & \dots & \dots & \dots & \dots \\ \dots & \dots & \dots & \dots & \dots & \dots & \dots \\ \dots & \dots & \dots & \dots & -1 & 2 & -1 \\ 0 & 0 & \dots & \dots & 0 & -1 & 2 \end{bmatrix}, \quad \text{and}$$

$$\mathbf{x} = \begin{bmatrix} x_1 \\ x_2 \\ \dots \\ \dots \\ x_m \end{bmatrix}, \qquad \mathbf{b} = h^2 \begin{bmatrix} f(x_1) \\ f(x_2) \\ \dots \\ \dots \\ f(x_n) \end{bmatrix}. \tag{2.10}$$

The matrix A in (2.10) has several interesting properties which we will study in later sections and chapters. One of these features is its *sparsity*, that so many of its entries are zero; special care is taken in the design of MATLAB to handle such matrices very efficiently, allowing us to carry out large calculations with relative ease, as we will see in several instances later in the text.

Returning to (2.5), we note that the equation $A\mathbf{x} = \mathbf{b}$ has a similar structure to the scalar equation $ax = b$, whose solution is $x = \frac{b}{a}$ when $a \neq 0$. This analogy will be the source of inspiration for seeking solutions to the matrix equation $A\mathbf{x} = \mathbf{b}$ in the form

$$\mathbf{x} = A^{-1}\mathbf{b}$$

once we have a proper definition of A^{-1}, the inverse of A, a concept we will take up later in this chapter.

We end this section with a few definitions of special matrices:

The *zero matrix*, denoted by Z, is a matrix with all entries equal to zero. The lines

```
n=10;
zmatrix=zeros(n);
```

generate a 10×10 zero matrix, which we have labeled `zmatrix`.

An $m \times n$ matrix is called a *square matrix* if $m = n$.

Given an $n \times n$ matrix A, the entries a_{ii}, $i = 1, ..., n$ constitute the *diagonal* of that matrix. Generally, when m is non-negative, the entries $a_{i,i+m}$, $i = 1, ..., n-m$, comprise the m-th *superdiagonal* of A. Similarly, the entries $a_{i+m,i}$, $i = 1, ..., n-m$, with m non-negative, form the m-th *subdiagonal* of A.

The matrix A in (2.10) has only three non-zero diagonals; the main diagonal, all of whose entries are 2, a superdiagonal and a subdiagonal of only -1's.

Mimicking MATLAB's notation, we may denote by

$$\text{diag}(\mathbf{a}, m)$$

to mean a matrix with zero entries everywhere, except on the m-th diagonal where the entries of the vector $\mathbf{a}$ are placed. For example,

$$A = \text{diag}(\langle 2, -7, 3, -5 \rangle, -2)$$

is the 6×6 matrix

$$A = \begin{bmatrix} 0 & 0 & 0 & 0 & 0 & 0 \\ 0 & 0 & 0 & 0 & 0 & 0 \\ 2 & 0 & 0 & 0 & 0 & 0 \\ 0 & -7 & 0 & 0 & 0 & 0 \\ 0 & 0 & 3 & 0 & 0 & 0 \\ 0 & 0 & 0 & -5 & 0 & 0 \end{bmatrix}.$$

Note that a negative m in this notation denotes a subdiagonal, a positive m a superdiagonal. Hence,

$$B = \text{diag}(\langle 2, 2 \rangle, 1)$$

is the 3×3 matrix

$$B = \begin{bmatrix} 0 & 2 & 0 \\ 0 & 0 & 2 \\ 0 & 0 & 0 \end{bmatrix}.$$

The following MATLAB lines reach the same conclusions:

```
a=[2 -7 3 -5];
A=diag(a,-2);
B=diag([2 2],1];
```

The *Identity* matrix, denoted by I, is an $n \times n$ matrix with ones on the diagonal and zeros elsewhere. The command `eye(n)` in MATLAB generates an $n \times n$ identity matrix.

A square matrix is called a *diagonal* matrix if $a_{ij} = 0$ if $i \neq j$; so a diagonal matrix is one whose nonzero entries may reside on the diagonal.

An $n \times n$ matrix is called *upper triangular* if $a_{ij} = 0$ with $i > j$. Similarly, an $n \times n$ matrix is called *lower triangular* if $a_{ij} = 0$ with $i < j$. The matrix A in (2.10), for example, may be written as the sum of three matrices, D, a diagonal matrix, and L and U, which are lower and upper triangular, respectively, as follows

$$A = D + L + U$$

where

$$\begin{aligned} D &= 2I \\ L &= \text{diag}(\mathbf{v}, -1) \\ U &= \text{diag}(\mathbf{v}, 1), \end{aligned}$$

where

$$\mathbf{v} = \langle -1, -1, ..., -1 \rangle$$

is a $1 \times n - 1$ vector with all entries equaling -1. The following lines in MATLAB generate a 10×10 version of A in (2.10):

```
n = 10;
a=-ones(n-1,1);
A=2*eye(n)+diag(a,-1)+diag(a,1);
```

Given a matrix A we construct A^T, called the *transpose* of A, by interchanging the rows and columns of A. For example, if $A = \begin{bmatrix} 1 & 2 \\ 3 & 4 \end{bmatrix}$, then $A^T = \begin{bmatrix} 1 & 3 \\ 2 & 4 \end{bmatrix}$. The transpose operation in MATLAB is accomplished by placing an ' after the matrix:

```
A=[1 2; 3 4];
B= A';      % B is the transpose of A
```

A square matrix A is said to be *symmetric* if $A = A^T$; the matrix A in (2.10), for example, is symmetric, while $A = \begin{bmatrix} a & 2 \\ 3 & b \end{bmatrix}$ is not. It is easy to show that the sum of two symmetric matrices is another symmetric matrix, while the product of two symmetric matrices may not in general be symmetric.

Given a complex-valued matrix A, the matrix $\bar{A}$ is the matrix one obtains by taking the complex conjugate of each entry of A. For example, if $A = \begin{bmatrix} 1+i & 2 \\ -3-i & 2+3i \end{bmatrix}$, then $\bar{A} = \begin{bmatrix} 1-i & 2 \\ -3+i & 2-3i \end{bmatrix}$. The *Conjugate Transpose* of a matrix A is defined by $\bar{A}^T$. A complex-valued matrix is called *Hermitian* if $A = \bar{A}^T$.

Problems 2.3

1. Use the `help` command in MATLAB to learn about `zeros, ones, eye, diag`.

2. Write down the following matrices:

   (a) $\text{diag}(\langle a, a\rangle, 1)$

   (b) $\text{diag}(\langle a, a\rangle, 0)$

   (c) $\text{diag}(\langle a, a, b\rangle, -1)$

   (d) `ones(2)`, `ones(2,1) zeros(3)`, `zeros(10,2)`; verify each answer in MATLAB.

   (e) `A=-2*diag(ones(3,1))`. How does the output differ from `A = -2*eye(3)`?

3. What will be the output of the following program in MATLAB? Write down the output of each line:

```
a=3;
b=-1;
A=a*diag(ones(2,1),1)+4*diag([b b],-1)+7*diag([b],2);
B=A';
```

4. Write the following systems of linear equations in matrix form:

$$\text{i)} \begin{cases} 2x - 3y &= 1 \\ 3x + y &= -2 \end{cases} \qquad \text{ii)} \begin{cases} ax + by &= \alpha_1 \\ cx + dy &= \alpha_2 \end{cases}$$

$$\text{iii)} \begin{cases} x + y + z &= 0 \\ x - y + z &= 1 \\ 2x + z &= 1 \end{cases} \qquad \text{iv)} \begin{cases} y &= 2x \\ x - y &= 0 \end{cases}$$

5. Let A and B two arbitrary 2×2 matrices. Show that the transpose of the product of A and B is the product of the transposes of B and A (note the change in order of multiplication), i.e.,

$$(AB)^T = B^T A^T.$$

6. Let A and B be two $n \times n$ symmetric matrices. Show that
   (a) $A + B$ is also symmetric.
   (b) $\alpha A + \beta B$ is symmetric for all α and β, where α and β are arbitrary real numbers.
   (c) Is the product of A and B necessarily symmetric? Either prove this statement or give a counterexample.

7. A square matrix A is called *skew symmetric* if $A^{\mathrm{T}} = -A$. Determine which of the following matrices is symmetric or skew-symmetric.

$$i) \begin{bmatrix} 2 & 1 \\ 1 & -1 \end{bmatrix} \qquad ii) \begin{bmatrix} 0 & 2 \\ -2 & 0 \end{bmatrix}.$$

8. Let A be an $n \times n$ skew-symmetric matrix. Show that $a_{ii} = 0$ for every i with $1 \leq i \leq n$.

9. Show that the zero matrix Z is the only matrix that is symmetric and anti-symmetric.

10. Let A be an $n \times n$ matrix. Show that A can be written as sum of two $n \times n$ matrices B and C, where B is symmetric and C an anti-symmetric matrix.

11. Consider $A = \begin{bmatrix} -1 & 1 & -1 \\ 0 & 1 & 2 \\ 3 & 2 & 1 \end{bmatrix}$. Find the B and C of the previous problem.

12. The vector $\mathbf{y} = \langle y_1, y_2, ..., y_n \rangle$ is related to $\mathbf{x} = \langle x_1, x_2, ..., x_n \rangle$ by the linear equations

$$y_i = \frac{x_{i+1} - x_{i-1}}{2h}, \qquad i = 1, 2, ..., n,$$

with $x_0 = x_n$ and $x_{n+1} = x_1$. Write these equations in matrix form $\mathbf{y} = A\mathbf{x}$; show that A is

$$A = \frac{1}{2h} \begin{bmatrix} 0 & 1 & 0 & 0 & ... & -1 \\ -1 & 0 & 1 & 0 & ... & 0 \\ ... & ... & ... & ... & ... & ... \\ ... & ... & ... & ... & ... & ... \\ 0 & 0 & ... & -1 & 0 & 1 \\ 1 & 0 & ... & 0 & -1 & 0 \end{bmatrix}.$$

This matrix is an example of a *Toeplitz* matrix, i.e., one that has constants only on each diagonal.

13. Do Project A in Section 2.12.

14. Consider the set of linear equations given by the formula

$$\frac{x_{i+1} - 2x_i + x_{i-1}}{h^2} + a\frac{x_{i+1} - x_{i-1}}{2h} + x_i = f(t_i),$$

where h and a are given, the index i ranges from 1 to 6, and $x_0 = x_7 = 0$.

(a) Write down this system as $A\mathbf{x} = \mathbf{b}$.

(b) Is A symmetric? Is it Skew-symmetric?

(c) Find D, L and U to write $A = D + L + U$ with D a diagonal matrix, L lower triangular, and U upper triangular.

2.4 Linear Spaces and Subspaces

Lines and planes of geometry are the fundamental objects in geometry for building and approximating other structures such as curves and surfaces. Lines and planes are examples of linear spaces and subspaces in three dimensions, which we develop here in the general setting of arbitrary dimensions.

We denote by R^n the collection of all n-tuples, i.e.,

$$R^n = \{(a_1, a_2, ..., a_n)|\, a_i \in R\}$$

where R is the set of real numbers. We note that we can add elements in R^n and multiply elements by scalars in the natural way:

$$\begin{aligned}(a_1, a_2, ..., a_n) + (b_1, b_2, ..., b_n) &= (a_1 + b_1, a_2 + b_2, ..., a_n + b_n),\\ c\,(a_1, a_2, ..., a_n) &= (c\,a_1, c\,a_2, ..., c\,a_n).\end{aligned}$$

Note that under the addition and scalar multiplication, the resulting n-tuples still belong to R^n. In this sense, we say that R^n is *closed* under the operations of addition and scalar multiplication and refer to R^n as the *n-dimensional Euclidean space*. Geometrically, R^2 is equivalent to the usual plane and R^3 to the three-dimensional physical space. The complex version of R^n, denoted by C^n, where

$$C = \{z|\, z = a + bi, a, b \in R\},$$

with $i = \sqrt{-1}$, is defined in the same way as R^n and is equipped with the natural operations of addition and scalar multiplication based on these operations with complex numbers.

Similar to R^n, we define E^n, the n-dimensional space of vectors, as the collection of vectors having n components

$$E^n = \{\langle a_1, a_2, ..., a_n\rangle \,|\, a_i \in R\}.$$

E^n is endowed with the vector addition and scalar multiplications defined earlier. There is, of course, the natural connection between R^n and E^n in that the n-tuple $A = (a_1, a_2, ..., a_n)$ in R^n can be viewed as the endpoint of the vector $\mathbf{A} = \langle a_1, a_2, ..., a_n\rangle$ in E^n, which begins at the origin of the coordinate system and ends up at the point A.

By a *linear subspace* (or *subspace* for short) of either R^n or E^n we mean a subset of these spaces that remains closed under the two operations of addition and scalar multiplications. For example, consider the space R^3 and the subspace

$$R_1 = \{(0, 0, a) \,|\, a \in R\}.$$

To see that R_1 is closed under addition and scalar multiplication, consider two arbitrary elements of R_1, $A_1 = (0, 0, a_1)$ and $A_2 = (0, 0, a_2)$. Their sum is $(0, 0, a_1) + (0, 0, a_2)$, which equals $(0, 0, a_1 + a_2)$ and is therefore another element of R_1. Similarly, with α an arbitrary scalar in R and $(0, 0, a)$ an arbitrary element of R_1, we note that $\alpha(0, 0, a)$ equals $(0, 0, \alpha a)$, again belonging to R_1. Thus, R_1 is closed under addition and scalar multiplication so it forms a subspace of R^3. Geometrically, this subspace is the z-axis of a typical rectangular coordinate system set up for the space R^3.

By contrast the set R_2 defined by

$$R_2 = \{(1, 0, a) \,|\, a \in R\}$$

is not a subspace of R^3 because this set is not closed under either addition or scalar multiplication: consider a typical element from this set, $(1, 0, a)$. With $\alpha \neq 1$, an otherwise arbitrary scalar, we have

$$\alpha(1, 0, a) = (\alpha, 0, \alpha a)$$

which does not belong to R_2 since $\alpha \neq 1$.

The subspace defined by R_1 is an example of a one-dimensional subspace of R^3, a straight line passing through the origin (we will give a precise definition of dimension shortly). In general any straight line passing through the origin forms a subspace of R^3. These subspaces constitute the only one-dimensional subspaces of R^3.

In addition to one-dimensional subspaces, R^3 also has two-dimensional subspaces. A typical one is R_3 defined by

$$R_3 = \{(a, b, 0) |\, a, b, \in R\}.$$

Geometrically R^3 is a plane passing through the origin. In fact any plane passing through the origin forms a two-dimensional subspace of R^3. As we will see later, because the definition of R_3 contains two free parameters, a and b, R_3 is a two-dimensional subspace of R^3. We will often refer to such parameters as the *degrees of freedom* of the subspace.

All in all R^3 has four types of subspaces, the one- and two-dimensional subspaces we have discussed already, the empty subspace (one that contains no element from R^3, and thus forms a zero-dimensional subspace), and R^3 itself.

Subspaces of E^n are generated in the same way as in R^n, by allowing a few parameters to vary. For example,

$$E_1 = \{\langle 0, 0, a\rangle |\, a \in R\}$$

and

$$E_2 = \{\langle a, a, 0\rangle |\, a \in R\}$$

each defines a one-dimensional subspace of E^3, while

$$E_3 = \{\langle a, b, a + b\rangle |\, a, b \in R\}$$

is a two-dimensional subspace of E^3.

We denote by $M_{m\times n}$ the set of all m by n matrices. This set is a linear space under the usual matrix addition and scalar multiplication of matrices. Subspaces in $M_{m\times n}$ come about in the same way that they are generated for R^n or E^n; all members of the subspace are $m \times n$ matrices with a fixed set their entries set to zero and the rest are allowed to be arbitrary. For example, consider $M_{2\times 2}$, the set of all 2 by 2 matrices. The set

$$M_1 = \{\begin{bmatrix} a & 0 \\ 0 & 0 \end{bmatrix} |\, a \in R\}$$

is a one-dimensional subspace of $M_{2\times 2}$, which the reader can verify easily by showing that this set is closed under matrix addition and scalar multiplication. Other one-dimensional subspaces of $M_{2\times 2}$ are constructed by placing the a in M_1 in the other three slots in the 2 by 2 matrix. Similarly, two-dimensional and three dimensional subspaces of $M_{2\times 2}$ are generated by placing two or three parameters in the various entry positions in the general 2 by 2 template.

We have taken some effort in this section to give examples of spaces and their subspaces of the kind that play fundamental roles in mathematical physics. We will see additional examples of linear spaces of special significance when we study methods to approximate functions and their applications to solving differential equations numerically. We will also see applications of subspaces when we study eigenvalues and eigenvectors of matrices, which constitute one of the most important tools in applied mathematics.

So far we have introduced the concepts of space and subspace in the context of vectors and matrices. These concepts, however, apply equally naturally to other structures in mathematics. For example, consider the set P_n

$$P_n = \{a_0 + a_1 x + a_2 x^2 + ... + a_n x^n |\, a_0, a_1, ..., a_n \in R\}, \tag{2.11}$$

which is the set of all polynomials of degree n, a positive integer, with real coefficients. If we impose the natural addition of polynomials and scalar multiplication of a polynomial by a scalar $\alpha \in R$ on P_n, then this set is closed under both addition and scalar multiplication, and hence forms a linear space. The similarity between P_n and R^{n+1} or E^{n+1} should be clear. We can associate to each typical element $a_0 + a_1 x + a_2 x^2 + ... + a_n x^{n+1}$ of P_n the point $(a_0, a_1, ...a_n)$ in R^{n+1} or the vector $\langle a_0, a_1, ..., a_n\rangle$ in E^{n+1}. In this precise sense, we can think of P_n to be equivalent to R^{n+1} and E^{n+1}, and when we discover properties of R^{n+1} or E^{n+1} we can ask if those same properties have analogs for P_n and vice versa.

A second important set of functions that can be viewed as a linear space is the set of trigonometric functions

$$T_n = \{a_0 + \sum_{i=1}^{n} a_i \cos\frac{i\pi x}{L} + \sum_{i=1}^{n} b_i \sin\frac{i\pi x}{L} |\, a_i, b_i \in R\}, \tag{2.12}$$

where L is a fixed real number and n a fixed positive integer. It is easy to see that the sum of two elements of T_n, and the scalar product of an element of T_n again belong to T_n, so T_n is a linear apace. Since there are $2n+1$ arbitrary coefficients in T_n (namely, the a_i's and b_i's), then T_n is equivalent to the $2n+1$-dimensional Euclidean space R^{2n+1} or E^{2n+1}. It is also relatively easy to see that sets such as

$$\{a_0 + \sum_{i=1}^{n} a_i \cos\frac{i\pi x}{L} |\, a_i \in R\} \qquad \text{and} \qquad \{\sum_{i=1}^{n} b_i \sin\frac{i\pi x}{L} |\, b_i \in R\}$$

are subspaces of T_n.

Problems 2.4

1. Verify whether the following sets form subspaces of R^3 or E^3. In case of a subspace, give the geometric interpretation:

(a) $\{(0, a, 0)|\, a \in R\}$
(b) $\{(a, a, 0)|\, a \in R\}$
(c) $\{(a, 1, 1)|\, a \in R\}$
(d) $\{\langle a, a, a\rangle|\, a \in R\}$
(e) $\{(a, b, 0)|\, a, b \in R\}$
(f) $\{\langle a, b, a+b\rangle|\, a, b \in R\}$
(g) $\{a, b, c)|\, a, b, c \in R\}$.

2. Write down all subspaces of $M_{2\times 2}$.

3. Show that $\{\begin{bmatrix} 0 & 0 & a \\ 0 & 0 & 0 \\ 0 & a & 0 \end{bmatrix} |\, a \in R\}$ is a subspace of $M_{3\times 3}$.

4. Find all subspaces of $M_{2\times 3}$.

5. Find a formula for the number of subspaces of $M_{n\times n}$.

6. Is the set $O_5 = \{a_1x + a_3x^3 + a_5x^5|\, a_i \in R\}$ a subset of P_n when $n \geq 5$?

7. Let A be an $n \times n$ matrix. the pair $(\lambda, \mathbf{x})$ is called an eigenvalue-eigenvector pair for A if

$$A\mathbf{x} = \lambda\mathbf{x}$$

with $\mathbf{x}$ having at least one non-zero entry (note that $\mathbf{x} = \mathbf{0}$ satisfies $A\mathbf{x} = \lambda\mathbf{x}$ trivially).

(a) Show that if $(\lambda, \mathbf{x})$ is an eigenvalue-eigenvector pair, so is $(\lambda, \alpha\mathbf{x})$ for any nonzero $\alpha \in R$.
(b) Show that if $\mathbf{x}_1$ and $\mathbf{x}_2$ are two eigenvectors associated with the same eigenvalue λ, then $\mathbf{x}_1 + \mathbf{x}_2$ is also an eigenvector associated with λ.
(c) Let $(\lambda_1, \mathbf{x}_1)$ and $(\lambda_2, \mathbf{x}_2)$, with $\lambda_1 \neq \lambda_2$, be two eigenvalue-eigenvector pairs of $A\mathbf{x} = \lambda\mathbf{x}$. Is it true that $\mathbf{x}_1 + \mathbf{x}_2$ is an eigenvector? If so, what is the eigenvalue?

2.5 Determinant and Inverse of Matrices

One of the key ideas we have introduced so far is related to how matrix algebra is used to write a system of algebraic equations compactly and what the consequence of this approach is in obtaining the solution to such a system. We now elaborate more on this point and discuss the theory of determinants of matrices and its implication in providing computational tools that lead to solutions of systems of algebraic equations.

In the previous sections several special matrices were introduced including the identity matrix. Recall that an $n \times n$ identity matrix, which we generally denote by I_n or by I, is a matrix with ones on the diagonal and zeros elsewhere; and let's recall that in MATLAB this matrix is accessed by entering

```
eye(n)
```

The identity matrix has the special property that it leaves a matrix A unchanged under multiplication, that is, $AI = IA = A$. In that sense this matrix acts as the multiplicative unity for matrices, similar to the role number one plays for the set of real numbers. For this reason, I also plays a significant role in the definition of A^{-1}, the inverse of A. We say a matrix B is an *inverse* of A if

$$AB = I. \tag{2.13}$$

It turns out that a matrix B that satisfies (2.13) commutes with A, that is $AB = BA$. Additionally, it turns out that B, when it exists, is the unique matrix that satisfies (2.13). The last two properties prompt us to denote by A^{-1} the (multiplicative) inverse to A so given a square matrix A, A^{-1}, when it exists, is the unique matrix that satisfies

$$AA^{-1} = A^{-1}A = I. \tag{2.14}$$

Not every square matrix A has an inverse, just as not every real number has a multiplicative inverse. But unlike scalars, where 0 is the only number that does not have an inverse, there are infinitely many matrices that do not have inverses. We illustrate this point for the class of 2 by 2 matrices. Let A be defined by

$$A = \begin{bmatrix} a_{11} & a_{12} \\ a_{21} & a_{22} \end{bmatrix}$$

where the entries are arbitrary real or complex numbers. Let B be a candidate for the inverse of A and write

$$B = \begin{bmatrix} b_{11} & b_{12} \\ b_{21} & b_{22} \end{bmatrix}.$$

Since $AB = I$, and keeping in mind that we want to compute the entries of B in terms of A's entries, we group the four equations we obtain from this relation as follows:

$$\begin{cases} a_{11}b_{11} + a_{12}b_{21} & = 1, \\ a_{21}b_{11} + a_{22}b_{21} & = 0, \end{cases} \quad \text{and} \quad \begin{cases} a_{11}b_{12} + a_{12}b_{22} & = 0, \\ a_{21}b_{12} + a_{22}b_{22} & = 1, \end{cases} \tag{2.15}$$

The above systems are simultaneous equations in the unknowns (b_{11}, b_{21}) and (b_{12}, b_{22}), respectively. Simple manipulations lead to

$$b_{11} = \frac{a_{22}}{D}, \quad b_{12} = -\frac{a_{12}}{D}, \quad b_{21} = -\frac{a_{21}}{D}, \quad b_{22} = \frac{a_{11}}{D}, \tag{2.16}$$

where D, called the *determinant* of the matrix A, is

$$D = a_{11}a_{22} - a_{12}a_{21}. \tag{2.17}$$

Hence,

$$A^{-1} = B = \frac{1}{a_{11}a_{22} - a_{12}a_{21}} \begin{bmatrix} a_{22} & -a_{12} \\ -a_{21} & a_{11} \end{bmatrix}. \tag{2.18}$$

Clearly if the determinant of the matrix A is zero, the formulas in (2.18) are not valid and A will not have an inverse. Such matrices are called *singular* and will play a significant role when we discuss eigenvalues and eigenvectors. By contrast *nonsingular*, or *invertible* matrices, are those with nonzero determinants, will have unique inverses, which are in turn used to determine the unique solution to the system of algebraic equations

$$A\mathbf{x} = \mathbf{b}. \tag{2.19}$$

To see this, multiply both sides of the above equation by A^{-1} to get

$$A^{-1}(A\mathbf{x}) = A^{-1}\mathbf{b}.$$

Since $A^{-1}A = I$, the left side reduces to $\mathbf{x}$ and we end up with

$$\mathbf{x} = A^{-1}\mathbf{b} \tag{2.20}$$

as the unique solution to (2.19). What we have illustrated is important enough that we state it as a theorem.

Theorem 2.5.1 (Existence and Uniqueness of Solutions)
Consider a system of linear algebraic equations in the form (2.19). Then (2.19) has the unique solution (2.20) if and only if A is nonsingular.

We have only illustrated this theorem in the context of 2 by 2 matrices. It turns out that its statement and conclusion are valid for $n \times n$ matrices, a claim that we can easily verify once we generalize the concept of determinant to these matrices. Before proceeding to that generalization, we note in passing that the set of 2×2 singular matrices, matrices with zero determinants, is neither finite (any matrix $\begin{bmatrix} a & a \\ a & a \end{bmatrix}$ is singular), nor is this set a linear subspace of $M_{2\times 2}$. For example, the two matrices

$$\begin{bmatrix} 1 & 1 \\ 1 & 1 \end{bmatrix}, \quad \begin{bmatrix} 1 & 2 \\ 2 & 4 \end{bmatrix}$$

are singular but their sum $\begin{bmatrix} 2 & 3 \\ 3 & 5 \end{bmatrix}$ is nonsingular. Similarly, the set of nonsingular matrices also does not form a linear space since, starting with a nonsingular matrix A, its additive inverse $B = -A$ is also nonsingular, yet their sum $A + B$ is the zero matrix, which is singular.

The computation of the determinant of an n by n matrix A is based on an iterative process, where each step of the process reduces the computation to evaluating the determinant of $i \times i$, $2 \leq i < n$, submatrices, each of which is constructed from A. As it turns out the algorithm we present here does not have a unique starting point — it is up to the user to decide which row or column of A is selected to start the algorithm to obtain the underlying submatrices. It is, however, the case that the determinant of a matrix is a unique scalar whose determination is independent of the starting strategy.

Given a_{ij}, the (i,j)-th entry of A, we define A_{ij} as the $(n-1) \times (n-1)$ submatrix constructed from A by eliminating its i-th row and j-th column. We define $\det(A)$ by

$$\det(A) = \sum_{j=1}^{n} (-1)^{i+j} a_{ij} \det(A_{ij}). \tag{2.21}$$

The formula in (2.21) is repeated until one reduces each $\det(A_{ij})$ to the computation of the determinant of a 2×2 matrix, whose formula we arrived at in (2.17). Although this algorithm is somewhat tedious, its iterative nature allows us to reduce the study of determinants of any matrix to smaller sized matrices induced by the original matrix.

Expression (2.21) is computed using the i-th row of A. A similar formula gives the same value when the computation is carried out in terms of a column of A instead:

$$\det(A) = \sum_{i=1}^{n} (-1)^{i+j} a_{ij} \det A_{ij}. \tag{2.22}$$

We summarize the determinant algorithm as follows:

Algorithm 2.5.1 (Computing the Determinant of an $n \times n$ matrix A by Row Evaluation):

1. (Step 1): *Select any of the rows of A. Label it the i-th row.*

2. (Step 2): *Construct the submatrices A_{ij}, $j = 1, ..., n$. Compute the determinant of each A_{ij}. Label it D_{ij}.*

   This is the recursive (iterative) step of the algorithm because computing the determinant of an $(n-1) \times (n-1)$ matrix may require repeating this algorithm until each A_{ij} becomes a 2×2 matrix.

3. (Step 3): *Compute the sum $\sum_{j=1}^{n} (-1)^{i+j} a_{ij} D_{ij}$, which is the determinant of A.*

We illustrate this algorithm for the 3 by 3 matrix

$$A = \begin{bmatrix} 1 & -2 & 3 \\ 0 & 1 & -2 \\ 3 & 2 & 1 \end{bmatrix}. \tag{2.23}$$

We compute the determinant of this matrix using its first row, hence, $i = 1$ in Step 1 of Algorithm 2.5.1. To accomplish Step 2, we need to compute D_{11}, D_{12} and D_{13}. They are

$$D_{11} = \det\left(\begin{bmatrix} 1 & -2 \\ 2 & 1 \end{bmatrix}\right) = 5, \qquad D_{12} = \det\left(\begin{bmatrix} 0 & -2 \\ 3 & 1 \end{bmatrix}\right) = 6,$$

and

$$D_{13} = \det\left(\begin{bmatrix} 0 & 1 \\ 3 & 2 \end{bmatrix}\right) = -3.$$

Step 3 leads to

$$\det(A) = \sum_{j=1}^{3} (-1)^{1+j} a_{1j} D_{1j} = a_{11}D_{11} - a_{12}D_{12} + a_{13}D_{13} =$$

$$1 \times 5 - (-2) \times 6 + 3 \times (-3) = 8.$$

This result can be checked in MATLAB by executing the following lines:

```
a=[1 -2 3; 0 1 -2;3 2 1];
det(a)
```

As pointed out earlier, the above algorithm may be implemented using columns of A, which is as follows:

Algorithm 2.5.2 (Computing the Determinant of an $n \times n$ matrix A by Column Evaluation):

1. (Step 1): *Select any column of* A. *Label it the* j*-th column.*

2. (Step 2): *Construct the submatrices* A_{ij}, $i = 1, ..., n$. *Compute the determinant of each* A_{ij}. *Label it* D_{ij}.

3. (Step 3): *Compute the sum* $\sum_{i=1}^{n} (-1)^{i+j} a_{ij} D_{ij}$, *which is the determinant of* A.

To illustrate this algorithm, instead of using the first row of A in (2.23), we use its third column, hence, $j = 3$ in Step 1 of Algorithm 2.5.2. Next we compute D_{i3}, $i = 1, 2$ and 3:

$$D_{13} = \det\left(\begin{bmatrix} 0 & 1 \\ 3 & 2 \end{bmatrix}\right) = -3, \qquad D_{23} = \det\left(\begin{bmatrix} 1 & -2 \\ 3 & 2 \end{bmatrix}\right) = 8,$$

and

$$D_{33} = \det\left(\begin{bmatrix} 1 & -2 \\ 0 & 1 \end{bmatrix}\right) = 1.$$

Step 3 leads to

$$\det(A) = \sum_{i=1}^{3} (-1)^{i+3} a_{i3} \det(A_{i3}) = a_{13} D_{13} - a_{23} D_{23} + a_{33} D_{33} =$$

$$3 \times (-3) - (-2) \times 8 + 1 \times 1 = 8.$$

In later sections we will describe other algorithms for computing the determinant of a matrix. The recursive algorithm we have described so far is a reasonable method to use when dealing with relatively small matrices, but it is quite inefficient when dealing with large matrices, principally because it requires on the order of n^3 algebraic operations (additions and multiplications) to determine the determinant of an $n \times n$ matrix. This algorithm, however, lends itself to deducing a few theoretical results about matrices, which we address next:

1. It follows directly from Step 3 of Algorithms 2.5.1 and 2.5.2 that if a row or a column of a matrix A consists of zero entries, then the det $(A) = 0$.

2. If two rows or two columns of a matrix are identical, then that matrix has zero determinant, which is easily arrived at for 2×2 matrices, and hence for any $n \times n$ matrix, by appealing to the recursive character of the two algorithms.

3. If a row of a matrix is a linear combination of two other rows of the same matrix, then the determinant of that matrix vanishes. This statement is readily proved for 2×2 matrices, and hence can be extended to all $n \times n$ matrices by induction.

Some of the exercises at the end of this section address these issues.

We end this section by briefly reviewing, at least in the context of simple examples, what we should expect for a solution to $A\mathbf{x} = \mathbf{b}$ when A is singular. Consider the following system of two equations

$$\begin{cases} ax + by &= e \\ ax + by &= f. \end{cases} \tag{2.24}$$

Note that the determinant of A is zero in this case, so we don't expect to be able to determine a unique solution. Geometrically, each equation in (2.24) is a straight line in the xy-plane. Since both equations have the same slope $-\frac{a}{b}$, the two lines are parallel. If the system in (2.24) is to have a solution, it will reside at the intersection of these lines. Two parallel lines intersect only if they coincide, i.e., only if $e = f$ in (2.24). Thus, evidently we may encounter two scenarios; if $e \neq f$, then (2.24) has no solutions. Alternatively, if $e = f$, then (2.24) has infinitely many solutions because now the two equations are identical, $ax + by = e$, in which case we are dealing with one equation in the two unknowns x and y. Hence, assuming $b \neq 0$, we solve y for x to get $y = \frac{1}{b}(e - ax)$, with x taking on any arbitrary real value. So, by letting $x = k$, with k any real number, any ordered pair of the form

$$\langle x, y\rangle = \langle k, \frac{1}{b}(e - ak)\rangle$$

is a solution of (2.24). We will elaborate on this point in a later section when we discuss the algorithm known as *Gaussian Elimination*, where we are able to generalize this discussion to systems of linear equations of any size. We summarize these observations in the following theorem:

Theorem 2.5.2 (Case of Singular Matrix of Coefficients)
Consider a system of linear algebraic equations in the form (2.19). Suppose the matrix A in (2.19) is singular. Then (2.19) either has no solutions or infinitely many solutions.

Problems 2.5

1. Complete the calculations that led to the formulas in (2.15).

2. Compute the determinant and inverse (if it exists) of the following matrices. Check the results in MATLAB: Here a and b are arbitrary real numbers.

$$(i) \quad \begin{bmatrix} 2 & 2 \\ 1 & -1 \end{bmatrix}, \qquad (ii) \quad \begin{bmatrix} 2 & -2 \\ -2 & 2 \end{bmatrix}, \qquad (iii) \begin{bmatrix} a & b \\ -b & a \end{bmatrix},$$

3. Compute the determinant of the following matrices. Check the results in MATLAB: Here a is an arbitrary real number.

$$\text{(i)} \begin{bmatrix} 2 & 2 & 0 \\ 0 & 1 & -1 \\ 0 & 0 & 1 \end{bmatrix}, \qquad \text{(ii)} \begin{bmatrix} a & a & a \\ -a & a & 0 \\ 0 & a & a \end{bmatrix},$$

$$\text{(iii)} \begin{bmatrix} 1 & 1/2 & 1/3 \\ 1/2 & 1/3 & 1/4 \\ 1/3 & 1/4 & 1/5 \end{bmatrix}.$$

4. Solve the following systems of equations and check the answers in MATLAB when appropriate:

   (a) $\begin{cases} 2x - y &= 1, \\ x + y &= 3, \end{cases}$

   (b) $\begin{cases} \alpha x - \beta y &= 1, \\ \alpha x + \beta y &= 3, \end{cases}$. Is there a solution to this problem for all values of $\alpha, \beta \in R$?

   (c) $\begin{cases} x - y &= 1, \\ 2x - 2y &= 3. \end{cases}$ How many solutions does this problem have?

   (d) $\begin{cases} x - y &= 1, \\ 2x - 2y &= 2. \end{cases}$ How many solutions does this problem have?

5. Solve the following systems of equations and check the answers in MATLAB when appropriate:

   (a) $\begin{cases} x + y + z &= 0, \\ 2x - y &= 1, \\ x - 2y + 3z &= -7. \end{cases}$

   (b) $\begin{cases} y + z &= 2, \\ x - y &= 0, \\ x + 3z &= 4. \end{cases}$

   (c) $\begin{cases} \alpha x + y + z &= 0, \\ x + \alpha y + z &= 0, \\ x + y + \alpha z &= 1. \end{cases}$ what happens when $\alpha = -2$? when $\alpha = 1$?

6. This problem concerns several properties of determinants.

   (a) Consider the two matrices $A = \begin{bmatrix} a_{11} & a_{12} \\ a_{21} & a_{22} \end{bmatrix}$ and $B =$

$\begin{bmatrix} a_{11} + \alpha a_{21} & a_{12} + \alpha a_{22} \\ a_{21} & a_{22} \end{bmatrix}$. Note that B is obtained from A by multiplying the second row of A by α and adding to its first row. Show that $\det A = \det B$. Show the same conclusion holds if B is obtained from A by replacing A's second row by adding α times its first row to the second row.

(b) Let A be an arbitrary 3×3 matrix. Let B be obtained from A by replacing its i-th row with the sum of the i-th and α times the j-th row, $i \neq j$. Then $\det A = \det B$.

(c) Show the above property holds for any $n \times n$ matrix.

(d) Let A and B two arbitrary 2×2 matrices. Show that $\det AB = \det A \det B$.

(e) Prove the conclusion of the above problem for arbitrary 3×3 matrices. This result is valid for arbitrary $n \times n$ matrices.

(f) Let A be an arbitrary $n \times n$ matrix. Let B be obtained from A by exchanging the i-th and the j-th rows. Show that $\det A = -\det B$. Prove the same results holds when columns of A are exchanged.

7. Let A and B be two arbitrary 2×2 matrices. Write the $\det(A+B)$ in terms of $\det A$ and $\det B$, if possible.

8. (*Hilbert Matrices*) The ij-th entry of an $n \times n$ Hilbert matrix H_n is

$$a_{ij} = \frac{1}{i+j-1}, \quad i, j = 1, ..., n. \tag{2.25}$$

Use the `Help` command in MATLAB and read about the Hilbert matrix and its properties. The command `hilb(n)` produces the $n \times n$ Hilbert matrix in MATLAB.

(a) Use MATLAB and compute the determinant and inverse of the $i \times i$ Hilbert matrix for $2 \leq i < 5$.

(b) The following program generates several Hilbert matrices, computes their determinant and then plots the graph of these values in the "semilog" scale (consult MATLAB's `Help` to read about `semilogx, semilogy, loglog` commands and their associated concepts):

```
b=[];
for i=1:10
    b=[b det(hilb(i+1))];
end
semilogy(b)
```

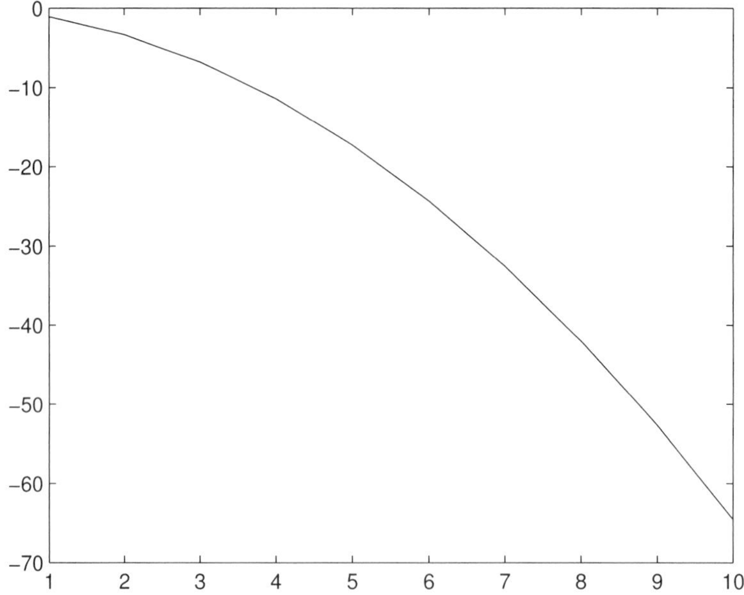

FIGURE 2.1: The graph of the determinant of the Hilbert matrix H_n as a function of n in semilog scale, illustrating the exponential decay of the value of the determinant with increasing n

See Figure 2.1 for the output. Note that although the determinant of H_n is nonzero, it rapidly approaches zero as n gets large. Hilbert matrices are examples of *ill-conditioned* matrices; while these matrices are nonsingular, their determinants are so close to zero that from a computational point of view, they almost behave like singular matrices. In particular, computing the inverse of an ill-conditioned nonsingular matrix turns out to be quite delicate.

9. This problem concerns several properties of inverses of matrices.

   (a) Show that inverse of a matrix A is unique, that is, if $BA = I$ and $CA = I$, then $B = C$.

   (b) Let A and B be two non-singular $n \times n$ matrices. Show that $(AB)^{-1} = B^{-1}A^{-1}$.

   (c) Let A be a nonsingular, but otherwise arbitrary, 2×2 matrix. Show that

$$\det\left(A^{-1}\right) = \frac{1}{\det A}.$$

(d) Let A be an arbitrary nonsingular 2×2 matrix. Let $B = A^{\mathrm{T}}$, the transpose of A. Show that B is also nonsingular and that $B^{-1} = (A^{-1})^{\mathrm{T}}$.

2.6 Computing $\mathbf{A^{-1}}$ Using Co-Factors

As in the case of 2×2 matrices, when the determinant of an $n \times n$ matrix A is nonzero a unique inverse A^{-1} exists. To compute it, we first construct a matrix, called the *cofactor matrix*, denoted by $\mathrm{cof}(A)$. The (i,j)-th entry of this matrix is $(-1)^{i+j}\det(A_{ij})$, i.e.,

$$\mathrm{cof}(A) = \begin{bmatrix} \det(A_{11}) & -\det(A_{12}) & \dots & \det(A_{1n}) \\ -\det(A_{21}) & \det(A_{22}) & \dots & -\det(A_{2n}) \\ \dots & \dots & \dots & \dots \\ \det(A_{n1}) & -\det(A_{n2}) & \dots & \det(A_{nn}) \end{bmatrix}, \tag{2.26}$$

when n is odd, and with appropriate changes with n is even – note that the sign of the entries in (2.26) alternate between plus and minus. The *adjoint* of A is defined as the transpose of $\mathrm{cof}(A)$:

$$\mathrm{Adj}(A) = \mathrm{cof}(A)^T. \tag{2.27}$$

The inverse of A^{-1} is now defined by the following formula:

$$A^{-1} = \frac{1}{\det(A)}\mathrm{adj}(A). \tag{2.28}$$

As an example, consider the 3 by 3 matrix

$$A = \begin{bmatrix} 1 & 1 & 0 \\ 0 & 1 & 1 \\ 1 & 0 & 1 \end{bmatrix}.$$

The cofactor matrix of A is

$$\mathrm{cof}(A) = \begin{bmatrix} 1 & 1 & -1 \\ -1 & 1 & 1 \\ 1 & -1 & 1 \end{bmatrix},$$

whose transpose gives us the adjoint of A:

$$\mathrm{adj}(A) = \begin{bmatrix} 1 & -1 & 1 \\ 1 & 1 & -1 \\ -1 & 1 & 1 \end{bmatrix}.$$

Since $\det(A) = 2$, the inverse of A is

$$A^{-1} = \frac{1}{2}\begin{bmatrix} 1 & -1 & 1 \\ 1 & 1 & -1 \\ -1 & 1 & 1 \end{bmatrix}.$$

The above calculation can easily be verified in MATLAB as follows:

```
A=[1 1 0; 0 1 1; 1 0 1];
inv(A)
```

MATLAB returns

```
ans =

    0.5000   -0.5000    0.5000
    0.5000    0.5000   -0.5000
   -0.5000    0.5000    0.5000
```

We have given no motivation for why the determinant of an $n \times n$ matrix is defined the way it was, why this scalar is uniquely determined, why the algorithm for computing the inverse of a matrix works, and why it leads to the unique inverse matrix. Some of these questions are explored in the exercises, but to gain more in-depth understanding of these concepts, the reader is strongly encouraged to consult the books on linear algebra that are listed at the end of this chapter, most notably the texts by Gilbert Strang, as well as the video lectures available at http://ocw.mit.edu/OcwWeb/Mathematics/18-06Spring-2005/CourseHome/.

Problems 2.6

1. Find the inverse of each matrix, if it exists. Verify answers in MATLAB when possible.

a) $\begin{bmatrix} 0 & 0 & 1 \\ 1 & 0 & 0 \\ 0 & 1 & 0 \end{bmatrix}$, b) $\begin{bmatrix} 1 & 0 & 1 \\ 1 & 1 & 0 \\ 0 & 1 & 1 \end{bmatrix}$, c) $\begin{bmatrix} 0 & 1 & 0 \\ 0 & 0 & 1 \\ -1 & -2 & -3 \end{bmatrix}$,

d) $\begin{bmatrix} a & 0 & a \\ b & 0 & 0 \\ 0 & c & 0 \end{bmatrix}$.

2. Let $A(n)$ stand for the $n \times n$ matrix defined in (2.10).

   (a) Find the inverse of $A(n)$ for $n = 2$ and 3 by hand.

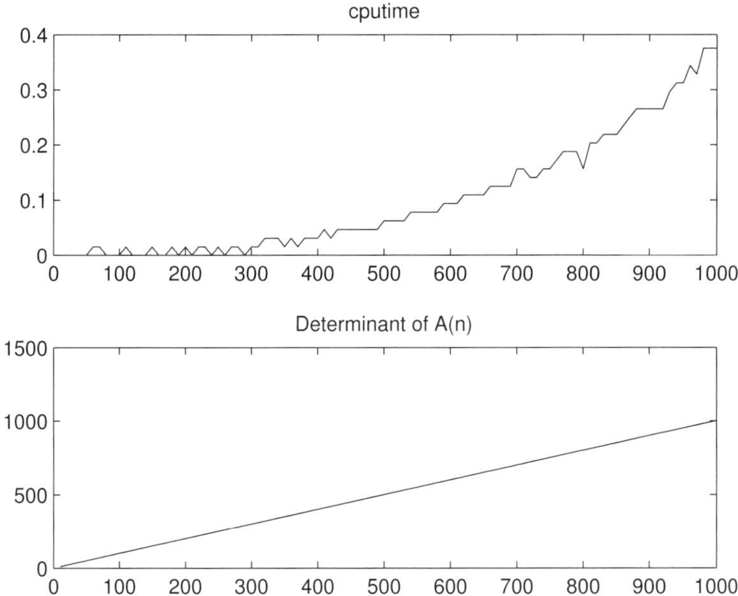

FIGURE 2.2: The first graph shows MATLAB's `cputime` when computing the inverse of $A(n)$, with n ranging between 10 and 1000 at increments of 10. The second graph shows the determinant of $A(n)$ over the same range of n.

(b) Find the inverse of $A(n)$ for $n = 10$ to 1000, at increments of 10, using MATLAB's `inv` function. For each inverse determine the cpu time, perhaps by running a script similar to the one below:

```
n=5;
A=-2*eye(n)+diag(ones(n-1,1),-1)+diag(ones(n-1,1),1);
start = cputime;  %This initializes 'start'
inverse=inv(A);
finish = cputime - start;
```

Figure 2.2 shows the cpu time for computing the inverse of $A(n)$ for n ranging from 10 to 1000 at increments of 10. The second graph shows how the determinant of $A(n)$ behaves as a function of n.

2.7 Linear Independence, Span, Basis, and Dimension

Intimately related to the concept of the inverse of a square matrix is the concept of linear independence of its rows and columns. We develop this concept in conjunction with the concepts of span, basis and the dimension of a linear space. We begin by presenting the concept of linear independence for vectors in E^n and relate it to the structure of general m by n matrices.

Consider a set of vectors $\{\mathbf{a}_1, \mathbf{a}_2, ..., \mathbf{a}_m\}$, each of which belongs to the linear space E^n. We define the *span* of these vectors as the set of all linear combinations of these vectors:

$$S = \{\mathbf{v} | \, \mathbf{v} = \alpha_1\mathbf{a}_1 + \alpha_2\mathbf{a}_2 + ... + \alpha_m\mathbf{a}_m, \text{ with } \alpha_i \in R, \, 1 \leq i \leq m\}. \quad (2.29)$$

As an example, consider the set consisting of a single vector $\mathbf{a}_1 = \langle 1, 0, 0 \rangle$. The span of this vector is

$$S_1 = \{\alpha\mathbf{a}_1\} = \{\alpha\langle 1, 0, 0\rangle\} = \{\langle \alpha, 0, 0\rangle\}, \quad \text{where } \alpha \in R.$$

Geometrically S_1 is equivalent to the traditional x-axis once we assign a set of coordinate axes to R^3. Similarly, consider the set $\{\mathbf{a}_1 = \langle 1, 0, 0\rangle, \mathbf{a}_2 = \langle 0, 0, 1\rangle\}$. The span of this set is

$$S_2 = \{\alpha_1\mathbf{a}_1 + \alpha_2\mathbf{a}_2 | \, \alpha_1, \alpha_2 \in R\}$$

which also has the familiar geometric interpretation of the xz-plane in R^3.

The reader is familiar with the three vectors $\mathbf{i} = \langle 1, 0, 0\rangle, \mathbf{j} = \langle 0, 1, 0\rangle$ and $\mathbf{k} = \langle 0, 0, 1\rangle$ and the role they play for the space E^3: these vectors span the entire space since given any vector $\mathbf{v} = \langle a, b, c\rangle \in E^3$, we can write $\mathbf{v}$ as a linear combination of $\{\mathbf{i}, \mathbf{j}, \mathbf{k}\}$:

$$\langle a, b, c\rangle = a\mathbf{i} + b\mathbf{j} + c\mathbf{k}.$$

Not only do these vectors span E^3, they form the smallest set of such vectors, that is, if we eliminate any of them from the set, we will lose valuable information. In contrast, the set of four vectors $\{\mathbf{i}, \mathbf{j}, \mathbf{i} + \mathbf{j}, \mathbf{k}\}$ still span E^3 because having access to $\mathbf{i} + \mathbf{j}$ has not presented us with any additional capabilities.

The latter set is an example of a linearly dependent set of vectors. In fact we have the following definition:

Definition 2.7.1 (Linear Independence)

We say a set of vectors $\{\mathbf{a}_1, \mathbf{a}_2, ..., \mathbf{a}_m\}$ *forms a linearly independent set if*

$$\alpha_1\mathbf{a}_1 + \alpha_2\mathbf{a}_2 + ... + \alpha_m\mathbf{a}_m = \mathbf{0} \quad \text{if and only if}$$
$$\alpha_1 = \alpha_2 = ... = \alpha_m = 0. \tag{2.30}$$

To test this definition on the set $\{\mathbf{i}, \mathbf{j}, \mathbf{k}\}$, we construct the linear combination in (2.30), namely, $\alpha_1\mathbf{i}+\alpha_2\mathbf{j}+\alpha_3\mathbf{k}$, which equals $\langle\alpha_1, \alpha_2, \alpha_3\rangle$. The only way this vector can equal the zero vector is for all three coefficients, α_1, α_2 and α_3 to vanish. Hence, the set of vectors $\{\mathbf{i}, \mathbf{j}, k\}$ is linearly independent. To see that the set $\{\mathbf{i}, \mathbf{j}, \mathbf{i}+\mathbf{j}, \mathbf{k}\}$ is dependent, it suffices to construct a linear combination in (2.30) for which some of the α's are nonzero. One such linear combination (with $\alpha_1 = \alpha_2 = 1$, $\alpha_3 = -1$ and $\alpha_4 = 0$) is

$$\mathbf{i} + \mathbf{j} - (\mathbf{i} + \mathbf{j})$$

which vanishes. Hence, this set of vectors is linearly dependent.

The next important concept in matrix theory is the concept of a *basis*, which is directly related to the concept of the *dimension* of a linear space.

Definition 2.7.2 (Basis and Dimension of a Linear Space)
A set of vectors forms a basis for a linear space if the set is linearly independent and it spans the space. The number of vectors in a basis constitutes the dimension of that space.

The set $\{\mathbf{i}, \mathbf{j}, \mathbf{k}\}$, for example, forms a basis for E^3. Similarly the set $\{\mathbf{i}+\mathbf{j}, \mathbf{j}+\mathbf{k}, \mathbf{k}+\mathbf{i}\}$ forms a basis for E^3 (see Problem (4) below) because this set is linearly independent and spans E^3.

By *components* of a vector $\mathbf{b} \in E^n$ in a basis $\{\mathbf{a}_1, \mathbf{a}_2, ..., \mathbf{a}_n\}$ we mean the set of scalars α_1, α_2, ..., α_n such that $\mathbf{b} = \alpha_1\mathbf{a}_1 + \alpha_2\mathbf{a}_2 + ... + \alpha_n\mathbf{a}_n$ and write

$$\mathbf{b} = \langle\alpha_1, \alpha_2, ..., \alpha_n\rangle, \tag{2.31}$$

and refer to (2.31) as the *representation* of $\mathbf{b}$ in terms of the basis $\{\mathbf{a}_1, \mathbf{a}_2, ..., \mathbf{a}_n\}$. We will refer to $\{\mathbf{e}_1 = \langle 1, 0, 0, ...0\rangle, \mathbf{e}_2 = \langle 0, 1, 0, ..., 0\rangle, ..., \mathbf{e}_n = \langle 0, 0, 0, ..., 1\rangle$ as the *standard* or the *Cartesian* basis in E^n.

So far we have concentrated almost exclusively on traditional vectors and linear spaces when defining the concepts of linear independence, span, basis, dimension and components. We saw in Section 2.4 that the concept of linear space can be naturally generalized to encompass sets whose elements may be polynomials or trigonometric functions, to give just two examples of linear spaces consisting of functions. We repeat here the two linear spaces P_n and T_n for reference:

$$P_n = \{a_0 + a_1x + a_2x^2 + ... + a_nx^n |\, a_0, a_1, ..., a_n \in R\}, \tag{2.32}$$

where x takes on values in an interval (a, b), and

$$T_n = \{a_0 + \sum_{i=1}^{n} a_i \cos \frac{i\pi x}{L} + \sum_{i=1}^{n} b_i \sin \frac{i\pi x}{L} | \, a_i, b_i \in R\}, \tag{2.33}$$

where $x \in (0, L)$.

As far as the linear space P_n is concerned, the set of "vectors" consisting of the functions

$$f_i(x) = x^i, \quad i = 0, 1, ..., n$$

forms a basis for P_n. To see this, we need to show that this set is linearly independent and that it spans P_n. That the set $\{x^i, i = 0, ..., n\}$ spans P_n follows trivially simply because of the way the set P_n is defined in (2.32). To see that this set is linearly independent, we refer to Definition 2.7.1 and construct the expression

$$c_0 + c_1 x + c_2 x^2 + ... + c_n x^n = 0, \text{ for any } x \in (a, b) \tag{2.34}$$

and now need to show that the only way this expression can hold is if all coefficients c_i vanish. We appeal to the fact that the expression in (2.34) holds for all $x \in (a, b)$, and hence we are allowed to differentiate this expression with respect to x to get

$$c_1 + 2c_2 x + 3c_3 x^2 + ... + nc_n x^{n-1} = 0, \text{ for any } x \in (a, b). \tag{2.35}$$

The expression in (2.35) also holds for all x in (a, b), again allowing us to differentiate it to obtain

$$2c_2 + 6c_3 x + 12c_4 x^2 + ... + n(n-1)c_n x^{n-2} = 0, \text{ for any } x \in (a, b). \tag{2.36}$$

Continuing with this line of argument, we end up differentiating the original expression in (2.34) $(n-1)$ times, and reach the conclusion that

$$n! \, c_n = 0, \tag{2.37}$$

which results in $c_n = 0$. Hence the expression in (2.34) reduces to

$$c_0 + c_1 x + c_2 x^2 + ... + c_{n-1} x^{n-1} = 0, \text{ for any } x \in (a, b). \tag{2.38}$$

It is easy to see now that the same argument of differentiation when applied to (2.38) results in $c_{n-1} = 0$; repeating this argument as many times as necessary eventually leads to the conclusion that all coefficients $c_i = 0$. Hence the set $\{x^i, i = 0, 1, ..., n\}$ is linearly independent. We summarize this discussion in the following theorem:

Theorem 2.7.1 (Linear Space P_n)
The set of polynomials $\{x^i, i = 0, 1, ..., n\}$ forms a basis for the linear space P_n. This space has dimension $n + 1$.

It follows from this theorem that the set $\{1, x, x^2\}$ forms a basis for P_2, the set of all second-order polynomials, $a + bx + cx^2$. In this sense P_2 is very similar to, and one could go as far as saying it is the same as, E^3 the vector space of all 3-tuples.

There is an alternative approach to proving that the set $\{1, x, x^2, ..., x_n\}$ is linearly independent. Let $\{x_0, x_1, ..., x_n\}$ be $n+1$ arbitrary but distinct values in the interval (a, b). Returning to the definition of linear independence, that the expression in (2.34) must hold for all x in the interval (a, b), this expression must be satisfied when x in (2.34) is replaced by any of the x_i's. We therefore obtain the following linear system of algebraic equations in the variables $\{c_0, c_1, ..., c_n\}$:

$$\begin{cases} c_0 + c_1x_0 + c_2x_0^2 + ... + c_{n-1}x_0^{n-1} + c_nx_0^n & = & 0, \\ c_0 + c_1x_1 + c_2x_1^2 + ... + c_{n-1}x_1^{n-1} + c_nx_1^n & = & 0, \\ c_0 + c_1x_2 + c_2x_2^2 + ... + c_{n-1}x_2^{n-1} + c_nx_2^n & = & 0, \\ \quad ... & = & ... \\ \quad ... & = & ... \\ c_0 + c_1x_n + c_2x_n^2 + ... + c_{n-1}x_n^{n-1} + c_nx_n^n & = & 0. \end{cases} \tag{2.39}$$

This system is equivalent to

$$A\mathbf{c} = \mathbf{0} \tag{2.40}$$

where

$$A = \begin{bmatrix} 1 & x_0 & x_0^2 & ... & x_0^{n-1} & x_0^n \\ 1 & x_1 & x_1^2 & ... & x_1^{n-1} & x_1^n \\ ... & ... & ... & ... & ... & .... \\ ... & ... & ... & ... & ... & .... \\ 1 & x_{n-1} & x_{n-1}^2 & ... & x_{n-1}^{n-1} & x_{n-1}^n \\ 1 & x_n & x_n^2 & ... & x_n^{n-1} & x_n^n \end{bmatrix}, \mathbf{c} = \begin{bmatrix} c_1 \\ c_2 \\ ... \\ ... \\ c_{n-1} \\ c_n \end{bmatrix}. \tag{2.41}$$

Hence if we could show that A is a nonsingular matrix, then the fact the right side of (2.40) is the zero vector will imply that $\mathbf{c}$ is the zero vector, thus establishing the desired result. The matrix A in (2.41) is nonsingular if we can show that its determinant does not vanish. It turns out that

$$\det A = (x_0 - x_1)(x_1 - x_2)...(x_{n-1} - x_n) \tag{2.42}$$

where (2.42) is the product of all possible combinations of the form $x_i - x_j$, $i \neq j$, where x_i and x_j take on all possible choices in the set

$\{x_0, x_1, ..., x_n\}$, exactly $(n+1)!/4$. Since we selected distinct x_i values, the determinant is nonzero and hence A is nonsingular.

A similar argument shows that the trigonometric functions in the set $\{1, \cos\frac{n\pi x}{L}, \sin\frac{n\pi x}{L}\}$, i ranging from 1 to n are linear independent and span T_n. Therefore the linear space T_n has dimension $2n+1$.

Problems 2.7

1. Show that the two vectors $\langle 1, 1\rangle$ and $\langle 1, -1\rangle$ are linearly independent, while the two vectors $\langle 1, 1\rangle$ and $\langle 2, 2\rangle$ are not.

2. Identify geometrically the span of the following set of vectors:

   (a) $S_1 = \{\langle 1, 0, 0\rangle\}$.

   (b) $S_2 = \{\langle 1, 0, 0\rangle, \langle 0, 1, 0\rangle\}$.

   (c) $S_3 = \{\langle 1, 0, 0\rangle, \langle 0, 0, 1\rangle\}$.

   (d) $S_4 = \{\langle 1, 1, 0\rangle, \langle 0, 1, 1\rangle\}$.

3. Show that set $\{\mathbf{a}, \mathbf{b}\}$, where $\mathbf{a} = \mathbf{i} + \mathbf{j}$ and $\mathbf{b} = \mathbf{i} - \mathbf{j}$, forms a basis for E^2. Consider the vector $\mathbf{c} \in E^2$ defined as $\mathbf{c} = \langle 2, -1\rangle$ in the standard basis (that is $\mathbf{c} = 2\mathbf{i} - \mathbf{j}$). Find the components of $\mathbf{c}$ in terms $\mathbf{a}$ and $\mathbf{b}$.

4. Show that the set of vectors $\{\mathbf{i}+\mathbf{j}, \mathbf{j}+\mathbf{k}, \mathbf{k}+\mathbf{i}\}$ forms a basis for E^3.

5. Consider the vector $\langle 1, 1, 1\rangle$ in the standard basis in E^3. Find the components of this vector in the basis defined in Problem 4.

6. By applying the definition of linear independence directly, show that the following sets of polynomials are linearly independent:

   (a) $\{1, x, x^2\}$

   (b) $\{1-x, 1+x, x^2\}$

   (c) $\{1, 3x-2, 2x^2+x-1\}$

7. Show that the trigonometric functions in the set $\{1, \cos\frac{n\pi x}{L}, \sin\frac{n\pi x}{L}\}$, i ranging from 1 to n are linearly independent in the interval $(0, L)$ and that they span T_n defined in (2.33). What is the dimension of T_n?

2.8 Linear Transformations

Up to this point we have presented matrices as a collection of arrays, and have defined various algebraic operations for them. We now view matrices from a slightly different point of view, namely as representations of certain functions which we will refer to as linear transformations, which operate between vector spaces.

Definition 2.8.1 (Linear Transformations)
A function T, with domain E^m and range E^n, is called a linear transformation if it satisfies the following the two properties:

1. *T is linear under vector addition, that is*

$$T(\mathbf{a} + \mathbf{b}) = T(\mathbf{a}) + (\mathbf{b})$$

for any vectors $\mathbf{a}$ *and* $\mathbf{b}$ *in* E^m.

2. *T is linear under scalar multiplication:*

$$T(\alpha\mathbf{a}) = \alpha T(\mathbf{a})$$

for any scalar α and any vector $\mathbf{a}$ *in* E^m.

Consider the example T, with domain E^3 and range in E^2, defined by

$$T(x\mathbf{i} + y\mathbf{j} + z\mathbf{k}) = (2x - z)\mathbf{i} + (y + z)\mathbf{j}. \tag{2.43}$$

In the usual component notation, we have

$$T(\langle x, y, z\rangle)) = \langle 2x - z, y + z\rangle. \tag{2.44}$$

Under the action of T, a vector such as $\langle 1, -1, 1\rangle$ in E^3 is mapped to the vector $\langle 1, 0\rangle$ in E^2.

The relation between linear transformations and matrices can be seen relatively easily from the above example: Each entry in the vector $\langle 2x - z, y + z\rangle$ on the right side of (2.44) can be viewed as the dot product of $\mathbf{x} = \langle x, y, z\rangle$ with a specific vector:

$$2x - z = \langle 2, 0, -1\rangle \cdot \langle x, y, z\rangle \quad \text{and} \quad y + z = \langle 0, 1, 1\rangle \cdot \langle x, y, z\rangle.$$

Hence, we can view $T(\mathbf{x})$ as

$$A\mathbf{x}$$

where A is the 2×3 matrix

$$A = \begin{bmatrix} 2 & 0 & -1 \\ 0 & 1 & 1 \end{bmatrix}, \tag{2.45}$$

and $\mathbf{x}$ the column vector $\begin{bmatrix} x & y & z \end{bmatrix}^{\mathrm{T}}$. In this viewpoint, every linear transformation T with domain and range in E^n and E^m, respectively, can be represented by an $m \times n$ matrix A. The entry a_{ij}, the ij-th entry of A, is obtained by computing the vector to which the i-th basis vector in E^n, the domain of T, is mapped. For example, the linear transformation T in the above example maps $\mathbf{i} = \langle 1, 0, 0 \rangle$ to the vector $\langle 2, 0 \rangle$ in E^2, which we note is the first column of A in (2.45). Similarly, the vector $\mathbf{j} = \langle 0, 1, 0 \rangle$ is mapped to $\langle 0, 1 \rangle$, which happens to be the second column of A, and so on. We deduce therefore that

the i-th column of the matrix representation of a linear transformation is the vector to which the i-th basis vector of the domain is mapped.

One of the consequences of representing a transformation T by a matrix A is that this matrix cannot be unique since it depends crucially on what bases we select for the domain and range. To appreciate this point, let's consider a different representation of the linear transformation T in the above example, by representing vectors

Problems 2.8

1. Let T be the transformation from E^3 to E^3 that transforms each vector $\mathbf{x}$ to $\lambda\mathbf{x}$, where λ is a fixed constant. Write down the matrix representation of T.

2. Let T be the transformation from E^2 to E^2 that takes $\mathbf{x}$ to $\mathbf{y}$, where $\mathbf{y}$ is the reflected image of $\mathbf{x}$ through the y-axis. Write down the matrix representation of T.

3. Let T be the transformation from E^2 to E^2 that takes $\mathbf{x}$ to $\mathbf{y}$, where $\mathbf{y}$ is the image of $\mathbf{x}$ after it has been rotated by 90 degrees in the counterclockwise direction. Write down the matrix representation of T.

4. Let T be the transformation from E^2 to E^2 that takes $\mathbf{x}$ to $\mathbf{y}$, where $\mathbf{y}$ is the image of $\mathbf{x}$ after it has been rotated by θ degrees in the counterclockwise direction. Write down the matrix representation of T.

2.9 Row Reduction and Gaussian Elimination

In Section 2.5 we discussed the system of algebraic equations $A\mathbf{x} = \mathbf{b}$ and its solution $\mathbf{x} = A^{-1}\mathbf{b}$ when A is invertible. It turns out that the process we described in the section on inverse of matrices is expensive numerically, especially when the size of this matrix is large. In this section we introduce an alternative method that is considerably more efficient and computationally economical.

From our experience with manipulating equations in a system of algebraic equations

$$\begin{cases} a_{11}x_1 + a_{12}x_2 + \ldots + a_{1n}x_n & = b_1, \\ a_{21}x_1 + a_{22}x_2 + \ldots + a_{2n}x_n & = b_2, \\ \ldots\ldots\ldots\ldots\ldots\ldots\ldots\ldots\ldots\ldots & = \ldots, \\ \ldots\ldots\ldots\ldots\ldots\ldots\ldots\ldots\ldots\ldots & = \ldots, \\ a_{n1}x_1 + a_{n2}x_2 + \ldots + a_{nn}x_n & = b_n, \end{cases} \tag{2.46}$$

we know that the following operations do not alter the solution to (2.46):

1. Exchanging two equations,

2. Multiplying an equation by a nonzero number,

3. Multiplying an equation by a number and adding it to another equation.

We now replace (2.46) with the augmented $n \times n+1$ matrix

$$\left[\begin{array}{ccccc|c} a_{11} & a_{12} & \cdots & \cdots & a_{1n} & b_1 \\ a_{21} & a_{22} & \cdots & \cdots & a_{2n} & b_2 \\ \cdots & \cdots & \cdots & \cdots & \cdots & \cdots \\ \cdots & \cdots & \cdots & \cdots & \cdots & \cdots \\ a_{n1} & a_{n2} & \cdots & \cdots & a_{nn} & b_n \end{array}\right] \tag{2.47}$$

where the $|$ is used to separate the matrix of coefficients $[a_{ij}]$ from the input vector $\mathbf{b}$. Our goal is to extend the above equation operations to the rows of (2.47) and *row reduce* this matrix to an *upper triangular* form

$$\left[\begin{array}{ccccc|c} a'_{11} & a'_{12} & \cdots & \cdots & a'_{1n} & b'_1 \\ 0 & a'_{22} & \cdots & \cdots & a'_{2n} & b'_2 \\ \cdots & \cdots & \cdots & \cdots & \cdots & \cdots \\ \cdots & \cdots & \cdots & \cdots & \cdots & \cdots \\ 0 & 0 & \cdots & \cdots & a'_{nn} & b'_n \end{array}\right]. \tag{2.48}$$

The latter system is equivalent to a system of algebraic equations

$$\left\{\begin{array}{ccccc} a'_{11}x_1+ & \cdots & a'_{1,n-1}x_{n-1} & a'_{1n}x_n & = b'_1, \\ & a'_{22}x_2+ & \cdots & a'_{2n}x_n & = b'_2, \\ & & \cdots & \cdots & = \dots, \\ & & \cdots & \cdots & = \dots, \\ & & a'_{n-1,n-1}x_{n-1}+ & a'_{n-1,n}x_n & = b'_{n-1}, \\ & & & a'_{nn}x_n & = b'_n, \end{array}\right. \tag{2.49}$$

Assuming $a'_{nn} \neq 0$ the last equation yields $x_n = b'_n/a'_{nn}$. Having the value of x_n in hand, we consider the next to last equation in (2.49) and solve for x_{n-1}:

$$a'_{n-1,n-1}x_{n-1} + a'_{n-1,n}x_n = b'_{n-1} \quad \Longrightarrow$$

$$x_{n-1} = \frac{1}{a'_{n-1,n-1}}(b'_{n-1} - \frac{a'_{n-1,n}}{a'_{nn}}b'_n).$$

Continuing with this **backward substitution** idea, we arrive at formulas for x_{n-2}, x_{n-3}, all the way back to x_1.

To see this method used in a concrete example, consider the system of linear equations

$$\left\{\begin{array}{rcl} x+y+z+w & = & 6, \\ 2x-3y+z+w & = & -1, \\ z-3x+w & = & 0, \\ 2x-y & = & w \end{array}\right. \tag{2.50}$$

which is equivalent to

$$\left[\begin{array}{rrrr|r} 1 & 1 & 1 & 1 & 6 \\ 2 & -3 & 1 & 1 & -1 \\ -3 & 0 & 1 & 1 & 0 \\ 2 & -1 & 0 & -1 & 0 \end{array}\right]. \tag{2.51}$$

Beginning with first row, we use row operations and convert every entry below a_{11} to zero: we multiply the first row by -2 and add it the second row, multiply the first row by 3 and add to the third row, and multiply the first row by -2 and add to the fourth row to get the equivalent matrix

$$\left[\begin{array}{rrrr|r} 1 & 1 & 1 & 1 & 6 \\ 0 & -5 & -1 & -1 & -13 \\ 0 & 3 & 4 & 4 & 18 \\ 0 & -3 & -2 & -3 & -12 \end{array}\right].$$

Next we move to the second row of the above matrix and make the entries below the $(2,2)$ entry zero: Multiply the second row by $\frac{3}{5}$ and add to the third row, and multiply the second row by $-\frac{3}{5}$ and add to the fourth row to get the equivalent matrix

$$\left[\begin{array}{cccc|c} 1 & 1 & 1 & 1 & 6 \\ 0 & -5 & -1 & -1 & -13 \\ 0 & 0 & \frac{17}{5} & \frac{17}{5} & \frac{51}{5} \\ 0 & 0 & -\frac{2}{5} & -\frac{12}{5} & -\frac{21}{5} \end{array}\right].$$

The latter system is equivalent to

$$\left[\begin{array}{cccc|c} 1 & 1 & 1 & 1 & 6 \\ 0 & -5 & -1 & -1 & -13 \\ 0 & 0 & 1 & 1 & 3 \\ 0 & 0 & -7 & -12 & -21 \end{array}\right].$$

To complete the row reduction phase of this method, we multiply the third row of the latter matrix by 7 and add to the fourth row to get

$$\left[\begin{array}{cccc|c} 1 & 1 & 1 & 1 & 6 \\ 0 & -5 & -1 & -1 & -13 \\ 0 & 0 & 1 & 1 & 3 \\ 0 & 0 & 0 & -5 & 0 \end{array}\right]. \tag{2.52}$$

We are now ready to apply the backward substitution part of the algorithm: The last row is equivalent to $-5w = 0$, or $w = 0$. The third row is equivalent to $z+w = 3$ which results in $z = 3$. Similarly, we arrive at $y = 2$ and $x = 1$.

The concepts we have introduced here, row reduction and backward substitution, provide us with a method, called *Gaussian Elimination*, for solving for solutions of $A\mathbf{x} = \mathbf{b}$. Gaussian Elimination has two significant properties. First, the steps involved in implementing this method typically require a reasonable number of additions and multiplications, especially when a large number of the entries of the matrix of coefficients A vanish, a scenario that happens often in the discretization of partial differential equations we encounter in the later chapters. The second property of this method is its iterative character, which results in relatively simple computer codes. Most modern software packages have internal commands that implement a version of this algorithm. MATLAB's `rref` command, for example, takes a matrix such as (2.51) and returns the *row reduced echelon form* (hence "rref") form of (2.51). This form, which is equivalent to (2.51), goes one step beyond (2.52) by converting the $n \times n$ block of (2.52) to the identity matrix. The extra steps that lead to such a form require that in each step of our Gaussian Elimination, we not only make the entries below the diagonal zero, but apply the same strategy to the entries above the diagonal. The following commands in MATLAB lead to the row reduced echelon form of (2.51):

```
A=[1 1 1 1 6;2 -3 1 1 -1;-3 0 1 1 0;2 -1 0 -1 0];
B=rref(A)
```

MATLAB returns

```
B =

     1     0     0     0     1
     0     1     0     0     2
     0     0     1     0     3
     0     0     0     1     0
```

Note that the solution to (2.51), namely, $x = 1$, $y = 2$, $z = 3$, and $w = 0$ appears in the last column of the `B` above.

The success of Gaussian Elimination (and by extension **rref**) is intimately related to invertibility of A. When A is invertible, where $A\mathbf{x} = \mathbf{b}$ has a unique solution, the row reduction methodology works extremely well. This technique, however, is also robust when it is applied to systems of algebraic equations where A is not invertible, or when A is not a square matrix. To better understand the variety of cases that appear in the general case, we introduce the concept of the *rank* of a matrix, which will enable us to write down a few results in compact form.

Given an $m \times n$ matrix A, let R be the span of the m row vectors of A. The rank of A, denoted by $\rho(A)$, is the dimension of R. Put in a slightly different way, $\rho(A)$ is the maximal number of linearly independent rows of A. In general, $\rho(A) \leq m$. When $\rho(A) = m$, we say A has *full rank*. When A is a square $m \times m$ matrix, it turns out that A is invertible if and only if $\rho(A) = m$.

When $\rho(A) < m$ the system of equations $A\mathbf{x} = \mathbf{b}$ may have infinitely many solutions or no solutions at all. For example, the 2×2 system

$$x + y = 1, \quad 2x + 2y = 2. \tag{2.53}$$

has infinitely many solutions ($x = a$, $y = 1 - a$, for any real-valued a) since clearly the two equations in (2.53) are the same. The matrix $A = \begin{bmatrix} 1 & 1 \\ 2 & 2 \end{bmatrix}$ is singular and has rank $\rho(A) = 1$.

The key idea that characterizes the multiplicity of solutions in (2.53), in particular, as well as for general systems, turns to be the relation between $\rho(A)$ and the rank of the *augmented* matrix $[A|\mathbf{b}]$. For (2.53) the augmented matrix is $\begin{bmatrix} 1 & 1 & 1 \\ 2 & 2 & 2 \end{bmatrix}$, and the ranks of both A and its augmented matrix are one, an important indicator of when a system with a deficient rank ends up having infinitely many solutions. In general, *when* $\rho(A) = \rho([A|\mathbf{b}])$, *the system* $A\mathbf{x} = \mathbf{b}$ *will have infinitely many solutions.* When, on the other hand, $\rho(A) < \rho([A, \mathbf{b}])$, the system will not have any solutions at all. An evidence of this feature can be seen in the system

$$x + y = 1, \quad x + y = 0. \tag{2.54}$$

where now the augmented matrix is $\begin{bmatrix} 1 & 1 & 1 \\ 2 & 2 & 0 \end{bmatrix}$, which has rank 2 while $\rho(A)$ is still one, and of course (2.54) does not have any solutions. The above discussion is significant enough that we state is as a theorem:

Theorem 2.9.1
Consider the system $A\mathbf{x} = \mathbf{b}$. *If*

1. A *is* $m \times m$ *and nonsingular, then* $\mathbf{x} = A^{-1}\mathbf{b}$ *is its unique solution.*

2. A *is* $m \times n$ *and* $\rho(A) = \rho([A|\mathbf{b}])$, *then the system has infinitely many solutions.*

3. A *is* $m \times n$ *and* $\rho(A) < \rho([A|\mathbf{b}])$, *then the system has no solutions.*

One of the interesting, and somewhat unintuitive, properties of the rank of a matrix is that $\rho(A) = \rho(A^{\mathrm{T}})$, i.e., the columns and rows of a matrix span the same linear space. In follows then that the rank of an $m \times n$ matrix A satisfies

$$\rho(A) = \min(m, n).$$

A word of caution about using `rref` with systems that do not have full rank. Note that when we apply `rref` to (2.53)

```
rref([1 1 1; 1 1 0])
```

MATLAB returns

```
ans =

     1     1     0
     0     0     1
```

In the absence of any other warning statements, the user must conclude from this output that (2.48) has no solutions since the second row of the output states $0 = 1$. While when we apply `rref` to (2.52)

```
rref([1 1 1;2 2 2])
```

we receive

```
ans =

     1     1     1
     0     0     0
```

The second row is equivalent to $0 = 0$, which indicates that the system is equivalent to a single equation ($x+y=1$), which of course has infinitely many solutions.

Problems 2.9

1. Consider the following systems of equations. Use Gaussian elimination to determine if each system has a unique solution, infinitely many solutions or no solution. Use `rref` to verify the results.

   (a) $\begin{cases} 2x+y &= 1, \\ x-3y &= -2. \end{cases}$

   (b) $\begin{cases} x+y+z &= 1, \\ x-3y+z &= 2, \\ z-3x &= 0 \end{cases}$

   (c) $\begin{cases} x+y+z &= 1, \\ x-y+3z &= 0, \\ 2x+4z &= 1. \end{cases}$

2. Consider the system $\begin{cases} ax+y &= b, \\ x+ay &= 0, \end{cases}$ where a and b are real numbers.

   (a) Determine all values of a for which this system has a unique solution, i.e., when the system is non-singular.

   (b) For each value of a for which this system is singular, determine all values of b for which i) the system has infinitely many solutions, and ii) no solution.

2.10 Eigenvalues and Eigenvectors

Eigenvalues and eigenvectors of matrices constitute some of the most important tools that matrix algebra offers in analyzing physical problems. By definition, given an $n \times n$ matrix A an eigenvalue-eigenvector pair $(\lambda, \mathbf{e})$, with $\mathbf{e} \neq \mathbf{0}$, of A satisfies

$$A\mathbf{e} = \lambda\mathbf{e}. \tag{2.55}$$

Geometrically, one can think of an eigenvector as a vector whose direction remains invariant under the action of A. To gain an appreciation of this point, consider the matrix A given by

$$A = \begin{bmatrix} 1 & 2 \\ 2 & 1 \end{bmatrix}. \tag{2.56}$$

Let $\mathbf{y} = A\mathbf{x}$. If $\mathbf{x} = \langle x_1, x_2 \rangle$, then $\mathbf{y}$, image of $\mathbf{x}$ under the action of A, is

$$\mathbf{y} = \langle x_1 + 2x_2, 2x_1 + x_2 \rangle.$$

Typically what A does to each vector $\mathbf{x}$ is to stretch (or compress) it as well as rotate it through an angle. In fact, if we consider the set of all $\mathbf{x}$'s having magnitude one, which trace a circle of radius 1 in the (x_1, x_2) plane, the images $A\mathbf{x}$ trace an ellipse (see Figure 2.3). To see this, note that the circle can be parameterized by

$$\mathbf{x} = \langle x_1, x_2 \rangle = \langle \cos t, \sin t \rangle, \qquad t \in [0, 2\pi).$$

The image of this curve is the ellipse

$$\mathbf{y} = \langle y_1, y_2 \rangle = \langle x_1 + 2x_2, 2x_1 + x_2 \rangle =$$

$$\langle \cos t + 2\sin t, 2\cos t + \sin t \rangle, \qquad t \in [0, 2\pi).$$

The following program in MATLAB graphs the above circle and ellipse:

```
clf
t=0:2*pi/100:2*pi;
ezplot('cos(t)','sin(t)');
hold on
ezplot('cos(t)+2*sin(t)','2*cos(t)+sin(t)');
```

Most of the vectors $\mathbf{x}$ and their images $A\mathbf{x}$ have different directions. For example, the vector $\langle 1, 0 \rangle$ has been mapped to $\langle 2, 1 \rangle$ by A. By contrast, two vectors, corresponding to the bisectors of the quadrants don't seem to have changed directions — the vector $\mathbf{e}_1 = \langle 1, 1 \rangle$ is mapped to $\langle 3, 3 \rangle$, and $\mathbf{e}_2 = \langle -1, 1 \rangle$ is mapped to $\langle -1, 1 \rangle$. To be more precise,

$$A \begin{bmatrix} 1 \\ 1 \end{bmatrix} = 3 \begin{bmatrix} 1 \\ 1 \end{bmatrix}, \qquad \text{and} \qquad A \begin{bmatrix} -1 \\ 1 \end{bmatrix} = - \begin{bmatrix} -1 \\ 1 \end{bmatrix}.$$

Therefore $(3, \mathbf{e}_1)$ and $(-1, \mathbf{e}_2)$ are the two eigenvalue-eigenvector pairs for A.

To compute eigenvalues and eigenvectors of any matrix we appeal to the definition (2.55). Note that $A\mathbf{e} = \lambda\mathbf{e}$ can be written as

$$(A - \lambda I)\mathbf{e} = \mathbf{0}, \tag{2.57}$$

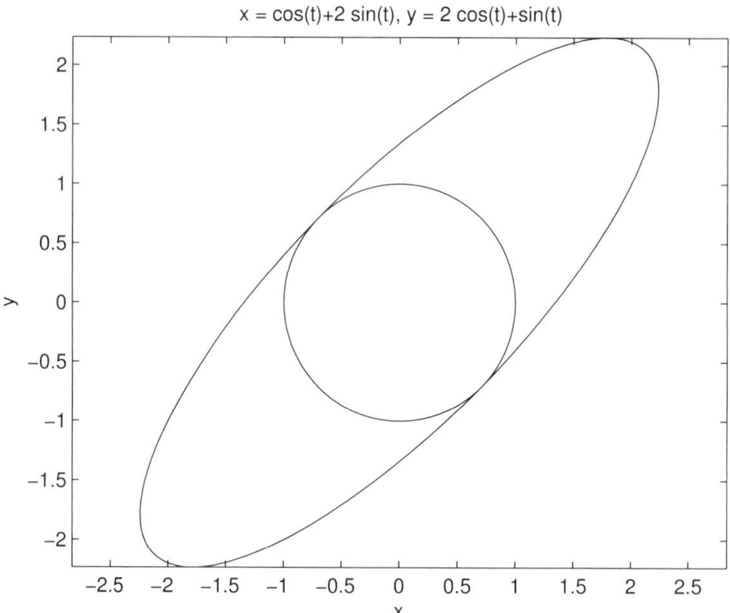

FIGURE 2.3: The circle and its image under the action of the matrix defined in (2.56).

where I is the $n \times n$ identity matrix. The above system is a special one in that if it has a unique solution $\mathbf{e}$, then $\mathbf{e}$ must be $\mathbf{0}$ since $\mathbf{0}$ trivially satisfies (2.57). On the other hand, if $\mathbf{e}$ is to be an eigenvector, it must be a nonzero vector. We conclude then that the solution to (2.57) cannot be unique. After appealing to Theorem 2.5.1, we see that $A - \lambda I$ must be singular, or equivalently, λ must be a solution to

$$\det(A - \lambda I) = 0. \tag{2.58}$$

The quantity $\det(A - \lambda I)$ is an n-th order polynomial in λ and since every such polynomial has n roots, counting multiplicity, any $n \times n$ matrix has n eigenvalues. These roots may be complex even if A has all real coefficients.

Returning to the earlier 2×2 example, $A = \begin{bmatrix} 1 & 2 \\ 2 & 1 \end{bmatrix}$, we determine A's eigenvalues by setting

$$\det(A - \lambda I) = \det(\begin{bmatrix} 1-\lambda & 2 \\ 2 & 1-\lambda \end{bmatrix}) = \lambda^2 - 2\lambda - 3$$

to zero. So $\lambda_1 = 3$ and $\lambda_2 = 1$ are the two eigenvalues of A, as was observed earlier.

We determine eigenvectors of a matrix by solving (2.57) using Gaussian elimination. For example, to obtain the eigenvector associated with $\lambda = 3$ in the above example we construct $(A - \lambda I)\mathbf{e} = \mathbf{0}$ as the augmented matrix

$$\left[\begin{array}{cc|c} -2 & 2 & 0 \\ 2 & -2 & 0 \end{array}\right]. \tag{2.59}$$

Note that $\rho(A) = \rho([A|\mathbf{0}]) = 1$, so by Theorem 2.9.1 the above system will have infinitely many solutions. In fact, it should be clear that the two rows in (2.59) correspond to identical algebraic equations. Concentrating on the first row, we see that, with $\mathbf{e}_1 = \langle x_1, x_2 \rangle$, x_1 and x_2 satisfy the linear equation $-2x_1 + 2x_2 = 0$. If we let $x_1 = c$, an arbitrary constant, then $x_2 = c$. Thus

$$\mathbf{e}_1 = c \begin{bmatrix} 1 \\ 1 \end{bmatrix}.$$

In a similar fashion we see that $\mathbf{e}_2$ satisfies $(A + I)\mathbf{e}_2 = \mathbf{0}$ or

$$\left[\begin{array}{cc|c} 2 & 2 & 0 \\ 2 & 2 & 0 \end{array}\right] \tag{2.60}$$

which results in

$$\mathbf{e}_2 = c \begin{bmatrix} 1 \\ -1 \end{bmatrix}.$$

The factor c in the above formulas for the eigenvectors is arbitrary, indicating geometrically that the entire lines defined by the equations $x_1 = x_2$ and $x_1 = -x_2$ remain invariant under the action of A. Often c is selected in such a way to render the eigenvector $\mathbf{e}$ a *unit* vector, i.e., c is chosen so that $||\mathbf{e}|| = 1$.

The MATLAB command `eig` computes eigenvalues and eigenvalues of this matrix.

```
[V,D]=eig([1 2;2 1])
```

returning

```
V =

   -0.7071    0.7071
    0.7071    0.7071


D =
```

```
   -1     0
    0     3
```

Note that columns of `V` are the eigenvectors of A (with $|c| = \frac{1}{\sqrt{2}}$, yielding a unit vector for each eigenvector), and that `D` contains the eigenvalues. The command `eig` is a very powerful command since computing eigenvalues and eigenvectors of a matrix could be quite tedious as some of the problems at the end of this section will demonstrate. To get a glimpse of the power of `eig` we apply this operation to compute the eigenvalues and eigenvectors of the 6×6 Hilbert matrix, namely, $A = [a_{ij}]$ where $a_{ij} = 1/(i + j - 1)$.

```
A=hilb(6)
[V,D]=eig(A)
```

which results in

```
A =


    1.0000    0.5000    0.3333    0.2500    0.2000    0.1667
    0.5000    0.3333    0.2500    0.2000    0.1667    0.1429
    0.3333    0.2500    0.2000    0.1667    0.1429    0.1250
    0.2500    0.2000    0.1667    0.1429    0.1250    0.1111
    0.2000    0.1667    0.1429    0.1250    0.1111    0.1000
    0.1667    0.1429    0.1250    0.1111    0.1000    0.0909


V =


   -0.0012   -0.0111    0.0622    0.2403   -0.6145    0.7487
    0.0356    0.1797   -0.4908   -0.6977    0.2111    0.4407
   -0.2407   -0.6042    0.5355   -0.2314    0.3659    0.3207
    0.6255    0.4436    0.4170    0.1329    0.3947    0.2543
   -0.6898    0.4415   -0.0470    0.3627    0.3882    0.2115
    0.2716   -0.4591   -0.5407    0.5028    0.3707    0.1814


D =


    0.0000         0         0         0         0         0
         0    0.0000         0         0         0         0
         0         0    0.0006         0         0         0
         0         0         0    0.0163         0         0
         0         0         0         0    0.2424         0
         0         0         0         0         0    1.6189
```

Although it may appear from `D` that the first two eigenvalues of A are zero, they are not. To see a more accurate representation of D, enter the following lines:

```
format long
D
```

which shows that the first eigenvalues are in fact $\lambda_1 = 0.00000010827995$ and $\lambda_2 = 0.00001257075712$.

Returning to the definition of an eigenvalue, the determinant of $A - \lambda I$ for a general $n \times n$ matrix will be an n-th order polynomial on λ. By the Fundamental Theorem of Algebra this polynomial will have n roots, $\lambda_1, \lambda_2, \ldots, \lambda_n$, counting multiplicity, which may be complex. Several of the problems below explore the connection between the eigenvalues of A and its entries a_{ij}.

Problems 2.10

1. Find the eigenvalues and eigenvectors of the following matrices. Verify the results in MATLAB when appropriate.

   (a) $\begin{bmatrix} 3 & 1 \\ 1 & 3 \end{bmatrix}$

   (b) $\begin{bmatrix} 3 & 2 \\ 1 & 3 \end{bmatrix}$

   (c) $\begin{bmatrix} 3 & -1 \\ 1 & 3 \end{bmatrix}$

   (d) The 2×2 identity and zero matrices.

   (e) $\begin{bmatrix} 0 & 1 \\ a & b \end{bmatrix}$

   (f) $\begin{bmatrix} a & 1 \\ 1 & a \end{bmatrix}$

   (g) $\begin{bmatrix} a & b \\ b & a \end{bmatrix}$

   (h) $\begin{bmatrix} a & -b \\ -b & a \end{bmatrix}$

2. Find the eigenvalues and eigenvectors of the following matrices. Verify the results in MATLAB when appropriate.

   (a) $\begin{bmatrix} 3 & 1 & 0 \\ 1 & 3 & 0 \\ 0 & 0 & 1 \end{bmatrix}$

(b) $\begin{bmatrix} 1 & 1 & 0 \\ 1 & 0 & 1 \\ 0 & 1 & 1 \end{bmatrix}$

(c) $\begin{bmatrix} 0 & 1 & 0 \\ 0 & 0 & 1 \\ 1 & 2 & 3 \end{bmatrix}$

(d) $\begin{bmatrix} a & 0 & 0 \\ 0 & b & 0 \\ 0 & 0 & c \end{bmatrix}$

3. Let $A = \begin{bmatrix} a & b \\ c & d \end{bmatrix}$. Let λ_1 and λ_2 be its eigenvalues. Show that $\lambda_1 + \lambda_2 = a + d$ and $\lambda_1 \lambda_2 = \det A$.

4. Let $A = \begin{bmatrix} a_{11} & a_{12} & a_{13} \\ a_{21} & a_{22} & a_{23} \\ a_{31} & a_{32} & a_{33} \end{bmatrix}$. Let λ_1, λ_2 and λ_3 be its eigenvalues. Let $\mathrm{tr}(A)$, called the trace of A, be the sum of the entries on A's diagonal. Show that $\lambda_1 + \lambda_2 + \lambda_3 = \mathrm{tr}(A)$ and $\lambda_1 \lambda_2 \lambda_3 = \det A$.

   It turns out the above result holds for general $n \times n$ matrices, that is, if

$$A = \begin{bmatrix} a_{11} & a_{12} & \cdots & \cdots & a_{1n} \\ a_{21} & a_{22} & \cdots & \cdots & a_{2n} \\ \cdots & \cdots & \cdots & \cdots & \cdots \\ \cdots & \cdots & \cdots & \cdots & \cdots \\ a_{n1} & a_{n2} & \cdots & \cdots & a_{nn} \end{bmatrix},$$

   and $\{\lambda_1, \lambda_2, ..., \lambda_n\}$ are its n eigenvalues, counting multiplicity, then $\mathrm{tr}(A) = \sum_{i=0}^{n} \lambda_i$, $\det A = \prod_{i=1}^{n} \lambda_i$.

2.11 Project A: Taylor Polynomials and Series

Recall that one way to define $y' = f'(a)$, the derivative of a function $y = f(x)$ at $x = a$, is by the following limit process:

$$f'(a) = \lim_{h \to 0} \frac{f(a+h) - f(a)}{h}.$$

Equally valid are the expressions

$$f'(a) = \lim_{h \to 0} \frac{f(a) - f(a-h)}{h},$$

and

$$f'(a) = \lim_{h \to 0} \frac{f(a+h) - f(a-h)}{2h}.$$

A good way to convince yourself of the validity of these formulas is to use Taylor Series formula, that

$$f(a+h) = f(a) + f'(a)h + \frac{f''(a)}{2!}h^2 + \frac{f'''(a)}{3!}h^3 + ... = \sum_{n=0}^{\infty} \frac{f^{(n}}{n!}h^n, \quad (2.61)$$

where $0! = 1$ and $f^{(n)}$ is the n-the derivative of f, with $f^0 = f$.

1. Use the Taylor series expression (2.61) to show that

   $$\frac{f(a+h) - f(a)}{h} = f'(a) + a_1 h + a_2 h^2 + a_3 h^3 + ...$$

   How do a_1, a_2 and a_3 depend on f?

2. Find a_1, a_2 and a_3 in each of the following expressions:

   $$\frac{f(a) - f(a-h)}{h} = f'(a) + a_1 h + a_2 h^2 + a_3 h^3 + \ldots,$$

   and

   $$\frac{f(a+h) - f(a-h)}{2h} = f'(a) + a_1 h + a_2 h^2 + a_3 h^3 + \ldots,$$

   and

   $$\frac{f(a+2h) - f(a+h)}{h} = f'(a) + a_1 h + a_2 h^2 + a_3 h^3 + \ldots,$$

3. The above formulas are all two-step approximations for $f'(a)$ since they all require only two function evaluations. The following formula is a multi-step approximations of $f'(a)$. Show that

   $$1/(12h)(f(a-2h) - 8f(a-2h) + 8f(a+h) - f(a+2h)) =$$

   $$f'(a) + a_1 h + a_2 h^2 + +a_3 h^3 + a_4 h^4 + \ldots \quad (2.62)$$

   for appropriate scalars a_i's.

4. Once we have developed formulas to approximate the first derivative of a function, these same formulas can readily be applied to develop appropriate approximations for the higher derivatives of a function.

(a) Since $f''(a) = \lim_{h\to 0} \frac{f'(a+h)-f'(a)}{h}$, show that $f''(a)$ is approximated by

$$\frac{f(a+2h) - 2f(a+h) + f(a)}{h^2}.$$

Apply the Taylor Series formula to this expression to determine the dependence of the first three terms of this expression on h.

(b) Apply the Taylor Series formula to each of the formulas below to show that they approximate $f''(a)$. For each formula determine the dependence on h.

$$\frac{f(a-2h) - 2f(a-h) + f(a)}{h^2}.$$

And

$$\frac{f(a+h) - 2f(a) + f(a-h)}{h^2}.$$

2.12 Project B: A Differentiation Matrix

The book by L. N. Trefethen, referenced in [6], is a monograph on a special numerical method for solving differential equations, the spectral method, and on the effective and optimal use of MATLAB. This project, as well as others in this text, are intended to encourage the reader to consult this book and thus benefit from its important content. In what follows, the reader is asked to read the first chapter of [6] and to derive some of its conclusions.

Given a function $y = f(x)$, its derivative y' can be approximated in many ways, among which is the *centered difference method* where $y'(x)$ is approximated by

$$y'(x) \approx \frac{y(x+h) - y(x-h)}{2h} \tag{2.63}$$

1. Consider the 2π-periodic function $y(x) = e^{\sin x}$, whose derivative is $y'(x) = \cos x\, e^{\sin x}$. Discretize the interval $(0, 2\pi)$ by $\{x_1, x_2, ..., x_{n-1}, x_n\}$ into n equal subintervals of length $h = \frac{2\pi}{n}$, where $x_1 = 0$ and $x_n = 2\pi - h$. Let w_i stand for the approximation of y' at $x = x_i$. Apply the formula on the right side of (2.63) to determine the following formulas for w_1, w_1, ..., w_n, taking advantage of the periodicity of the function $y(x)$ (recall that the function

y is 2π-periodic, so $y_0 = y_{n-1}$ and $y_{n+1} = y_0$):

$$\begin{array}{rcl} w_1 & = & \frac{1}{2h}(y_2 - y_{n-1}), \\ w_2 & = & \frac{1}{2h}(y_3 - y_1), \\ \ldots & = & \ldots \\ \ldots & = & \ldots \\ w_n & = & \frac{1}{2h}(y_1 - y_{n-1}). \end{array} \tag{2.64}$$

2. Let $\mathbf{y} = \langle y_1, y_2, ..., y_n \rangle^{\mathrm{T}}$ and $\mathbf{w} = \langle w_1, w_2, ..., w_n \rangle^{\mathrm{T}}$ and show that (2.64) can be written as $\mathbf{w} = D\mathbf{y}$ where D is

$$D = \frac{1}{2h} \begin{bmatrix} 0 & 1 & 0 & 0 & \ldots & \ldots & 0 & -1 \\ -1 & 0 & 1 & 0 & \ldots & \ldots & 0 & 0 \\ 0 & -1 & 0 & 1 & \ldots & \ldots & 0 & 0 \\ \ldots & \ldots & \ldots & \ldots & \ldots & \ldots & \ldots & \ldots \\ \ldots & \ldots & \ldots & \ldots & \ldots & \ldots & \ldots & \ldots \\ 0 & 0 & 0 & 0 & \ldots & -1 & 0 & 1 \\ 1 & 0 & 0 & 0 & \ldots & \ldots & -1 & 0 \end{bmatrix}. \tag{2.65}$$

Note that this matrix consists of relatively few nonzero entries and is ana example of a *sparse* matrix. We will encounter sparse matrices later in the text and will explore how to use MATLAB's efficient handling of sparse matrices to carry out calculations involving quite large matrices.

3. Write a MATLAB program to plot the graph of the function $\mathbf{w}$ and the exact derivative $\cos x \, e^{\sin x}$ over the interval $(0, 2\pi)$ with $n = 10, 100$ and 1000. See Figure 2.4 for a similar figure. The MATLAB internal function `diag` is the ideal tool for constructing D.

4. Apply the above approach to the following functions:

   (a) $y(x) = \sin x + 0.1 \cos x$,

   (b) $y(x) = \sin x + \sin 10x$,

   (c) $y(x) = e^{\sin x \cos x}$.

5. Apply the formula in (2.62) to compute the derivative $y'(x)$ if $y(x) = \sin x + 2 \sin 3x \cos x$ on the interval $(0, 10\pi)$. Experiment with the discretization of $(0, 10\pi)$ to arrive at a reasonable approximation to y'.

6. The matrix D defined in (2.65) is the operator that allows us to compute the first derivative of a function. It seems reasonable then that D^2 should provide an approximation to y''. Let $\mathbf{z} = D^2\mathbf{y}$ and

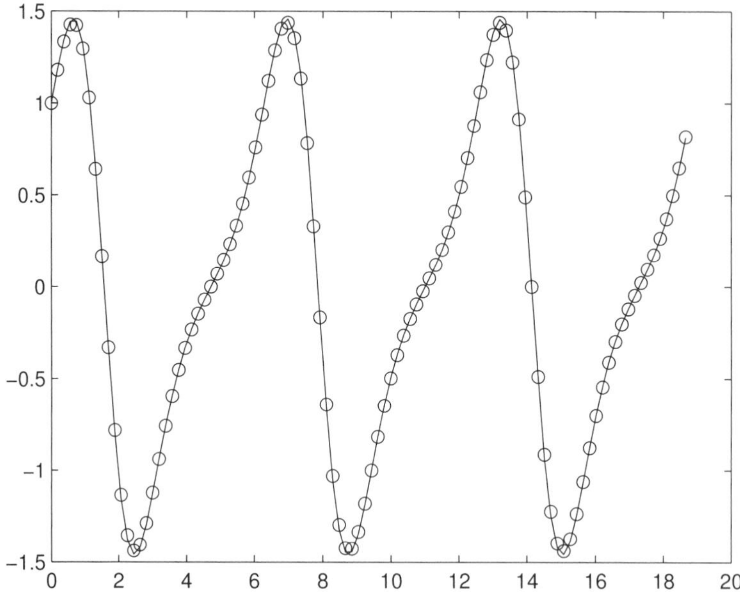

FIGURE 2.4: The exact derivative of the function $y(x) = e^{\sin x}$ and its approximation, using the centered difference method with $n = 100$ in the interval $[0, 6\pi)$.

test this idea on the function $y = e^{\sin x}$ by comparing $\mathbf{z}$ to the second derivative of y. Compare the structure of D^2 to D and comment on the number of nonzero diagonals of the former.

7. Explore further the connection between the operation of matrix multiplication and differentiation by considering A^n, where A is defined in (2.65), as a proxy for $y^{(n)}$, the n-th derivative of y. Write a MATLAB program that computes the first four derivatives of a function such as $y(x) = \sin x + 0.3 \sin 2x - 0.4 \sin 3x$. Of particular importance is to understand the relation between h, the step-size, and the accuracy of the approximation as the order of differentiation increases. See Figure 2.5.

8. Denote the differentiation matrix (2.65) by D_N, acknowledging the fact that this matrix, its size and content, depends on N, the number of subintervals in the discretization. Explore the eigenvalues of D_N as a function of $N = 2^i$ as i ranges from 3 to 12. Do any patterns emerge? For example, is D_N nonsingular for any N? Are there any eigenvalues of D_N with nonzero real part?

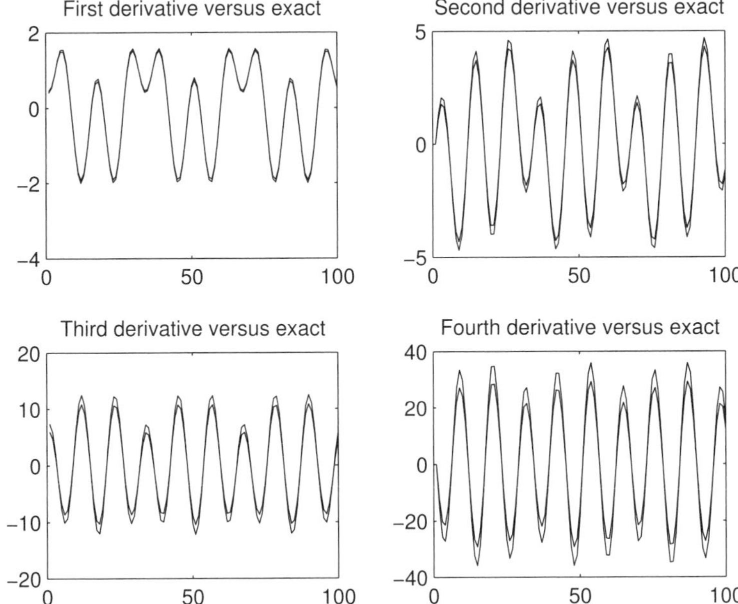

FIGURE 2.5: A comparison of the first four derivatives of the function $y(x) = \sin x + 0.3\sin 2x - 0.4\sin 3x$ and its approximation, using the centered difference method with $n = 100$ in the interval $[0, 6\pi)$.

2.13 Project C: Spectral Method and Matrices

The approach of the previous project allows us to compute a relatively good approximation of the derivative of a function $y(x)$ by considering function evaluations at the two neighboring points immediately to the left and right of $x = a$. We can easily generalize this method to other finite difference approximations of the derivative that we have already discussed, including for instance approximating $y'(a)$ by considering four neighbors of $x = a$, namely, by considering evaluating y at $x = a+h$, $x = a+2h$, $x = a-h$, and $x = a-2h$. It is not difficult to see, and perhaps not surprising, that having access to more data about y in the neighborhood of $x = a$ will lead to more accurate approximation of $y'(a)$.

A natural question arises as to the accuracy of the approximation we would achieve if we used all of the information available to us as far as the domain of y is concerned to approximate the value $y'(a)$ (and equivalently $y''(a)$ or higher derivatives of y at a). This idea is the essence of the *Spectral Method* for generating differentiation matrices. Recalling from Project B that the more nearby points we used to approximate the derivative of y at $x = a$, the more nonzero diagonals appeared in the differentiation matrix A, we can expect that the diagonal entries of the spectral differentiation matrix will almost all be nonzero.

Spectral methods have received considerable amount of attention in the past few decades and several outstanding texts are available that treat this material. We have already alluded to the text by L. Trefethen in Project B. Another text worthy of special note is the one by J. Hesthaven, S. Gottlieb, and D. Gottlieb, (see Reference [7]), from which we have taken the formula (2.66) below.

1. Consider a 2π-periodic function $y(x)$. Discretize the domain $[0, 2\pi)$ as before, with a uniformly distributed set of points at step-size h – Let $x_i = (i-1)h$, $i = 1, ..., n$. Define the matrix A as follows:

$$A_{ij} = \frac{(-1)^{i+j}}{2} \cot\left(\frac{x_i - x_j}{2}\right), \tag{2.66}$$

   and $A_{i,j} = 0$ if $i = j$. Write a MATLAB program to generate A for any n.

2. With $y(x) = e^{\sin x}$, apply the spectral differentiation matrix A defined above to compute an approximation to y', y'' and y''' on the interval $[0, 10\pi)$. Compare the graphs of the approximations to the exact values. See Figure 2.6.

2.14 Concluding Remarks

The discussion in this chapter on matrix algebra is just a brief introduction to this important topic, arguably the most important area of applied mathematics, especially when considering its applications to numerical computations. The interested reader should continue consulting the texts listed below for more complete treatment of linear algebra. In addition to the books referenced there, there are now numerous resources for how one takes advantage of MATLAB's unique capabilities in implementing matrix algebra to solve fundamental problems in applied mathematics. Some of these resources, such as Lloyd Trefethen's text on the Spectral Methods, have already reached the status of a classic text. For others, the reader is strongly encouraged to consult the World Wide

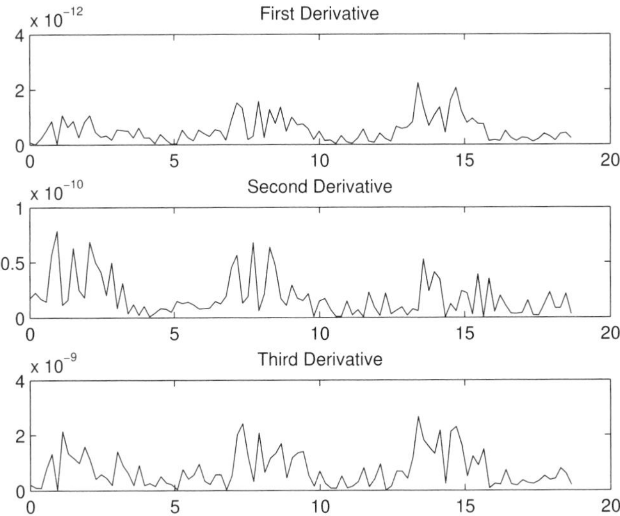

FIGURE 2.6: A comparison between the exact and approximate values of the first three derivatives of the function $y(x) = e^{\sin x}$ in the interval $[0, 6\pi)$ using the Spectral Method with $n = 100$.

Web regularly in search of new materials that become available almost on a daily basis and are often written for beginner or intermediate users of MATLAB.

2.15 References and Further Reading

1. Strang, G., *Linear Algebra and Its Application*, 3rd edition, Brooks/Cole Publishing, 1988.

2. Strang, G., *Introduction to Linear Algebra*, 3rd edition, Wellesley Cambridge Press, 2003.

3. Nakos, G., Joyner, D., *Linear Algebra with Applications*, Brooks/Cole Publishing, 1998.

4. Malek-Madani, R., *Advanced Engineering Mathematics with Mathematica® and MATLAB®*, Addison-Wesley, 1998.

5. Golub, G., Van Loan, C., *Matrix Computations*, The Johns Hopkins University Press, 3rd edition, 1996.

6. Trefethen, L., *Spectral Methods in MATLAB*, Society for Industrial and Applied Mathematics, 2000.

7. Hesthaven, J., Gottlieb, S., Gottlieb, D., *Spectral Methods for Time-Dependent Problems*, Cambridge University Press, 2007.

Chapter 3

Differential and Integral Calculus

In this chapter we develop the essential concepts from differential and integral calculus and discuss the role they play in this text in the context of geophysical fluid dynamics. We will also use this opportunity to hint at the issues we will face when we need to approximate the typical rates of change that appear in the governing equations of motion.

3.1 Derivative

The standard definition of the derivative of f, a function of one variable, at the point $x = a$ is

$$f'(a) = \lim_{h\to 0} \frac{f(a+h) - f(a)}{h}, \tag{3.1}$$

when that limit exists. Alternative ways of defining the same quantity are

$$f'(a) = \lim_{h\to 0} \frac{f(a) - f(a-h)}{h}, \tag{3.2}$$

$$f'(a) = \lim_{h\to 0} \frac{f(a+h) - f(a-h)}{2h}, \tag{3.3}$$

or

$$f'(a) = \lim_{h\to 0} \frac{f(a+2h) + f(a+h) - 2f(a)}{3h}, \tag{3.4}$$

which constitute just a few formulas, out of infinitely many such formulas, that lead to determining $f'(a)$. We use the concept of derivative primarily to relate the rates of growth of various variables in a physical process. In this context it is not significant which of the definitions in (3.1)–(3.4) we use to develop our arguments. This choice becomes quite significant, however, in our second application of the definition of derivative, namely when we need to approximate $f'(a)$ by one of the many "rise-over-run" ratios on the right side of (3.1)–(3.4). In the context of solving differential equations, a subject we will take up in the next chapter, which one of the representations of $f'(a)$ in (3.1)–(3.4) is selected could have a significant impact on the accuracy of the numerical schemes one develops.

Higher order derivatives of f are defined analogously, by applying the formulas in (3.1)–(3.4) to lower order derivatives. For example, $f''(a)$ is determined as

$$f''(a) = \lim_{h \to 0} \frac{f'(a+h) - f'(a)}{h}. \tag{3.5}$$

We may apply any of the formulas in (3.1)–(3.4) to the right side of (3.5) to arrive at formulas for $f''(a)$ that involve evaluation of the function f at a and its neighboring values and not any of its derivatives. For instance

$$f''(a) = \lim_{h \to 0} \frac{f(a+h) - 2f(a) + f(a-h)}{h^2} \tag{3.6}$$

and

$$f''(a) = \lim_{h \to 0} \frac{f(a+2h) - 2f(a+h) + f(a)}{h^2} \tag{3.7}$$

result from applying (3.1) and (3.2) (see the Problem 8 at the end of this section).

Partial derivatives of a function f, when f depends on several independent variables, are defined precisely as laid out in (3.1) and its equivalent forms because the partial derivative of f with respect to one of its independent variables, say x, is the rate of change of f when x is allowed to vary while all other independent variables are kept constant. For simplicity, let us assume f is a function of three variables, denoted by x, y and z, in a domain D, a subset of R^3. Let $P = (a, b, c)$ be a point in the domain at which we are interested in determining f's rate of change in the x direction. This quantity, which we denote by $\frac{\partial f}{\partial x}$ or by f_x, is obtained as follows:

$$\frac{\partial f}{\partial x}|_P = \lim_{h \to 0} \frac{f(a+h, b, c) - f(a, b, c)}{h}, \tag{3.8}$$

if the limit exists. We obtain $\frac{\partial f}{\partial y}$ and $\frac{\partial f}{\partial z}$ in a similar fashion:

$$\frac{\partial f}{\partial y}|_P = \lim_{h \to 0} \frac{f(a, b+h, c) - f(a, b, c)}{h},$$

$$\frac{\partial f}{\partial z}|_P = \lim_{h \to 0} \frac{f(a, b, c+h) - f(a, b, c)}{h}. \tag{3.9}$$

Higher order derivatives of f are obtained by repeated application of the definition of differentiation. So, $f_{xx}|_P$ is obtained by first computing f_x

at an arbitrary point (x, y, z) in the domain and then applying (3.8) to f_x:

$$f_{xx}|_P = \frac{\partial^2 f}{\partial x^2} = \lim_{h\to 0} \frac{f_x(a+h,b,c) - f_x(a,b,c)}{h}. \tag{3.10}$$

Similarly,

$$f_{yx} = \frac{\partial^2 f}{\partial y \partial x}|_P = \lim_{h\to 0} \frac{f_x(a,b+h,c) - f_x(a,b,c)}{h}. \tag{3.11}$$

As the reader may suspect, the order of differentiation in expressions such as f_{xy} and f_{yx} do not matter, at least for the large class of functions f that we encounter in typical applications in this text. The standard texts on Advanced Calculus, several of which are listed at the end of this chapter, devote a substantial amount of effort in developing the right mathematical theorems that ensure the well-posedness of the topics we have discussed; for example, under what conditions on f does the limit in (3.1) exist, so that we can be assured that the function f is *differentiable* at P? And when can we be sure that the order of differentiation in (3.11) is immaterial? While this assertion seems natural and apparently should hold, it is somewhat surprising that there are plenty of counterexamples (see Problem 14 for one such example), although these examples are relatively pathological and may not appear very often in nature.

Problems 3.1

1. Consider the function $f(x) = x^3 + 2x^2 - 1$. Use (3.1) and (3.5) to show that $f'(a) = 3a^2 + 4a$ and $f''(a) = 6a$.

2. Consider the function $f(x) = x^2 + bx + c$, where a, b and c are constants. Use (3.3) and (3.4) to show that $f'(d) = 2d + b$.

3. Use (3.1) to show that nx^{n-1} is the derivative of x^n, where n is a positive integer. Also, show that $f''(a) = n(n-1)a^{n-2}$.

4. Use (3.1) to show that $\cos x$ is the derivative of $\sin x$.

5. Consider the function $f(x) = |x|$. Compute $f'(1)$, $f'(-1)$ and $f'(0)$, if they exist.

6. Consider the function $f(x) = |x|x$. Determine $f'(x)$.

7. Consider the function $f(x) = x \sin \frac{1}{x}$ when $x \neq 0$. Define $f(0) = 0$. Is f continuous at $x = 0$? (Remark: Recall that a function f is continuous at $x = a$ if and only if $\lim_{x\to a} f(x) = f(a)$.) Is f differentiable at $x = 0$?

8. Derive the formulas in (3.6) and (3.7) from (3.1) and (3.2).

9. The *composition* of two functions f and g, denoted by $f \circ g$ is defined by
$$(f \circ g)(x) = f(g(x)) \tag{3.12}$$
as long as the range of g and the domain of f are compatible enough for (3.12) to make sense. The *Chain Rule* of differentiation provides a formula for differentiating $f \circ g$. It states
$$(f \circ g)'(x) = f'(g(x))g'(x). \tag{3.13}$$
Use the Chain Rule to differentiate $\sin 2x$, $\sin(x^2)$, $\sqrt{x^3+1}$ and $\ln \frac{1}{\cos x}$.

10. Consider the function $f(x, y) = x^2 + 3y^2 - xy$. Determine
$$\text{i) } f_x(1,2), \quad \text{ii) } f_y(1,2), \quad \text{iii) } f_{xx}(x,y), \quad \text{iv) } f_{yy}(a,b), \quad \text{v) } \frac{\partial^2 f}{\partial x \partial y}.$$

11. Find all a and b, both constants, so that $f(x, y) = ax^2 + by^2$ satisfies the equation
$$f_{xx} + f_{yy} = 0.$$

12. Find all a and b, both constants, so that $u(x, y) = \sin ax \cos by$ satisfies the equation
$$u_{xx} - u_{yy} = 0.$$

13. Consider the complex-valued function $h(t, x) = e^{i\omega t + kx}$ where $i = \sqrt{-1}$ and ω and k are constants. Determine h_t, h_{tt}, h_{tx} and h_{xx}.

14. Let $f(x, y) = \frac{xy(x^2-y^2)}{x^2+y^2}$ if $(x, y) \neq (0, 0)$ and 0 otherwise. Show that $f_{xy}(0, 0) = -1$ but $f_{yx}(0, 0) = 1$.

3.2 Taylor Polynomial and Series

The Taylor polynomial is one of the main tools in developing approximate formulas to represent functions. This concept is typically applied when one has local information about a function, the information consisting of knowledge of the functional value and several of its derivatives at a single point $x = x_0$ in the domain. Since the first derivative of a function f represents the slope of the tangent line to the graph of f, this information can be used to build a linear approximation to the function, thus obtaining a formula that serves as a reasonable approximation as long as one applies it only near x_0. Repeating this procedure for the various derivatives of f at x_0, one obtains a polynomial approximation to f.

To see this procedure in a concrete setting, consider a function f in a domain (a, b) and a point $x_0 \in (a, b)$. Assume that $f(x_0)$, $f'(x_0)$, $f''(x_0)$, ..., $f^{(n)}(x_0)$ are known. Then the n-th order polynomial

$$P_n(x) = f(x_0) + f'(x_0)(x - x_0) + \frac{f''(x_0)}{2!}(x - x_0)^2 + ... + \frac{f^{(n)}(x_0)}{n!}(x - x_0)^n \tag{3.14}$$

is the *n-th order Taylor Polynomial* approximation of f. As the reader can easily verify, the function f and all of its derivatives of up to order n agree with the corresponding values of P_n. That is

$$f(x_0) = P_n(x_0),\ f'(x_0) = P_n'(x_0),\ f''(x_0) = P_n''(x_0),\ ...,\ f^{(n)}(x_0) = P_n^{(n)}(x_0).$$

A standard theorem from Calculus gives an excellent estimate on the error one incurs when one considers P_n in place of f. This error is proportional to the $n+1$ derivative of f but evaluated at a point ξ, somewhere between x and x_0:

$$f(x) = P_n(x) + \frac{f^{(n+1)}(\xi)}{(n+1)!}(x - x_0)^{n+1}, \quad \xi \in (x_0, x). \tag{3.15}$$

The term $\frac{f^{(n+1)}(\xi)}{(n+1)!}(x - x_0)^{n+1}$ is referred to as the *remainder* and its absolute value in the domain (a, b) provides insight on the amount of error one makes when approximating f by P_n.

As an example, consider the 7-th order Taylor polynomial approximation of the function $f(x) = \sin x$ about $x_0 = 0$. The formula in (3.14) gives

$$P_7(x) = x - \frac{x^3}{3!} + \frac{x^5}{5!} - \frac{x^7}{7!}$$

with the remainder $\frac{\sin\xi}{8!}x^8$, with $\xi \in (0, x)$. Since $|\sin\xi| \leq 1$ the maximum error we sustain by replacing f with P_7 in the interval $(0, a)$ is $\frac{a^8}{8!}$. When $a = 1$, say, this error is less than 0.00002. The following MATLAB program plots f and P_7 on the interval $(0, 2\pi)$, showing how well P_7 approximates f on the interval $(0, 1)$ and beyond, but that the approximation begins to deteriorate when $x > 5$ or so (see Figure 3.1). This figure is obtained by running the following code in MATLAB:

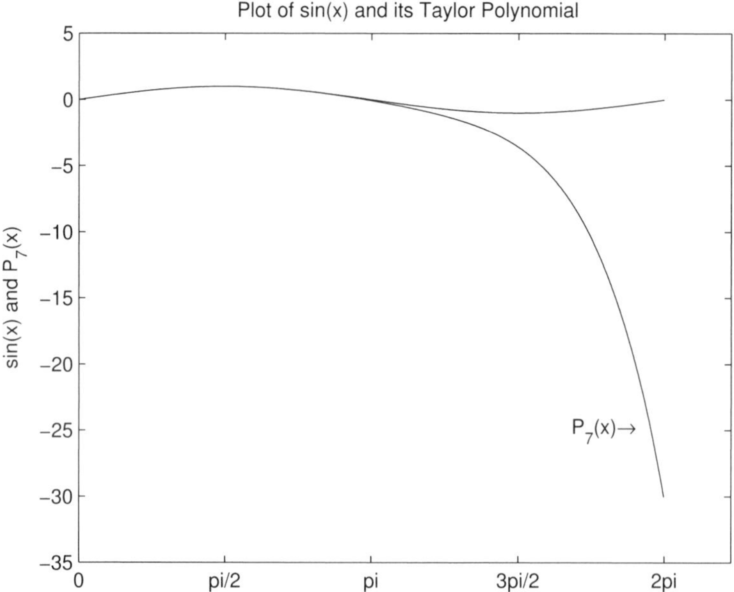

FIGURE 3.1: The function $\sin x$ and its 7-th order Taylor Polynomial approximation.

```
clf
x=0:0.01:2*pi;
plot(x,sin(x));
hold on
z=zeros(size(x));
for i=1:4
     z=z+(-1)^(i-1)*x.^(2*i-1)/factorial(2*i-1);
end
plot(x,z)
set(gca,'XTick',0:pi/2:2*pi)
set(gca,'XTickLabel',{'0','pi/2','pi','3pi/2','2pi'})
ylabel('sin(x) and P_7(x)')
title('Plot of sin(x) and its Taylor Polynomial')
text(2*pi-1,-25,'P_7(x)\rightarrow', ...
            'HorizontalAlignment','left')
```

One of the important applications of the Taylor polynomial is in computing the *truncation error* of *finite difference* schemes, which we will study in detail later, when we approximate the derivative of a function by any of the several right sides in formulas (3.1)–(3.4). For example, formula (3.1) suggests that we replace $f'(a)$ by the ratio

$$\Delta f(a,h) = \frac{f(a+h) - f(a)}{h}. \tag{3.16}$$

The expression in (3.16) converges to $f'(a)$ as h approaches zero but now we can find out the rate of this convergence as a function of h. Define the function g by $g(h) = f(a+h)$ and expand g about $h = 0$ to get

$$g(h) = g(0) + g'(0)h + ... = f(a) + f'(a)h + \frac{f''(a)}{2!}h^2 + \text{h.o.t} \tag{3.17}$$

where h.o.t. stands for higher order terms in h. Returning to the definition of Δf in (3.16), we see that

$$\Delta f(a,h) - f'(a) = \frac{f''(a)}{2!}h + \text{h.o.t.}$$

where now h.o.t. stands for terms in h with powers equal or higher than two. Since the remainder $\Delta f(a,h) - f'(a)$ is proportional to h (and assuming that $f''(a)$ does not vanish), we say the difference scheme given by Δf is *first order*. Some of the problems at the end of this section deal with computing the truncation errors in the other definitions of $f'(a)$ in (3.2)–(3.4), as well as for higher-order differential operators. See also Project A in Chapter 2.

The process of obtaining an n-th order polynomial approximation of a function f can of course be implemented for any n as long as f is sufficiently differentiable. If it turns out that the function f is in fact infinitely many times differentiable, we can push this process to its limit by allowing n approach to infinity and obtain an infinite series representation of f:

$$f(x) = \sum_{n=0}^{\infty} \frac{f^{(n)}(a)}{n!}(x-a)^n. \tag{3.18}$$

This series representation behaves very well for a large class of functions, including the familiar ones such as exponential, trigonometric and rational functions, but one must be considerably more careful when using (3.18), as compared with (3.14), since we now must deal with the prospect of the convergence of the right side of (3.18), a rich subject of analysis treated in several of the books cited in the references. Since we do not make much use of the Taylor series in the applications we encounter in this text, we will not pursue further the discussion of that subject.

Problems 3.2

1. Compute the 3rd, 5th and 7th order Taylor polynomial approximations of the function $f(x) = e^{-x}\sin x$ about $x = 0$ and plot the graphs of each approximation against the original f over the interval $(0, 4)$.

2. Apply the Taylor polynomial method to determine the order and the truncation error of each of the following finite-difference approximations for $f'(a)$:

   (a) $\Delta f(a,h) = \frac{f(a)-f(a-h)}{h}$,

   (b) $\Delta f(a,h) = \frac{f(a+h)-f(a-h)}{2h}$,

   (c) $\Delta f(a,h) = \frac{f(a+2h)+f(a+h)-2f(a)}{3h}$.

3. Apply the Taylor polynomial method to determine the order and the truncation error of the finite-difference approximation

$$\frac{f(a+h) - 2f(a) + f(a+h)}{h^2}$$

   for $f''(a)$.

3.3 Functions of Several Variables and Vector Fields

The typical physical quantities we study are represented mathematically by functions or vector fields that depend on several variables. Pressure, salinity, density and temperature are examples of physical entities that are scalars that depend on several space dimensions and on time, while velocity, acceleration and wind stress are examples of vector quantities that typically vary with space and time. The results and theorems we develop in this section relate rates of change of these quantities and provide information about their local behavior.

We begin by considering f, a function of several variables, and address questions about the various rates of change of f and how they relate to surfaces along which f remains constant. For simplicity let f depend on only two independent variables x and y; the results we discuss readily generalize to higher dimensions. The *gradient* of f, denoted by ∇f, is defined by

$$\nabla f = \langle \frac{\partial f}{\partial x}, \frac{\partial f}{\partial y} \rangle. \tag{3.19}$$

The *directional derivative* of f at the point $P = (a, b)$ in the direction $\mathbf{e}$, a unit vector, is denoted by $\frac{df}{d\mathbf{e}}$ and defined by the relation

$$\frac{df}{d\mathbf{e}} = \nabla f|_P \cdot \mathbf{e}. \tag{3.20}$$

Recall that the dot product of two vectors $\mathbf{a}$ and $\mathbf{b}$ provides information about the angle θ between the two vectors because $\mathbf{a} \cdot \mathbf{b}$ and $\cos\theta$ are related by

$$\mathbf{a} \cdot \mathbf{b} = ||\mathbf{a}||\,||\mathbf{b}|| \cos\theta. \tag{3.21}$$

Returning to (3.20) we note that $|\frac{df}{d\mathbf{e}}| = |\nabla f|_P \cdot \mathbf{e}| = ||\nabla f|_P||\,||\mathbf{e}||\,|\cos\theta|$. Keeping P fixed for the time being, and noting that $||\mathbf{e}|| = 1$, we deduce that the directional derivative of f at P in the direction of $\mathbf{e}$ reaches its maximum when $|\cos\theta| = 1$, or when∇f and $\mathbf{e}$ are parallel, i.e., when

$$\mathbf{e} = \frac{1}{||\nabla f||}\nabla f.$$

with this expression for $\mathbf{e}$, the directional derivative $\frac{df}{\mathbf{e}}$ is $||\nabla f||$ by (3.20). This result is significant enough that we summarize it in a theorem.

Theorem 3.3.1 (Direction of Steepest Ascent)
Let f be a differentiable function of its arguments. Let P be a point in its domain. Then the quantity $\frac{df}{d\mathbf{e}}$ achieves its largest value, which is $||\nabla f||$, when $\mathbf{e} = \frac{1}{||\nabla f||}\nabla f$. Similarly, $\frac{df}{d\mathbf{e}}$ achieves its minimum $-||\nabla f||$ when $\mathbf{e} = -\frac{1}{||\nabla f||}\nabla f$, which we refer to as the direction of steepest descent.

The *contour level* (or the *level set*) of a function f is the set of all points in the domain of f at which f remains constant. Assuming, again without loss of generality, that f depends on only two independent variables, the contour level of f is defined by

$$\{(x, y)|\, f(x, y) = \text{const.}\}. \tag{3.22}$$

The following argument shows that ∇f and its contour level must be orthogonal: Let C be a contour of f with k the constant such that $f(x, y) = k$. Let $(x(t), y(t))$, $t \in (a, b)$, be the parametrization of this curve, that is,

$$f(x(t), y(t)) = k, \quad \text{for all } t \in (a, b).$$

Differentiate both sides of the above expression to arrive at

$$\frac{\partial f}{\partial x}\frac{dx}{dt} + \frac{\partial f}{\partial y}\frac{dy}{dt} = 0, \quad \text{for all } t \in (a, b),$$

which can be rewritten as

$$\nabla f \cdot \langle \frac{dx}{dt}, \frac{dy}{dt} \rangle = 0,$$

signifying that ∇f and $\langle \frac{dx}{dt}, \frac{dy}{dt} \rangle$ are orthogonal. Since the vector $\langle \frac{dx}{dt}, \frac{dy}{dt} \rangle$ is tangential to the contour level C, we have shown that ∇f and C are orthogonal. We summarize this discussion as a theorem.

Theorem 3.3.2 (Contour Levels and Gradients)
Level curves (surfaces) of a function f and its gradient are orthogonal.

The observation we have made about the gradient of f and its contours is easily captured in MATLAB by the following example. Here $f(x, y) = e^{-x} \sin y$.

```
clf;
[x,y]=meshgrid(0:0.2:3,0:0.15:pi);
z=exp(-x).*sin(y);
[qx,qy]=gradient(z,0.2,0.15);
contour(x,y,z);
hold on
quiver(x,y,qx,qy);
```

See Figure 3.2 for the output of the above program. The color bar on the right side of the figure shows the range of contour values, in this case ranging from 0 to 1; applying MATLAB's `surf` command will shed more light on why the contours appear as they are. Also note that the gradient vectors are clearly orthogonal to the associated contour levels.

The distribution of the gradient vectors in Figure 3.2 is an example of a *vector field*, namely, an assignment of vectors to positions P in a domain. A substantial part of this text will be dedicated to computing *velocity vector fields*. A velocity vector field is a special collection of vectors $\mathbf{v}(x, y, z, t)$ expressing the tendency a particle, which is located at position (x, y, z) at time t, possesses in order to move in the direction designated by $\mathbf{v}$. Since position and velocity are intimately related through the expressions

$$\frac{dx}{dt} = u(x, y, z, t), \quad \frac{dy}{dt} = v(x, y, z, t), \quad \frac{dz}{dt} = w(x, y, z, t), \qquad (3.23)$$

where u, v and w are the components of $\mathbf{v}$, these equations provide us with a system of ordinary differential equations whose solution will lead to particle trajectories of the flow induced by the velocity field. Solving equations like (3.23) is the subject of the next chapter.

MATLAB's `quiver` command, which was used in the code listed above, is the appropriate tool for displaying vectors in a vector field.

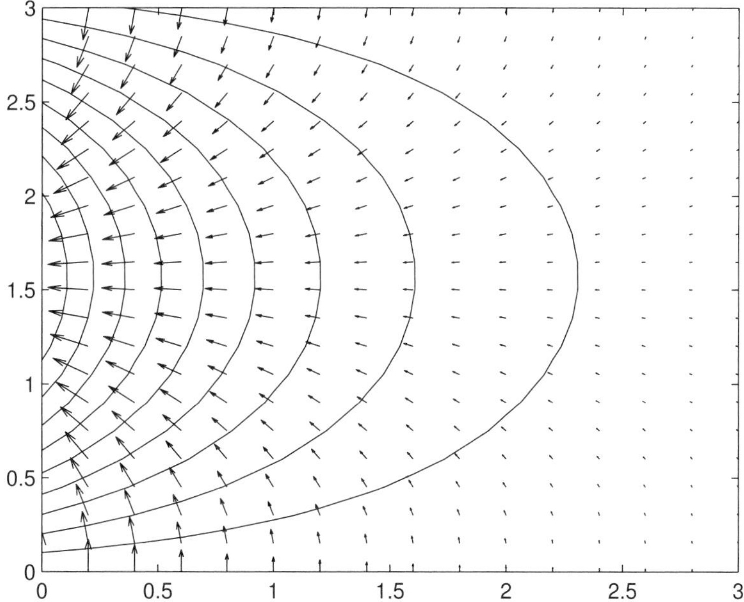

FIGURE 3.2: The contour and gradient vectors of the function $f(x, y) = e^{-x} \sin y$.

Problems 3.3

1. For each function listed, plot the equivalent of Figure 3.2. In each case use `surf` to plot the graph of the surface. Experiment with the domain of each function to display regions where the function undergoes substantial change.

   (a) $x^2 + y^2$,

   (b) $\sin(x^2 + y^2)$,

   (c) $x^2 - y^2$,

   (d) $\tan(x^2 - y^2)$,

   (e) $x^2 + 2y^2$,

   (f) $x^2 + 10y^2$,

   (g) $\sin \pi x \sin \pi y$,

   (h) $\sin \pi x \cos \pi y$,

   (i) $\ln(x^2 + y^2)$,

   (j) $1 + \ln(x^2 + y^2)$,

(k) $x + \ln(x^2 + y^2)$.

2. For each function listed, plot the equivalent of Figure 3.2.

   (a) $\frac{x^2+y^2}{x^2+y^2+1}$,
   (b) $\frac{x}{x^2+y^2}$,
   (c) $\frac{xy}{x^2+y^2} + \epsilon \sin(||\mathbf{r}||)$ where $\mathbf{r} = \langle x, y \rangle$. Let $\epsilon = -1$, 0 and 1.
   (d) $\frac{\sin r}{r}$ where $r = ||\langle x, y \rangle||$.

3. Determine the direction of steepest descent at the designated points for each of the following functions.

   (a) $x^2 + y^2$ at $P = (1, 1)$ and at $P = (1, 2)$.
   (b) $x^2 + 3y^2$ at $P = (a, b)$.
   (c) $\ln(x^2 + y^2)$ at $P = (-1, 1)$.

4. Consider the function $f(x, y) = x^2 + 3y^2 - 2x$ and the set of points (x, y) on the contour level 1 (i.e., $f(x, y) = 1$). Determine the magnitude of steepest descent at each one of these points, and find the point on this set where this rate of change is minimized.

5. Verify the following identities. Here f and g are arbitrary differentiable functions of x, y and z.

   (a) $\nabla(f + g) = \nabla f + \nabla g$.
   (b) $\nabla(cf) = c\nabla f$, where c is a constant.
   (c) $\nabla(fg) = g\nabla f + f\nabla g$.
   (d) $\nabla(\frac{f}{g}) = \frac{1}{g^2}(g\nabla f - f\nabla g)$.

6. *(Basis in Polar Coordinates)* Show that $\{\mathbf{e}_r, \mathbf{e}_\theta\}$ defined by

$$\mathbf{e}_r = \langle \cos\theta, \sin\theta \rangle, \quad \mathbf{e}_\theta = \langle -\sin\theta, \cos\theta \rangle \tag{3.24}$$

   forms a basis for E^2.

7. *(Gradient in Polar Coordinates)* Let $F(r, \theta)$ be the representation of $f(x, y)$ in polar coordinates, that is,

$$F(r, \theta) = f(r\cos\theta, r\sin\theta).$$

   Show that

$$\nabla f = \frac{\partial F}{\partial r}\mathbf{e}_r + \frac{1}{r}\frac{\partial F}{\partial \theta}\mathbf{e}_\theta, \tag{3.25}$$

   where $\mathbf{e}_r$ and $\mathbf{e}_\theta$ are defined in (3.24).

8. *(Basis in Spherical Coordinates)* Show that $\{\mathbf{e}_\rho, \mathbf{e}_\theta, \mathbf{e}_\phi\}$ defined by

$$\begin{aligned} \mathbf{e}_\theta &= \langle -\sin\theta, \cos\theta, 0\rangle, \\ \mathbf{e}_\phi &= \langle \cos\theta\cos\phi, \sin\theta\cos\phi, -\sin\phi\rangle, \\ \mathbf{e}_\rho &= \langle \cos\theta\sin\phi, \sin\theta\sin\phi, \cos\phi\rangle, \end{aligned} \tag{3.26}$$

   forms a basis for E^3.

9. *(Gradient in Spherical Coordinates)* Let $F(\rho, \theta, \phi)$ be the representation of $f(x, y, z)$ in polar coordinates, that is,

$$F(\rho, \theta, \phi) = f(\rho\cos\theta\sin\phi, \rho\sin\theta\sin\phi, \rho\cos\phi).$$

   Show that

$$\nabla f = \frac{\partial F}{\partial \rho}\mathbf{e}_\rho + \frac{1}{\rho}\frac{\partial F}{\partial \theta}\mathbf{e}_\theta + \frac{1}{\rho\sin\theta}\frac{\partial F}{\partial \phi}\mathbf{e}_\phi, \tag{3.27}$$

   where $\mathbf{e}_\rho$, $\mathbf{e}_\theta$ and $\mathbf{e}_\phi$ are defined in (3.26).

3.4 Divergence

Since vector fields vary with position over their domains we need mathematical tools, similar to the concept of the derivative of a function of a single variable, to analyze their local behavior. Divergence and Curl are two such tools.

Consider a vector field $\mathbf{v}$ in E^3. If we represent $\mathbf{v}$ by

$$\mathbf{v} = \langle u(x, y, z), v(x, y, z), w(x, y, z)\rangle \tag{3.28}$$

in the standard basis, we define the *divergence* of $\mathbf{v}$, denoted by div $\mathbf{v}$ or by $\nabla \cdot \mathbf{v}$, as

$$\text{div } \mathbf{v} = \frac{\partial u}{\partial x} + \frac{\partial v}{\partial y} + \frac{\partial w}{\partial z}. \tag{3.29}$$

Note that if we define the *del operator*, denoted by ∇, as

$$\nabla = \langle \frac{\partial}{\partial x}, \frac{\partial}{\partial y}, \frac{\partial}{\partial z}\rangle,$$

then (3.29) is equivalent to

$$\nabla \cdot \mathbf{v},$$

that is, the dot product of the "vector" ∇ and the vector $\mathbf{v}$.

The divergence of a vector field provides information about stretching and compression of space under the action of $\mathbf{v}$. We will make this point precise in the context of conservation of mass in a later chapter. Here we will bring up one important application of divergence in the context of two-dimensional vector fields whose divergence vanishes. Consider the vector field

$$\mathbf{v} = \langle u(x, y, t), v(x, y, t) \rangle \tag{3.30}$$

endowed with the property div $\mathbf{v} = 0$, or $\frac{\partial u}{\partial x} + \frac{\partial v}{\partial y} = 0$. This property is automatically satisfied if the components of $\mathbf{v}$ are related to a single function ψ by

$$u = \frac{\partial \psi}{\partial y}, \quad v = -\frac{\partial \psi}{\partial x}. \tag{3.31}$$

Under reasonable conditions (see theorem below) on the smoothness of u and v, it turns out that when div $\mathbf{v} = 0$ then there exists a function ψ where (3.31) holds. Such a function is called a *Stream Function* in fluid dynamics and a *Hamiltonian* in mathematics. One of the important features of having a stream function in hand is that ψ remains *invariant* under the action of the system of differential equations (3.23) when the right side of (3.30) is time-independent. To see this, consider (3.30) in combination with (3.31):

$$\frac{dx}{dt} = u(x, y) = \frac{\partial \psi}{\partial y}, \quad \frac{dy}{dt} = v(x, y) = -\frac{\partial \psi}{\partial x}. \tag{3.32}$$

Then, with $(x(t), y(t))$ a solution of (3.23) and $\psi = \psi(x(t), y(t))$, we have

$$\frac{d\psi}{dt} = \frac{\partial \psi}{\partial x}\frac{dx}{dt} + \frac{\partial \psi}{\partial y}\frac{dy}{dt} = \frac{\partial \psi}{\partial x}\frac{\partial \psi}{\partial y} - \frac{\partial \psi}{\partial y}\frac{\partial \psi}{\partial x} = 0, \tag{3.33}$$

which indicates that ψ remains constant along trajectories $(t, x(t), y(t))$ of the velocity field (3.30).

Another significant feature of having a stream function in hand is that we can go far in determining the orbits $(x(t), y(t))$ of (3.30). This property follows from the special relations in (3.32) and the fact that the gradient of any function is perpendicular to its contour levels. To see this, note that

$$\nabla \psi \cdot \mathbf{v} = \langle \frac{\partial \psi}{\partial x}, \frac{\partial \psi}{\partial y} \rangle \cdot \langle v_1, v_2 \rangle = \langle -v_2, v_1 \rangle \cdot \langle v_1, v_2 \rangle = 0. \tag{3.34}$$

Hence $\mathbf{v}$ is perpendicular to $\nabla \psi$, or, equivalently, parallel to $\psi = k$, k a constant. Since $\mathbf{v}$ is instantaneously tangential to the orbits of (3.30), we have valuable information about this system of differential equations in the special case when the system is two-dimensional, its velocity field is divergence-free, and its right side is independent of t. In this setting the contours of ψ end up being the orbits of (3.30). We summarize the above discussion as a theorem.

Theorem 3.4.1 (Stream Functions)
Let $\mathbf{v}$ *be a continuously differentiable two-dimensional vector field with div* $\mathbf{v} = 0$. *Then*

1. *There is a continuously differentiable function* $\psi(x, y, t)$ *such that the relations in (3.30) hold.*

2. *When* $\mathbf{v}$ *(and by extension,* ψ*) is time-independent, then the orbits of the system of differential equations in (3.32) and contour levels of* ψ *coincide.*

A point regarding this theorem is worth emphasizing. We have used the term "orbits" of (3.32), rather than "trajectories," to call attention to the fact that the statement of the theorem involves the set of points $(x(t), y(t))$ (and not $(t, x(t), y(t))$) — it turns out that the uniqueness property of trajectories of systems such as (3.32) extend to the orbits of that system when $\mathbf{v}$ does not explicitly depend on t. This property plays a crucial role in the proof of the contention about the coincidence of orbits and contour levels.

The task of determining ψ in a concrete setting reduces to integrating (3.31). Consider the example $\mathbf{v} = \langle y, -x\rangle$, which satisfies the requisite condition div $\mathbf{v} = 0$. To determine ψ we need to integrate (3.31), which in this example reduces to

$$y = \frac{\partial \psi}{\partial y}, \quad -x = -\frac{\partial \psi}{\partial x}. \tag{3.35}$$

Integrating the first equation with respect to y yields

$$\psi(x, y) = \frac{1}{2}y^2 + f(x),$$

where f is the constant of integration (with respect to y). Differentiating the latter with respect to x gives us $\frac{\partial \psi}{\partial x} = f'(x)$, which when compared with (3.35)b, yields $f'(x) = x$. Hence $f(x) = \frac{1}{2}x^2 + c$, with c a universal constant. Hence the stream function for $\mathbf{v} = \langle y, -x\rangle$ is

$$\psi(x, y) = \frac{1}{2}(x^2 + y^2) + c.$$

Since contours of this stream function are concentric circles about the origin, we conclude, following Theorem 3.4.1, that the orbits of the system of differential equations

$$\frac{dx}{dt} = y, \quad \frac{dy}{dt} = -x$$

are concentric circles.

Problems 3.4

1. Verify the following identities. All vector fields are assumed smooth enough to allow differentiations of all orders needed.

   (a) div $(\mathbf{v} + \mathbf{w}) = \text{div } \mathbf{v} + \text{div } \mathbf{w}$.

   (b) div $(c\mathbf{v}) = c \,\text{div } \mathbf{v}$, where c is a constant.

   (c) div $(\rho\mathbf{v}) = \nabla\rho \cdot \mathbf{v} + \rho \,\text{div } \mathbf{v}$, where ρ is a smooth function.

   (d) div $(\nabla f) = \Delta f$, where Δ, the *Laplace Operator*, is $\frac{\partial^2}{\partial x^2} + \frac{\partial^2}{\partial y^2} + \frac{\partial^2}{\partial z^2}$.

   (e) div $(f\nabla g) = f\Delta g + \nabla f \cdot \nabla g$.

   (f) div $(f\nabla g) -$ div $(g\nabla f) = f\Delta g - g\Delta f$.

   (g) div $(\nabla f \times \nabla g) = 0$.

2. Consider the stream function $\psi(x, y) = \frac{1}{\sqrt{2x^2 - y^2}}$. Determine its associated vector field $\mathbf{v}$.

3. Show that if ψ is a stream function and $\mathbf{v}(x, y)$ its attendant vector field, then

$$\text{div } (\psi\mathbf{v}) = 0. \tag{3.36}$$

4. Consider the velocity field $\mathbf{v}(x, y) = \langle \frac{y}{\sqrt{x^2+y^2}}, -\frac{x}{\sqrt{x^2+y^2}} \rangle$. Show that this velocity field has a stream function, determine it, and use the result of Theorem 3.4.1 and MATLAB to plot $\mathbf{v}$'s orbits.

5. *(Divergence in Polar Coordinates)*

   (a) Let $\mathbf{v}$ be a vector field in E^2 with components (u, v) in Cartesian coordinates and (u_r, u_θ) in polar coordinates, i.e.,

$$\mathbf{v} = u\mathbf{i} + v\mathbf{j} = u_r\mathbf{e}_r + u_\theta\mathbf{e}_\theta,$$

   where $\mathbf{e}_r$ and $\mathbf{e}_\theta$ were defined in (3.24).

   i. Show that

$$u_r = u\cos\theta + v\sin\theta, \quad u_\theta = -u\sin\theta + v\cos\theta.$$

   ii. Verify that the following formula holds for the divergence of a vector field in polar coordinates.

$$\text{div } \mathbf{v} = \frac{\partial u_r}{\partial r} + \frac{1}{r}u_r + \frac{1}{r}\frac{\partial u_\theta}{\partial \theta}.$$

(b) (*Cylindrical Coordinates*) Let $\mathbf{v} = u\mathbf{i} + v\mathbf{j} + w\mathbf{k} = u_r\mathbf{e}_r + u_\theta\mathbf{e}_\theta + w\mathbf{k}$. Show that

$$\text{div}\ \mathbf{v} = \frac{\partial u_r}{\partial r} + \frac{1}{r}u_r + \frac{1}{r}\frac{\partial u_\theta}{\partial \theta} + \frac{\partial w}{\partial z},$$

where w is the z-component of $\mathbf{v}$.

6. Apply the results in Problem 5 and compute the divergence of each $\mathbf{v}$, first in Cartesian coordinates and then in polar coordinates.

(a) $\mathbf{v} = x\mathbf{i} - 2y\mathbf{j}$.

(b) $\mathbf{v} = \frac{x}{x^2+y^2}\mathbf{i} + \frac{y}{x^2+y^2}\mathbf{j}$.

(c) $\mathbf{v} = \frac{y}{\sqrt{x^2+y^2}}\mathbf{i} - \frac{x}{\sqrt{x^2+y^2}}\mathbf{j}$.

(d) $\mathbf{v} = \frac{1}{r^3}\mathbf{e}_\theta$.

7. In this problem we determine the relationship between $\psi(x, y)$, the stream function of a vector field, represented in the standard Cartesian basis, and $\Psi(r, \theta)$, the representation of the same stream function in polar coordinates. Note that

$$\psi(x, y) = \Psi(r, \theta) \tag{3.37}$$

where $x = r\cos\theta$ and $y = r\sin\theta$. Show that the relations in (3.31) take the form

$$u_r = \frac{1}{r}\frac{\partial \Psi}{\partial \theta}, \quad u_\theta = -\frac{\partial \Psi}{\partial r}, \tag{3.38}$$

where u_r and u_θ are defined in Problem 5.

8. (*Laplace Operator in Polar Coordinates*) The Laplace operator in Cartesian coordinates has the form

$$\frac{\partial^2 \psi}{\partial x^2} + \frac{\partial^2 \psi}{\partial y^2} + \frac{\partial^2 \psi}{\partial z^2}.$$

Show that this operator has the form

$$\frac{\partial^2 \Psi}{\partial r^2} + \frac{1}{r}\frac{\partial \Psi}{\partial r} + \frac{1}{r^2}\frac{\partial^2 \Psi}{\partial \theta^2} + \frac{\partial^2 \Psi}{\partial z^2}$$

in polar (cylindrical) coordinates.

9. (*Divergence in Spherical Coordinates*) Consider the vector field $\mathbf{v} = u_\rho\mathbf{e}_\rho + u_\theta\mathbf{e}_\theta + u_\phi\mathbf{e}_\phi$. Show that the divergence of $\mathbf{v}$ is given by the following formula:

$$\frac{\partial u_\rho}{\partial \rho} + \frac{2}{\rho}u_\rho + \frac{1}{\rho}\frac{\partial u_\theta}{\partial \theta} + \frac{\cot\theta}{\rho}u_\theta + \frac{1}{\rho\sin\theta}\frac{\partial u_\phi}{\partial \phi}. \tag{3.39}$$

Hint: If you have access to the symbolic manipulator *Mathematica®*, you may try the following code and compare your result to its output:

```
Clear[v1, v2, v3, e1, e2, e3, r, theta, phi, u, v, w]
r = Sqrt[x^2 + y^2 + z^2]; theta = ArcTan[y/x];
phi = ArcSin[z/Sqrt[x^2 + y^2 + z^2]];
v1 = u[r, theta, phi]; v2 = v[r, theta, phi];
     v3 = w[r, theta, phi];
e1 = Simplify[{-Sin[theta], Cos[theta], 0},
      {x > 0, y > 0, z > 0}];
e2 = Simplify[{(-Cos[theta])*Sin[phi],
     (-Sin[theta])*Sin[phi], Cos[phi]},
         {x > 0, y > 0, z > 0}];
e3 = Simplify[{Cos[theta]*Cos[phi],
       Sin[theta]*Cos[phi], Sin[phi]},
         {x > 0, y > 0, z > 0}];
conds1 = {Sqrt[x^2 + y^2 + z^2] -> r1,
            ArcTan[y/x] -> th1,
              ArcSin[z/Sqrt[x^2 + y^2 + z^2]] -> ph1};
conds2 = {x -> r1*Cos[th1]*Cos[ph1],
    y -> r1*Sin[th1]*Cos[ph1],
       z -> r1*Sin[ph1]};
vector = Simplify[v1*e1 + v2*e2 + v3*e3,
       {x > 0, y > 0, z > 0}];
divergence = Simplify[D[vector[[1]], x]+
     D[vector[[2]], y] + D[vector[[3]], z],
           {x > 0, y > 0, z > 0}];
div1 = divergence /. conds1;
div2 = div1 /. conds2;
answer = Expand[FullSimplify[div1 /. conds2,
    {r1 > 0, Cos[ph1] > 0}]];
answer
```

10. *(Laplace Operator in Spherical Coordinates)* See (3.26) for definitions. Show that the Laplace operator (see Problem 8 for definitions) has the form

$$\frac{\partial^2\Psi}{\partial\rho^2} + \frac{2}{\rho}\frac{\partial\Psi}{\partial\rho} + \frac{1}{\rho^2}\frac{\partial^2\Psi}{\partial\theta^2} + \frac{\cot\theta}{\rho^2}\frac{\partial\Psi}{\partial\theta} + \frac{1}{\rho^2\sin^2\theta}\frac{\partial^2\Psi}{\partial\phi^2}$$

in spherical coordinates.

3.5 Curl and Vector Fields

Given a vector field $\mathbf{v}$ in E^3 we define the *curl* of $\mathbf{v}$, denoted

$$\text{curl } \mathbf{v}$$

or more commonly by

$$\nabla \times \mathbf{v},$$

as the following vector

$$\nabla \times \mathbf{v} = \langle \frac{\partial w}{\partial y} - \frac{\partial v}{\partial z}, \frac{\partial u}{\partial z} - \frac{\partial w}{\partial x}, \frac{\partial v}{\partial x} - \frac{\partial u}{\partial y} \rangle. \tag{3.40}$$

Symbolically, this definition can be written as the determinant of the 3×3 matrix

$$\nabla \times \mathbf{v} = \det \begin{bmatrix} i & j & k \\ \frac{\partial}{\partial x} & \frac{\partial}{\partial y} & \frac{\partial}{\partial z} \\ u & v & w \end{bmatrix}. \tag{3.41}$$

The curl of a vector field provides information about the tendency for rotation and spin in particles whose motion is influenced by $\mathbf{v}$. A typical example to keep in mind is provided by $\mathbf{v} = \langle y, -x \rangle$ where as we saw in the last section, it induces a motion in the shape of concentric circles about the origin. Its curl

$$\nabla \times \langle y, -x, 0 \rangle = \langle 0, 0, -2 \rangle = -2\mathbf{k} \tag{3.42}$$

reinforces this point because $-2\mathbf{k}$ points to a clockwise rotation in the xy-plane (clockwise because of the coefficient of $\mathbf{k}$ is negative). When $\mathbf{v}$ is a velocity vector field, $\nabla \times \mathbf{v}$ is called the **vorticity** of $\mathbf{v}$ and is often denoted by ω.

The curl operation is one of three analytical tools, counting gradient and divergence as the other two, at our disposal to study the local behavior of a function of several variables or a vector field. When these operations combine they often provide detailed information about the structure of vector fields. For example, when a two-dimensional vector field $\mathbf{v}$ has a stream function ψ and is *irrotational*, that is, its curl vanishes, then ψ must satisfy *Laplace's equation*

$$\Delta\psi = 0. \tag{3.43}$$

To see this, let $\mathbf{v} = \langle \frac{\partial \psi}{\partial y}, -\frac{\partial \psi}{\partial x} \rangle$ and note that $\nabla \times \mathbf{v} = \langle 0, 0, -\Delta\psi \rangle$. Thus, the curl of $\mathbf{v}$ vanishes if and only if (3.43) holds. In general, however, the vorticity of a typical flow is non-zero, and flows are rotational. When a flow is two-dimensional, its vorticity takes the form

$$\omega = f(x, y, t)\mathbf{k}.$$

If, in addition, this flow is divergence-free, then its stream function $\psi(x, y, t)$ satisfies the *Poisson Equation*

$$-\Delta\psi = f. \tag{3.44}$$

The curl of a vector field $\mathbf{v}$ also determines whether $\mathbf{v}$ is endowed with a scalar potential or a vector potential. A function ϕ is called a *(scalar) potential* for $\mathbf{v}$ if

$$\mathbf{v} = \nabla\phi. \tag{3.45}$$

Alternatively, a vector-valued function Ψ is called a *(vector) potential* for $\mathbf{v}$ if

$$\mathbf{v} = \nabla \times \Psi. \tag{3.46}$$

Generally speaking, vector fields have neither scalar nor vector potentials. But when a vector field possesses such a potential, its study is often reduced to analyzing less complicated mathematical equations. The conditions for the existence of such potentials are explored in the problems at the end of this section.

Problems 3.5

1. Show that equations (3.31), which define the relationship between $\mathbf{v}$ and its stream function ψ, when ψ exists, are equivalent to

$$\mathbf{v} = \nabla \times (\psi\mathbf{k}).$$

2. For each vector field determine whether $\mathbf{v}$ is divergence-free, and compute its curl:

   (a) $\mathbf{v} = \langle y - z, x - z, x + y \rangle$.

   (b) $\mathbf{v} = \langle 2y, -3x + xy \rangle$.

   (c) $\mathbf{v} = \langle ax, by \rangle$, where a and b are constants.

   (d) $\mathbf{v} = \langle \frac{\alpha y}{x^2+y^2}, -\frac{\beta x}{x^2+y^2} \rangle$, α and β are constants.

   (e) $\mathbf{v} = \langle y, -\sin x \rangle$.

   (f) $\mathbf{v} = \langle x^2 - y^2, xy \rangle$.

3. Assume that all of the functions below are sufficiently differentiable to allow the necessary differentiation operations to go through. Verify the following identities:

(a) div(curl $\boldsymbol{\Psi}$) $= 0$. Hence, returning to the definition of a vector potential (see (3.46)), that $\mathbf{v}$ is divergence-free is a necessary condition for the existence of a vector potential Ψ.

(b) $\nabla \times (\nabla\phi) = \mathbf{0}$. According to this identity the necessary condition for $\mathbf{v}$ to have a scalar potential ϕ is $\nabla \times \mathbf{v} = \mathbf{0}$.

(c) $\text{curl}(\phi\mathbf{v}) = \text{grad}\ \phi \times \mathbf{v} + \phi\,\text{curl}\ \mathbf{v}$.

(d) $\text{div}\ (\mathbf{v} \times \mathbf{w}) = \mathbf{w} \cdot \text{curl}\ \mathbf{v} - \mathbf{v} \cdot \text{curl}\ \mathbf{w}$.

(e) $\text{curl}\ (\phi\mathbf{v}) = \phi\,\text{curl}\ \mathbf{v} + \nabla\phi \times \mathbf{v}$.

(f) $\nabla \times (\nabla \times \mathbf{v}) = \nabla(\nabla \cdot \mathbf{v}) - \Delta\mathbf{v}$ where $\Delta\mathbf{v} = \langle \Delta u, \Delta v, \Delta w\rangle$.

4. *(Curl in Cylindrical Coordinates)* Following the approach of Problem 5 of Section 3.4, let $\mathbf{v} = u_r\mathbf{e}_r + u_\theta\mathbf{e}_\theta + w\mathbf{k}$ and show that

$$\nabla \times \mathbf{v} = (\frac{1}{r}\frac{\partial w}{\partial \theta} - \frac{\partial u_\theta}{\partial z})\mathbf{e}_r + (\frac{\partial u_\rho}{\partial z} - \frac{\partial w}{\partial r})\mathbf{e}_\theta + (\frac{\partial u_\theta}{\partial r} + \frac{1}{r}u_\theta - \frac{1}{r}\frac{\partial u_r}{\partial \theta})\mathbf{k}$$

is the curl of $\mathbf{v}$ in polar-cylindrical coordinates.

5. *(Curl in Spherical Coordinates)* Following the approach of Problem 9 of Section 3.4, let $\mathbf{v} = u_\rho\mathbf{e}_\rho + u_\theta\mathbf{e}_\theta + u_\phi\mathbf{e}_\phi$ and show that

$$\nabla \times \mathbf{v} = \frac{1}{\rho}(\frac{\partial u_\phi}{\partial \theta} + \cot\theta\, u_\phi - \frac{1}{\sin\theta}\frac{\partial u_\theta}{\partial \phi})\mathbf{e}_\rho +$$

$$+(\frac{1}{\rho\sin\theta}\frac{\partial u_\rho}{\partial \phi} - \frac{\partial u_\phi}{\partial \rho} - \frac{1}{\rho}u_\phi)\mathbf{e}_\theta + (\frac{\partial u_\theta}{\partial \rho} + \frac{1}{\rho}u_\theta - \frac{1}{\rho}\frac{\partial u_\rho}{\partial \theta})\mathbf{e}_\phi$$

is the curl of $\mathbf{v}$ in spherical coordinates.

3.6 Integral Theorems

Integration and differentiation are inverse operations. Each operation has its own utility and place in analysis. The derivative of a function provides local behavior information about that function — this information is precise but is generally confined to a small neighborhood of the point at which the derivative is computed. By contrast, an integral of a function over an interval provides information that is global but only in an averaged sense. In the context of functions of several variables and vector fields, when integration and differentiation are combined properly, the result is often quite powerful.

The *line integral* of a vector field $\mathbf{v}$ over a curve C, denoted by $\int_C \mathbf{v} \cdot d\mathbf{r}$, is defined by

$$\int_C \mathbf{v} \cdot \mathbf{r} = \int_a^b \mathbf{v}|_C \cdot \mathbf{r}'(t)\, dt, \tag{3.47}$$

where $\mathbf{r}$ is a parametrization of C, that is, points P on C are end-points of vectors $\mathbf{r}(t)$ as t ranges over the interval (a,b):

$$C = \{(x(t), y(t), z(t)) |\, \mathbf{r}(t) = \langle x(t), y(t), z(t)\rangle, \quad t \in (a,b)\}.$$

Since the integrand in (3.47) is the dot product of two vectors, one of which, $\mathbf{r}'$, is tangential to C, the line integral in (3.47) measures to what extent the vector field $\mathbf{v}$ deviates from being tangential to C — if $\mathbf{v}$ and $\mathbf{r}'$ are orthogonal, say, the contribution of $\mathbf{v} \cdot \mathbf{r}'$ to the integrand is zero, while if $\mathbf{v}$ is parallel with $\mathbf{r}'$, this contribution is maximized. On the whole, the integral in (3.47) gives some information about the disposition of $\mathbf{v}$ relative to C. When $\mathbf{v}$ represents a force field, (3.47) measures *work*, while when $\mathbf{v}$ represents a velocity field and C is a closed curve, which will be defined shortly, (3.47) measures *circulation.*

As an example, consider the vector field $\mathbf{v} = \langle \frac{y}{\sqrt{x^2+2y^2}}, -\frac{x}{\sqrt{x^2+2y^2}}, 0\rangle$ and the curve C given by $r(t) = \langle \cos t, \sin t, 0\rangle$, $t \in (0, \frac{\pi}{2})$, a quarter circle traversed in the counterclockwise direction. The line integral (3.47) of $\mathbf{v}$ over C is

$$\int_C \mathbf{v} \cdot\, d\mathbf{r} =$$

$$\int_0^{\frac{\pi}{2}} \langle \frac{\sin t}{\sqrt{\cos^2 t + 2\sin^2 t}}, -\frac{\cos t}{\sqrt{\cos^2 t + 2\sin^2 t}}, 0\rangle \cdot \langle -\sin t, \cos t, 0\rangle\, dt =$$

$$= \int_0^{\frac{\pi}{2}} \frac{1}{\sqrt{\cos^2 t + 2\sin^2 t}}\, dt = -1.3110.$$

The last integration was carried in MATLAB:

```
F=@(t) -1./(cos(t).^2+ 2*sin(t).^2).^(1/2);
quad(F,0,pi/2)
```

Continuing with this example, let C be the completed curve with $t \in (0, 2\pi)$. This curve is an example of a smooth *simple closed* curve, for which the tangent vector at every point exists and where the curve intersects itself only once, in this case when its beginning touches its end. For such a curve we use the notation $\oint_C$ to emphasize that C is a closed curve. As stated earlier, the quantity $\oint_C \mathbf{v} \cdot d\mathbf{r}$, which in this case equals -5.2441, is the *circulation* of $\mathbf{v}$ about C.

We note in passing that `quad`, which performs numerical integration (quadrature) in MATLAB, is quite efficient with integrands that are not highly oscillatory. By contrast, the functions `quadl`, `quadv` and `quadgk` are the suitable integrators to use when the integrand rapidly oscillates.

The *surface* integral of a vector field $\mathbf{v}$ over a surface S, denote by $\int\int_S \mathbf{v}\cdot d\mathbf{S}$, is defined by

$$\int\int_S \mathbf{v}\cdot d\mathbf{S} = \int\int_D \mathbf{v}\cdot(\mathbf{r}_u \times \mathbf{r}_v)\, dudv \tag{3.48}$$

where $\mathbf{r}(u,v)$ is the parametrization of the surface S, and hence $\mathbf{r}_u \times \mathbf{r}_v$ is normal to the surface, and D is the domain of this parametrization. For example, when S is the surface of the northern hemisphere of radius 1, then $D = \{(u,v)|\, 0 \le u < 2\pi, 0 \le v < \frac{\pi}{2}\}$, i.e., u and v are the standard longitude and co-latitude and $\mathbf{r}$ is

$$\mathbf{r}(u,v) = \langle \cos u \sin v, \sin u \sin v, \cos v\rangle.$$

Note that $\mathbf{r}_u \times \mathbf{r}_v = \langle \cos u \sin v, \sin u \sin v, \cos v\rangle$, a radial vector, which is of course normal to the surface of the sphere.

The computation of a surface integral always reduces to computing a double integral, as described in (3.48). The physical interpretation of this quantity is that of *flux* when $\mathbf{v}$ is a velocity vector field – since $\mathbf{v}$ has dimensions of length over time, and $du\,dv$ has dimensions of length squared, the combination $\mathbf{v}\cdot\mathbf{r}_u \times \mathbf{r}_v$ has dimensions of volume over time, or flux (note that $\mathbf{r}_u$ and $\mathbf{r}_v$ are dimensionless). As an example consider $\mathbf{v} = x^2\mathbf{k}$, which, as a velocity field, describes a flow in the z-direction whose strength varies with the horizontal component of the position of each particle. To compute the flux of this flow through the unit northern hemisphere, we compute

$$\int_0^{2\pi}\int_0^{\frac{\pi}{2}} \langle 0, 0, \cos^2 u \sin^2 v\rangle \cdot \langle \cos u \sin v, \sin u \sin v, \cos v\rangle\, du\, dv =$$

$$= \int_0^{2\pi}\int_0^{\frac{\pi}{2}} \cos^2 u \sin^2 v \cos v\, du\, dv = 1.0472,$$

a conclusion we reach by either computing this integral analytically or by using MATLAB:

```
F=@(u,v) (cos(u).^2.*sin(v).^2.*cos(v));
dblquad(f,0,2*pi,0,pi/2)
```

One of the main applications of the line and surface integrals is in the context of the generalization of the "Fundamental Theorem of Calculus" in higher dimensions. This theorem in one-space dimension relates f and its derivative f' in the familiar identity

$$\int_a^b f'(x)\, dx = f(b) - f(a), \tag{3.49}$$

relating the average value of f' over the domain (a, b) to the net "flow" of f through the boundary. There are two analogs of this theorem in higher dimensions, the *Stokes Theorem* and the *Divergence Theorem* or *Gauss's Theorem*, which we now state.

Theorem 3.6.1 (Stokes's Theorem)
Let $\mathbf{v}$ *be a smooth vector field defined in a domain* $D \subset R^3$*. Let* S *be a surface contained in* D *with boundary* C*. Then the following identity holds:*

$$\oint_C \mathbf{v} \cdot d\mathbf{r} = \int\int_S \nabla \times \mathbf{v} \cdot d\mathbf{S}, \tag{3.50}$$

where the parametrization of C *and* S *need to be compatible in the following sense: The curve* C *and* S *are parametrized according to the right-handed rule so that when the curve* C *is traversed in the direction of the parametrization, the normal to* S *always points to the left.*

Note that this identity involves a double integral of a derivative of $\mathbf{v}$, in this case the curl of $\mathbf{v}$, balanced by the integral of $\mathbf{v}$ itself, where the latter integration is over the boundary of S. Its similarity to (3.49) cannot be overemphasized.

As stated earlier when $\mathbf{v}$ is a velocity vector field of a fluid flow, the quantity $\nabla \times \mathbf{v}$ is called the vorticity of the fluid flow. The Stokes Theorem relates the "flux of the vorticity," the quantity on the right side of (3.50), to its circulation of the flow, the quantify on the right side of (3.50). This identity plays a crucial role in providing insight into "vortex lines" and their dynamics, an important concept in rotating fluid flows.

The second theorem involves a surface integral as well.

Theorem 3.6.2 (Divergence or Gauss's Theorem)
Let $\mathbf{v}$ *be a smooth vector field defined in a domain* $D \subset R^3$*. Then the following identity holds:*

$$\int\int\int_D \operatorname{div} \mathbf{v} \, dV = \int\int_{\partial D} \mathbf{v} \cdot d\mathbf{S}, \tag{3.51}$$

where ∂D *is parameterized in such a way that its normal always points to the outside of* D*.*

As was the case with the Stokes Theorem, the Divergence Theorem relates the integral of a derivative of $\mathbf{v}$, in this case its divergence, to the net change of $\mathbf{v}$ on the boundary. This theorem plays a key role in the development of the governing equations of motion because it will show us how to establish conservation laws of mass, linear momentum and energy by relating the internal changes in physical quantities to their net influx of flow through the boundary.

In addition to the applications already alluded to, the Stokes and

Divergence theorems give intuitive interpretations of the curl and divergence operations when they are combined with the Mean Value Theorem. We consider (3.50) first. Let P be a fixed point in the domain of $\mathbf{v}$ with S a surface passing through P, which for simplicity we assume it to be a plane. Consider a square in that plane centered at P, and remove the rest of the plane for the remainder of this discussion. We are now in the setting of Theorem 3.6.1 with S a square centered at P and C consisting of four edges that constitute the square. Consider a sequence of squares S_n, concentric at P, and shrinking to P as n approaches infinity. Divide both sides of (3.50) by the area of S_n and take the limit of both sides as the area of S_n approaches zero (with n approaching infinity):

$$\lim_{n\to\infty} \frac{1}{\Delta S_n} \oint_{C_n} \mathbf{v} \cdot d\mathbf{r} = \lim_{n\to\infty} \frac{1}{\Delta S_n} \int\!\!\int_{S_n} \nabla \times \mathbf{v} \cdot d\mathbf{S}. \tag{3.52}$$

We note that the integral on the right side in (3.52) is equivalent to

$$\int\!\!\int_{D_n} \nabla \times \mathbf{v} \cdot \mathbf{N}_n \, du \, dv$$

where D_n is the domain of parametrization of S_n and $\mathbf{N}_n$ the normal (i.e., $\mathbf{r}_u \times \mathbf{r}_v$) to the square S_n. Note that $\mathbf{N}_n$'s direction is fixed, although its length depends on n. By the Mean Value Theorem

$$\frac{1}{\Delta D_n} \int\!\!\int_{D_n} \nabla \times \mathbf{v} \cdot \mathbf{N}_n \, du \, dv = (\nabla \times \mathbf{v} \cdot \mathbf{N}_n)|_{P_n} \tag{3.53}$$

where P_n is a point in D_n. But $\lim_{n\to\infty} P_n = P$ and $\lim_{n\to\infty} N_n = \mathbf{N}$, a unit normal to the original surface S. Returning to (3.52) we have

$$(\nabla \times \mathbf{v} \cdot \mathbf{N})|_P = \lim_{n\to\infty} \frac{1}{\Delta D_n} \oint_{C_n} \mathbf{v} \cdot dr. \tag{3.54}$$

The above result gives a precise relationship between $\oint_C \mathbf{v} \cdot d\mathbf{r}$, the circulation of $\mathbf{v}$, and its vorticity $\nabla \times \mathbf{v}$.

A similar application to (3.51) results in the following identity:

$$(\text{div } \mathbf{v})|_P = \lim_{n\to\infty} \frac{1}{\Delta V_n} \int\!\!\int_{S_n} \mathbf{v} \cdot d\mathbf{S} \tag{3.55}$$

where V_n is a sequence of regions, all containing the point P and shrinking to P as n approaches infinity. This result gives a geometric interpretation of how the divergence of a vector field at a point P is related to flux per unit volume of that flow in a small neighborhood of P.

Problems 3.6

1. Verify the Divergence Theorem in the following setting: Let $\mathbf{v} = \langle x, y, z \rangle$ and D a hemisphere of radius 1 centered at the origin, i.e., $D = \{(x, y, z) |\, x^2 + y^2 + z^2 \leq 1, z \geq 0\}$.

2. Verify the Stokes Theorem in the following setting: Let $\mathbf{v} = x^2 k$ and S the surface of the northern hemisphere given by $x^2+y^2+z^2 = 1$ and $z > 0$.

3. (*Leibniz's formula*) An important extension of the Fundamental Theorem of Calculus, (3.49) and its variation $\frac{d}{dt}(\int_c^t f(\eta)\, d\eta) = f(t)$, is to the case when the integrand and the limits of integration vary with respect to the parameter of differentiation. Consider the function g defined as

$$g(t) = \int_{a(t)}^{b(t)} f(t, \eta)\, d\eta. \tag{3.56}$$

Using the Chain Rule of Differentiation, compute $g'(t)$ and show that

$$\frac{\partial}{\partial t}\left(\int_{a(t)}^{b(t)} f(t, \eta)\, d\eta\right) = \int_{a(t)}^{b(t)} \frac{\partial f}{\partial t}(t, \eta)\, d\eta + b'(t) f(t, b(t)) - a'(t) f(t, a(t)). \tag{3.57}$$

(Hint: Write g as $g(t) = \int_c^{b(t)} f(t, \eta)\, d\eta - \int_c^{a(t)} f(t, \eta)\, d\eta$, where c is an arbitrary but fixed constant.)

3.7 References and Further Reading

1. Malek-Madani, Reza, *Advanced Engineering Mathematics with Mathematica® and MATLAB®*, Addison-Wesley, 1998.

2. Marsden, Jerrold, E., Trobma, A. *Vector Calculus*, 5th edition, W. H. Freeman, 2003.

Chapter 4

Ordinary Differential Equations

Our approach to understanding the capabilities as well as the limitations of a physical model often hinges on computing solutions to ordinary differential equations (ODEs). In this chapter we will review some of the fundamental concepts associated with ODEs and introduce a few basic approximate techniques for obtaining these solutions when exact solutions are not available in analytic form. An added goal in this section is to provide enough background to motivate the use of MATLAB's powerful suite of ODE solvers, including `ode45`. Some of the techniques discussed here are then extended to solving Partial Differential Equations (PDEs).

4.1 Linear Independence and Space of Functions

Before we proceed to describe the general techniques for solving ODEs, we pause to generalize the two concepts of linear independence and linear space from matrix algebra to sets of functions. Consider a set C of functions $\{\phi_1, \phi_2, ..., \phi_n\}$. We say C (or equivalently, the functions ϕ_i's) is *linearly independent* on the interval (a, b) if

$$c_1\phi(x) + c_2\phi_2(x) + ... + c_n\phi_n(x) = 0 \quad \text{for all } x \in (a, b)$$

$$\text{if and only if } c_1 = c_2 = ... = c_n = 0. \tag{4.1}$$

The key phrase to keep in mind in this definition is "for all x," that is, the linear combination of the functions ϕ's must vanish, no matter what value x takes in the specified domain. For example, the set of functions $\{\sin x, \sin 2x, \sin 3x\}$ is linearly independent on the interval $(0, 1)$, while $\{\sin x, \sin 2x, 2\sin x - \sin 2x\}$ is not. To see that the latter violates (4.1) we note that if we identify the functions ϕ_i's as

$$\phi_1(x) = \sin x, \quad \phi_2(x) = \sin 2x, \quad \text{and} \quad \phi_3(x) = 2\sin x - \sin 2x,$$

then the linear combination

$$-2\phi_1 + \phi_2 + \phi_3$$

is identically zero, which violates (4.1) for $c_1 = -2$, $c_2 = 1$ and $c_3 = 1$. By contrast the first set of functions is linear independent. To see that, let F stand for any linear combination of the functions ϕ's in this example, i.e.,

$$F(x) = c_1 \sin x + c_2 \sin 2x + c_3 \sin 3x.$$

According to (4.1) this combination must vanish for all $x \in (a, b)$. Since F is a smooth function, the fact that F vanishes identically implies that its first and second derivatives also vanish identically. This observation leads to the following three equations for the variables c_1, c_2 and c_3:

$$\begin{cases} c_1 \sin x + c_2 \sin 2x + c_3 \sin 3x & = 0 \\ c_1 \cos x + 2c_2 \cos 2x + 3c_3 \cos 3x & = 0 \\ -c_1 \sin x - 4c_2 \sin 2x - 9c_3 \sin 3x & = 0, \end{cases}$$

or in matrix form

$$A\mathbf{c} = \mathbf{0}$$

where

$$A = \begin{bmatrix} \sin x & \sin 2x & \sin 3x \\ \cos x & 2\cos 2x & 3\cos 3x \\ -\sin x & -4\sin 2x & -9\sin 3x \end{bmatrix}. \tag{4.2}$$

We recall from Chapter 2 that a system of algebraic equations $A\mathbf{x} = \mathbf{b}$ has a unique solution when A is nonsingular. The determinant of the matrix A in (4.2) is $-16\sin^6 x$, which can be readily verified in a symbolic manipulator such as *Mathematica*®. Since A is a nonsingular matrix for a typical value of x in the interval (a, b), we conclude that $\mathbf{c} = A^{-1}\mathbf{0} = \mathbf{0}$, thus verifying (4.1). See Problem 1 for further discussion on linear independence.

We say C, a set of functions, forms a *linear space* if

1. *for every $f \in C$ and $g \in C$ we have $f + g \in C$; when this property holds, we say C is closed under the operation of addition. And if*

2. *for every $f \in C$ and c a scalar (real or complex) we have $cf \in C$. Similarly, when this property holds, we say that C is closed under the operation of scalar multiplication.*

The concepts of *span*, *basis* and *dimension* generalize verbatim to space of functions. We say the span S of a set $C = \{\phi_1, \phi_2, ..., \phi_n\}$ is

$$S = \{f |\, f(x) = c_1\phi_1(x) + c_2\phi_2(x) + ... + c_n\phi_n(x)\}$$

where the constants $c_1, c_2, \ldots$ take arbitrary real or complex values. If the set C is linearly independent, the information content of the function ϕ_i will be truly different from the information content of ϕ_j, with $i \neq j$, and therefore to characterize the span S we will need every function ϕ_i in C. In this sense, we say that C forms a basis for S and that the dimension of S is the cardinality n of the set C. These concepts are identical with the equivalent concepts that we discussed in Chapter One.

On the other hand, when there is redundancy in the set C, when one of the ϕ_i's is actually a linear combination of other ϕ_i's, then the dimension of S will be generally less n and one needs to work a bit more to identify an appropriate basis. Some of the problems at the end of this section address this issue.

As an example, consider the set of functions $C = \{\sin \pi x, \sin 2\pi x, \sin 3\pi x, \ldots, \sin n\pi x\}$, which forms a linearly independent set. Its span

$$S = \{\sum_{i=1}^{N} c_i \sin i\pi x |\, c_i \in R\},$$

is an N-dimensional space for which the functions in C form a basis. In a sense C is equivalent to the linear space of N-dimensional vectors in E^N, where we can think of each element of C, $c_1 \sin \pi x + c_2 \sin 2\pi x + \ldots + c_n \sin n\pi x$, as the vector $\langle c_1, c_2, \ldots, c_n \rangle$. With this perspective the space of functions spanned by the basis $\{\sin i\pi x\}$, $i = 1, \ldots, N$, is in one-to-one and onto correspondence with the space E^N.

Problems 4.1

1. Consider a set of smooth functions $C = \{\phi_1(x), \phi_2(x), \ldots, \phi_n(x)\}$. Show that the functions ϕ_i are linearly independent if the matrix

$$\begin{bmatrix} \phi_1 & \phi_2 & \cdots & \cdots & \phi_n \\ \phi_1' & \phi_2' & \cdots & \cdots & \phi_n' \\ \cdots & \cdots & \cdots & \cdots & \cdots \\ \cdots & \cdots & \cdots & \cdots & \cdots \\ \phi_1^{(n-1)} & \phi_2^{(n-1)} & \cdots & \cdots & \phi_n^{(n-1)} \end{bmatrix}$$

is nonsingular. The notation $f^{(i)}$ stands for $\frac{d^i f}{dx^i}$. The determinant of the above matrix is often called the *Wronskian* of the functions ϕ.

2. Show that the following set of functions are linearly independent. In each case write down the linear space they span and identify this space with E^n for an appropriate n.

   (a) $\phi_1(x) = 1$, $\phi_2(x) = x$.

(b) $\phi_1(x) = x$, $\phi_2(x) = x^2$.
(c) $\phi_1(x) = 1$, $\phi_2(x) = x$, $\phi_3(x) = x^2$.
(d) $\phi_1(x) = 1$, $\phi_2(x) = x$, $\phi_3(x) = x^2$, ..., $\phi_n(x) = x^n$.

3. Consider the set C consisting of the functions $\phi_1(x) = 1$, $\phi_2(x) = x$, $\phi_3(x) = x^2$. Let S be the span of C.

   (a) What is the dimension of S?

   (b) Consider the set of functions $D = \{\psi_1, \psi_2, \psi_3\}$ defined by

   $$\psi_1(x) = \phi_1(x) - \phi_2(x), \quad \psi_2(x) = \phi_2(x) - \phi_3(x),$$
   $$\psi_3(x) = \phi_3(x) - \phi_1(x).$$

   i. Show that the set D is linearly independent.
   ii. Write ϕ_1 as a linear combination of ψ_i's, i.e., determine the coefficients c_1, c_2 and c_3 so that $\phi_1(x) = c_1\psi_1(x) + c_2\psi_2(x) + c_3\psi_3(x)$.
   iii. Let f be an arbitrary function in the span of C, i.e., there are constants c_1, c_2 and c_3 such that

   $$f(x) = c_1\phi_1(x) + c_2\phi_2(x) + c_3\phi_3(x).$$

   The function f also belongs to the span of D, that is, there are constants d_1, d_2 and d_3 such that

   $$f(x) = d_1\psi_1(x) + d_2\psi_2(x) + d_3\psi_3(x).$$

   Find the relationship between c_i's and d_j's. Moreover, write down the 3×3 matrix A that relates the vectors $\mathbf{c} = \langle c_1, c_2, c_3 \rangle$ and $\mathbf{d} = \langle d_1, d_2, d_3 \rangle$.

4. Show that the following sets of functions are linearly independent. Write down the linear space they span. What is the dimension of each space?

   (a) $f_1(x) = \sin x$, $f_2(x) = \sin 2x$
   (b) $f_0(t) = 1$, $f_1(t) = \cos \omega t$, $f_2(t) = \sin \omega t$, $f_3(t) = \sin 2\omega t$, $f_4(t) = \cos 2\omega t$, where ω is a nonzero real number.

5. Consider the set of functions $\{e^{-t}, e^{-2t}, e^{-3t}, ..., e^{-nt}\}$ on the interval $(0, 1)$. Is this set linearly independent?

6. Consider the set of functions $C = \{\sin 2x, \sin 4x, 2\sin 2x - 3\sin 4x\}$.

   (a) Is this set linearly independent?
   (b) What is the span S of C? Is the function $\sin 3x$ in S? if not, why not?
   (c) Write down two sets of bases for S.

4.2 Linear ODEs

Consider the differential equation

$$y'' + ay' + by = f(x) \tag{4.3}$$

where a and b are constants, and f is a known function. This equation is *second order*, because the highest order derivative present in (4.3) is second order, is *linear* because all terms involving the unknown function y enter linearly in this equation, and *nonhomogeneous* if f is nonzero. Typically equation (4.3) is supplemented by the two *initial conditions*

$$y(x_0) = y_0, \quad y'(x_0) = y_1, \tag{4.4}$$

or by two *boundary conditions*

$$y(c) = \alpha, \quad y(d) = \beta. \tag{4.5}$$

In either case one first typically finds the general solution to (4.3) and then applies the initial or boundary conditions to find the exact solution.

The general solution to (4.3) is obtained in two steps: first one determines the general solution (sometimes called the *complementary* solution) of the *homogeneous* part

$$y'' + ay' + by = 0, \tag{4.6}$$

and second a *particular* solution y_p of (4.3), by any method available. The general solution of (4.3) is then the sum of y_c and y_p:

$$y(x) = y_c(x) + y_p(x). \tag{4.7}$$

Because (4.6) is linear and has constant coefficients, its general solution is a linear combination of exponential functions $e^{\lambda x}$. Substituting this expression into (4.6) leads to

$$e^{\lambda x}(\lambda^2 + a\lambda + b) = 0$$

from which we infer that λ must be the roots of the second order polynomial

$$\lambda^2 + a\lambda + b$$

which we refer to as the *characteristics polynomial.* When the two roots, λ_1 and λ_2, of the characteristic polynomial are distinct, we obtain two linearly independent solutions of the homogeneous equation (4.6),

$y_1(x) = e^{\lambda_1 x}$ and $y_2(x) = e^{\lambda_2 x}$, from which we construct the general solution y_c by simply combining a linear combination of y_1 and y_2:

$$y_c(x) = c_1 e^{\lambda_1 x} + c_2 e^{\lambda_2 x}, \tag{4.8}$$

where c_1 and c_2 are arbitrary constants. In the special circumstance when the two roots of the characteristic polynomial are identical, i.e., when $\lambda_1 = \lambda_2 = \lambda$, we are able to construct two linearly independent solutions $e^{\lambda x}$ and $xe^{\lambda x}$, and the general solution to (4.6) is

$$y_c(x) = c_1 e^{\lambda x} + c_2 x e^{\lambda x}. \tag{4.9}$$

We note that λ_1 and λ_2 in (4.8) may be complex numbers. When a and b in (4.3) are real-valued constants, and λ_1 and λ_2 have ended up being complex-valued, the latter must be complex conjugates of each other, i.e., $\lambda_1 = \gamma + \delta i$ and $\lambda_2 = \bar{\lambda}_1$. In that case the two functions $e^{\gamma x}\cos\delta x$ and $e^{\gamma x}\sin\delta x$ form two linearly independent and real-valued solutions of (4.6) so that

$$y_c(x) = c_1 e^{\gamma x}\cos\delta x + c_2 e^{\gamma x}\sin\delta x \tag{4.10}$$

is the general solution to (4.6).

As mentioned earlier, we obtain a particular solution y_p to (4.3) by any means possible, including a judicious guess. Often this forcing term f in (4.3) ends up being of sinusoidal type, as will be the case in the forcing terms induced by the prevailing winds in most of the models we will consider. In those cases we will apply a simple ansatz, as we will show shortly by an example, based on the general form of the forcing term itself, to arrive at y_p. Once (4.7) is determined, we apply the initial conditions (4.4) or the boundary conditions (4.5) to compute c_1 and c_2 in any of the formulas (4.8), (4.9) or (4.10).

We summarize these findings in the following theorem.

Theorem: (IVP and BVP for Linear ODEs)
Consider the second-order ordinary differential equation (4.3)

$$y'' + ay' + by = f(x),$$

where a and b are real numbers, supplemented by either initial conditions (4.4), which we will refer to as an Initial Value Problem (IVP), or boundary conditions (4.5), referred to as a Boundary Value Problem (BVP). The general solution of (4.3) is of the form (4.7), that is

$$y(x) = y_c(x) + y_p(x)$$

where y_c is the solution of the homogeneous equation (4.6)

$$y'' + ay' + by = 0$$

(see (4.8), (4.9) and (4.10)) and y_p is any particular solution of the full equation. When the roots of the characteristic polynomial

$$\lambda^2 + a\lambda + b$$

are real and distinct, the general solution of (4.3) is

$$y(x) = c_1 e^{\lambda_1 x} + c_2 e^{\lambda_2 x} + y_p(x),$$

and the constants c_1 and c_2 are chosen appropriately so that the side conditions (4.4) or (4.5) are satisfied. When the roots of the characteristic polynomial are repeated or are complex, the solution to the homogeneous part takes the form (4.9) and (4.10), respectively.

While determining y_c, the solution to the homogeneous part (4.3), is straightforward, finding a particular solution y_p of (4.3) requires our attention. There are several techniques for coming up with a particular solution y_p, some of which we will encounter in the exercises at the end of this section; here we outline only one of these techniques, a relatively simple one, called the *Method of Undetermined Coefficients*,. This technique is suited particularly well for the case when the forcing term f in (4.3) is a linear combination of simple trigonometric functions, $\sin \omega x$ and $\cos \omega x$, or exponential functions; we will come across this type of a forcing term when we study the Stommel solution to the Gulf-Stream problem later in the text.

As an example to illustrate the Method of Undetermined Coefficients, consider the initial value problem

$$y'' + 0.1y' + 3y = 2\sin 3x, \quad y(0) = y'(0) = 0. \tag{4.11}$$

The homogeneous part of (4.11) is

$$y'' + 0.1y' + 3y = 0, \tag{4.12}$$

which, after substituting $e^{\lambda x}$, leads to the characteristic polynomial equation $\lambda^2 + 0.1\lambda + 3 = 0$. Its roots are $\lambda_1 = -0.05 + 1.7313i$ and $-0.05 - 1.7313i$ (which is MATLAB's output to `roots([1 0.1 3])`). We therefore conclude that y_c, the complementary solution, is

$$y_c(x) = e^{-0.05x}(c_1 \cos 1.7313x + c_2 \sin 1.7313x). \tag{4.13}$$

Next, to find a particular solution y_p, we try the template

$$y_p(x) = A\sin 3x + B\cos 3x, \tag{4.14}$$

which consists of the forcing term $\sin 3x$ itself and all of its derivatives (i.e., $\cos 3x$ and $\sin 3x$). We substitute the template (4.14) into (4.11) and equate the coefficients of $\sin 3x$ and $\cos 3x$ on either side of (4.11) to get the following set of algebraic equations in A and B:

$$-6A - 0.3B = 2, \quad -6B + 0.3A = 0, \tag{4.15}$$

resulting in $A = -0.332502$ and $B = -0.0166251$. Hence, the general solution to (4.11) is

$$y(x) = e^{-0.05x}(c_1 \cos 1.7313x + c_2 \sin 1.7313x)$$

$$-0.332502 \sin 3x - 0.0166251 \cos 3x. \tag{4.16}$$

The constants c_1 and c_2 are now found by applying the initial data in (4.11): Since $y(0) = 0$ and $y'(0) = 0$, we have

$$0 = y(0) = c_1 - 0.0166251, \quad 0 = y'(0) = -0.05c_1 +$$

$$1.7313c_2 - 0.997506$$

or $c_1 = 0.016625$ and $c_2 = 0.57664$. Hence, the exact solution to (4.11) is

$$y(x) = e^{-0.05x}(0.016625 \cos 1.7313x + 0.57664 \sin 1.7313x) +$$

$$-0.332502 \sin 3x - 0.0166251 \cos 3x. \tag{4.17}$$

The following commands in MATLAB lead to Figure 4.1.

```
f=@(x)exp(-0.05*x).*(0.016625*cos(1.7313*x)+ ...
0.57664*sin(1.7313*x))-0.332502*sin(3*x)-...
+0.0166251*cos(3*x);
ezplot(f,[0,2*pi])
```

Most of what we have outlined above easily generalizes to higher-order differential equations, especially when initial data are specified. We will not pursue the development of that topic since its utility in the other topics we intend to develop is limited.

Problems 4.2

1. Consider the following pair of functions y_1 and y_2. In each case show that y_1 and y_2 are linearly independent on any arbitrary interval (a, b).

   (a) $y_1(x) = e^{\lambda_1 x}$ and $y_2(x) = e^{\lambda_2 x}$, where $\lambda_1 \neq \lambda_2$.

   (b) $y_1(x) = e^{\lambda x}$ and $y_2(x) = xe^{\lambda x}$.

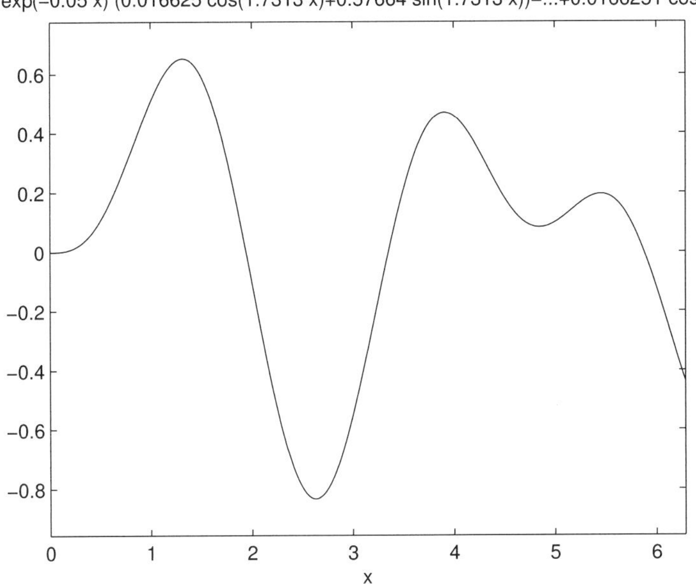

FIGURE 4.1: The solution to (4.11).

(c) $y_1(x) = e^{\lambda x} \cos \omega x$ and $y_2(x) = e^{\lambda x} \sin \omega x$, where $\omega \neq 0$.

2. Find the general solution of the following ODEs:
   (a) $y'' + 3y' + 2y = 0$.
   (b) $y'' - 4y' + 4y = 0$.
   (c) $y'' + 16y = 0$.
   (d) $y'' + 4y = 3 \sin x$.
   (e) $y'' - 3y' + 4y = 2e^x + 3 \cos x$.

3. (*WolframAlpha*) If you have access to the internet, access the link
   http://www.wolframalpha.com
   to call up this powerful information engine on the web. Having access to the capabilities of *Mathematica* is one of the features of this web site. Try

   $$\text{solve } y'' + 4y = 3\sin(2x)$$

   in the input window to see this site's capability to provide information about this ODE or any of the differential equations in the above problem.

4. Find the solution of the following IVPs:

   (a) $y'' + 2y' + y = 5\cos 2x, \quad y(0) = y'(0) = 0.$
   (b) $x'' + x = \sin 2t, \quad x(0) = 0,\ x'(0) = 1.$

5. (*The Phenomenon of Resonance*) When the forcing term f in (4.3) is of the trigonometric or exponential type and f itself is a solution of the homogeneous equation $y'' + ay' + by = 0$, the approach of trying a template based on f and all its derivatives fails to yield a particular solution y_p. Instead, we modify this method by seeking particular solutions of the form $xf(x)$ and its derivatives. As an example, consider $y'' + y = \sin x$, noting that the $\sin x$ is already a solution to $y'' + y = 0$. Hence seeking a particular solution of the form $y_p(x) = A\sin x + B\cos x$ fails. The suggested modification of the Method of Undetermined Coefficients, that one should seek particular solutions of the form $y_p(x) = x(A\sin x + B\cos x)$, results in the desired solution. We note, however, that a term such as $x\sin x$ becomes unbounded in x so that the energy imparted to the system by the forcing term $\sin x$ is amplified by x, leading to solutions that grow unboundedly. This phenomenon of resonance is observed in many systems, especially in mechanical and electrical systems and often leads to undesirable outcomes, typically to system failure. Resonance is possibly at the heart of the unusual enhancement of wind-driven waves in harbors and estuaries.

   In each problem below determine a particular solution to the ODE:

   (a) $y'' + y = \sin x$. (Try $y_p(x) = x(A\sin x + B\cos x)$ and find A and B.)
   (b) $y'' + 4y' + 4y = e^{-2x}$. (Try $y_p(x) = Axe^{-2x}$.)
   (c) $y'' + 6y' + 9y = \sin x + e^{-3x}$.

6. Find the solution of the following BVPs:

   (a) $y'' + 4y' + 3y = 1, \quad y(0) = 1,\ y(2) = -1.$
   (b) $y'' - y = \sin t, \quad y'(0) = 0,\ y(3) = 0.$

4.3 General Systems of ODEs

The equation we discussed in the previous, Equation (4.11), is an example of a larger class of ordinary differential equations of the form

$$\mathbf{y}' = \mathbf{f}(x, \mathbf{y}), \quad \mathbf{y}(x_0) = \mathbf{y}_0. \tag{4.18}$$

Higher order equations such as (4.3) can be converted to a first-order system like (4.18) by simply renaming the various derivatives in (4.3): To illustrate, consider the n-th equation

$$\frac{d^n z}{dx^n} = f(x, z, z', z'', ..., z^{(n-1)}), \tag{4.19}$$

subject to the initial conditions

$$z(x_0) = z_0, \quad z'(x_0) = z_1, \quad ..., \quad z^{(n-1)} = z_{n-1}.$$

Define a new variable $\mathbf{y} = \langle y_1, y_2, ..., y_n \rangle$ by

$$y_1 = z, \quad y_2 = z', \quad y_3 = z'', \quad ..., \quad y_n = \frac{d^{n-1}z}{dx^{n-1}}. \tag{4.20}$$

Differentiate the first expression, $y_1 = z$, to arrive at

$$y_1' = z'.$$

The second expression in (4.20), however, relates z' to y_2. Hence, $y_1' = z' = y_2$ and we obtain the first equation in the new variables:

$$y_1' = y_2.$$

Similarly

$$y_2' = z'' = y_3,$$

and

$$y_3' = z''' = y_4,$$

and so forth. Finally we arrive at $y_n' = \frac{d^n z}{dz^n}$ which reduces to

$$y_n' = f(x, y_1, y_2, ..., y_n)$$

once we invoke the original differential equation (4.19). We have thus demonstrated that the original n-th order differential equation is equivalent to the following n-th system of first-order equations:

$$\begin{cases} y_1' &= y_2, \\ y_2' &= y_3, \\ \cdots & \cdots \\ y_{n-1}' &= y_n, \\ y_n' &= f(x, y_1, y_2, ..., y_n), \end{cases} \tag{4.21}$$

with the initial data

$$y_1(x_0) = z_0, \quad y_2(x_0) = z_1, \quad ..., \quad y_n(x_0) = z_{n-1}. \tag{4.22}$$

This scheme, when applied to (4.3), leads to the following first-order system: Let $y_1 = y$ and $y_2 = y'$. Then

$$y_1' = y_2, \quad y_2' = -ay_1 - by_2 + f(x), \tag{4.23}$$

or in matrix form

$$\mathbf{y}' = \mathbf{f}(x, \mathbf{y}), \text{ where } \mathbf{y} = \begin{bmatrix} y_1 \\ y_2 \end{bmatrix}, \mathbf{f} = \begin{bmatrix} y_2 \\ -ay_1 - by_2 + f(x) \end{bmatrix}, \tag{4.24}$$

with $\mathbf{y}_0 = \begin{bmatrix} y_0 \\ y_1 \end{bmatrix}$.

Initial-value problems for a general system of the form (4.18) are extremely difficult to solve analytically and we often have no choice but to try to seek approximate solutions. It would be very helpful, however, if we had some guarantee that there is a solution to the problem posed in (4.18) and that this solution is unique — once the existence of a *unique* solution is guaranteed, it does not matter then which approximate or numerical method we attempt, since all of these approximate solutions should be close to the one and only solution of (4.18). To answer the question of existence and uniqueness for (4.18), and its natural extension to the partial differential equations that we will study later, has constituted one of the cornerstones of modern mathematical analysis. In particular, a comprehensive qualitative theory of differential equations has been developed to identify conditions on $\mathbf{f}$ to guarantee existence and uniqueness of solutions to (4.18) and the asymptotic behavior of these solutions as x approaches infinity. Here we deal with the set of sufficient conditions that guarantee existence and uniqueness.

Let the function $\mathbf{f}$ in (4.18) be continuous in x and $\mathbf{y}$ in domain $D \subset R \times E^n$, and *Lipschitz* continuous in $\mathbf{y}$, that is, there is a constant M, perhaps depending on x, such that

$$||\mathbf{f}(x, \mathbf{y}_2) - \mathbf{f}(x, \mathbf{y}_1)|| \leq M||\mathbf{y}_2 - \mathbf{y}_1||, \tag{4.25}$$

for $\mathbf{y}_1$ and $\mathbf{y}_2$ in the domain of f. Then a fundamental theorem of ordinary differential equations (see Ref. [1]) states that a unique solution of (4.18) exists for x in a neighborhood of x_0. Moreover, either this solution exists for all x or the solution $\mathbf{y}$ will blow-up in finite time, that is, there is an L such that

$$\lim_{x \to L} ||\mathbf{y}(x)|| = \infty. \tag{4.26}$$

Interestingly, since solutions of linear equations such as (4.11) are combinations of exponential functions, finite-time blow-up is not an option for them. For nonlinear equations, however, this behavior is common, as seen in the example

$$y' = -y^2, \; y(0) = -1, \tag{4.27}$$

a Riccati-type equation, whose exact solution is $y(x) = \frac{1}{x-1}$, which blows up as x approaches 1. Note that the function $-y^2$ is smooth in all of R and there is no hint of the blow-up behavior by examining the right side of (8.127).

The Lipschitz property quoted above seems to be necessary for the uniqueness property. As a counterexample, consider the equation $y' = \sqrt{y}$ with $y(0) = 0$, which has the two distinct solutions

$$y(x) \equiv 0 \quad \text{and} \quad y(x) = \frac{x^2}{4}. \tag{4.28}$$

The problem with this equation is that its right side, $\sqrt{y}$, has infinite slope at $y = 0$, the initial value, and cannot support a bound such as (4.25) for any M.

4.4 MATLAB's `ode45`

We digress momentarily from our development of the theory of ODEs to present the syntax for MATLAB's `ode45` because, unlike the linear ODEs we discussed earlier, in this text we will often be dealing with nonlinear ODEs whose analytical solutions are intractable. We will be discussing in some detail in the next several sections how to generate approximate numerical solutions to nonlinear ODEs, but before addressing the mathematical development of numerical schemes, we describe here how to use `ode45`, an accurate and efficient ODE-solver, because we will resort to this tool in this chapter as a benchmark when we introduce some elementary numerical schemes for solving ODEs. In a later section we will also see how to use `ode45` with the *method of lines* to solve partial differential equations.

Consider the initial-value problem (IVP)

$$y' = f(t, y), \quad y(t_0) = y_0.$$

Here we may have a single equation, i.e., y may be a scalar, or a system of ODEs where y is vector in E^n. The syntax for using `ode45` on this IVP is

```
[TOUT,YOUT] = ODE45(ODEFUN,TSPAN,YO)
```

where `ODEFUN` defines the right side of the ODE (more on this later), `TSPAN` defines the domain of t, typically (t_0, T), and `YO` contains the initial data. The quantities TOUT and YOUT contain the output of ode45. As an example consider the IVP

$$y' = -y^2 + t, \quad y(0.1) = -0.3.$$

The following lines in MATLAB will result in an approximate solution of this problem in the interval $(0.1, 4)$:

```
%
%  Use the inline command to define f
%
f=inline('-y.^2 + t', 't', 'y');
%
% Apply ode45
%
[t, y] = ode45(f, [0.1 4], -0.3);
%
```

The output in stored in t and in y. What ode45 has done is to use its technique to break up the t values in the interval (0.1 4) into smaller subintervals, with endpoints that we label (t_i, t_{i+1}) and then proceeded to compute y_i and each grid point t_i, where y_i is an excellent approximation of y at t_i. The values of t_i and y_i are stored in t and y. These vectors have equal length (try length(t) and length(y) to see the size of these vectors) and can be plotted against each other using the plot command:

```
plot(t,y)
```

A close look at t shows that these grid points are not uniform, that is, the distance $t_{i+1} - t_i$ varies with i. This grid-size adaptivity is actually one of the special features of ode45, to which we will return later in the chapter.

Applying ode45 to a system of ODEs is similar. The one difference arises in the difficulty with using the inline command when defining the right side of the ODEs, which is rather cumbersome when it comes to concatenating expressions. Instead we will use the M-file utility of MATLAB as described in the following example. Consider the system

$$x' = y, \quad y' = -0.1y - \sin x + \cos t, \qquad x(0) = 1, \quad y(0) = 2.$$

We intend to solve this IVP and obtain an approximation to the solution $(x(t), y(t))$ for $t \in (0, 3)$. To that end, we first define the right side of the ODEs in an M-file (called rhs.m for later reference) in MATLAB:

```
function yprime=rhs(t,y);
%
yprime = [y(2); -0.1*y(2)-sin(y(1))+cos(t)];
```

We apply `ode45` to `rhs.m` as follows:

```
[t,y]=ode45('rhs',[0 3],[1 2]);
```

or by entering

```
[t,y]=ode45(@rhs,[0 3],[1 2]);
```

The output is stored in `t` and in `y`. We can choose to plot each component of `y` versus `t` by entering

```
plot(t,y(:,1))  % or
plot(t,y(:,2))
```

or plot the phase-plane diagram (x versus y components) by

```
plot(y(:,1),y(:,2))
```

A particularly useful option within `ode45` is `odeplot` (which is used in combination with `ODEset`). This tool generates the graphs of the time plots of the output variables and places circles at the evaluated points. The output of

```
[t,y]=ode45(@rhs,[0 3],[1 2],odeset('OutputFcn',@odeplot));
```

is shown in Figure 4.2. Equally useful option in `ode45` is `odephas2` which generates the phase-plane portrait of a two-dimensional system of ODEs. The output of

```
[t,y]=ode45(@rhs,[0 30],[1 2],odeset('OutputFcn',@odephas2));
```

is shown in Figure 4.3. Note the t-domain in the latter figure.

Problems 4.4

1. Use `ODE45` to solve the following IVPs:

   (a) $y'' + 2y' + y = \sin t$, $y(0) = 1$, $y'(0) = 0$ in the interval $(0, 10)$.

   (b) $y'' + 2ty' + y^2 = \cos t$, $y(0) = 0$, $y'(0) = 0$ in the interval $(0, 4)$.

2. Plot the graphs of the solutions of the following systems of ODEs:

   (a) $x_1' = x_2 - x_1$, $x_2' = -2x_1 - 0.3x_2$, $x_1(0) = 1$, $x_2(0) = -1$.

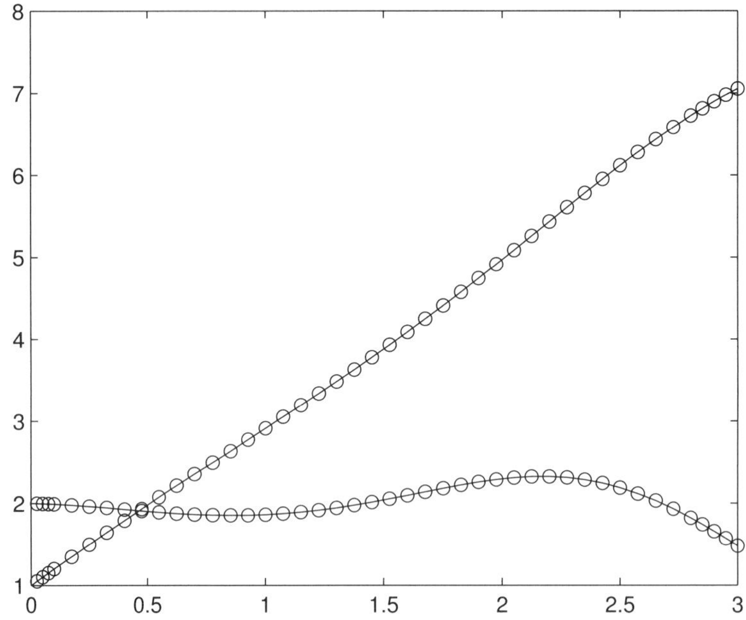

FIGURE 4.2: The output of `odeplot` when used within `ode45`.

(b) $x_1' = \frac{x_2 - x_1^2}{x_1^2 + x_2^2}$, $x_2' = \frac{x_1 + x_2}{x_1^2 + x_2^2}$, $x_1(0) = 0$, $x_2(0) = 1$.

3. The Lorenz system of equations, which displays complex and chaotic behavior, is

$$x' = \sigma(y - x),\ y' = x(\rho - z) - y,\ z' = xy - \beta z,$$

with initial data

$$x(0) = x_0,\ y(0) = y_0,\ z(0) = z_0,$$

where σ, ρ and β are physical constants. Plot the graphs of the solutions for the parameter values and initial positions listed:

(a) $\sigma = \rho = 1, \beta = 0.1$ and $x_0 = y_0 = 0$, $z_0 = 1$ with $t \in (0, 100)$.

(b) $\sigma = \rho = 1, \beta = -0.1$ and $x_0 = y_0 = 0$, $z_0 = 1$ with $t \in (-100, 100)$.

(c) Let $\sigma = 10, \rho = 28$, $x_0 = y_0 = 0$, $z_0 = 1$ and $t \in (0, 100)$. Plot the graph of the solution when

i. $\rho = 20$.

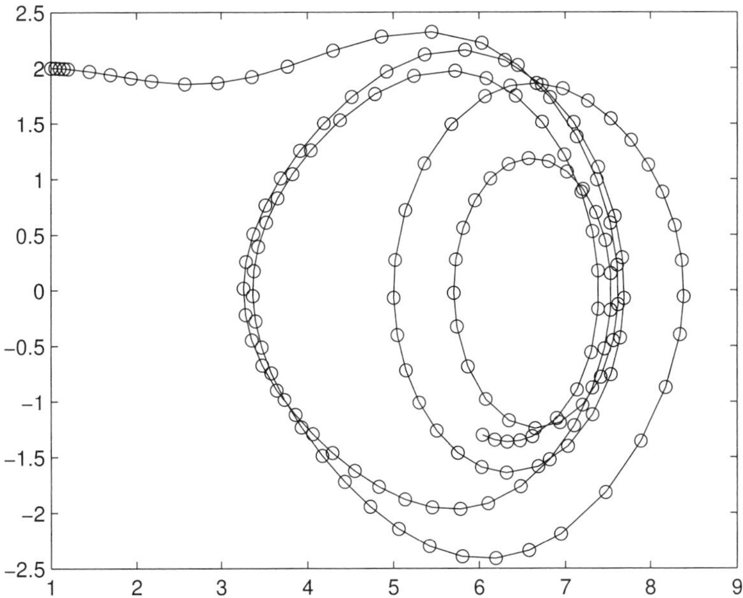

FIGURE 4.3: The output of `odephas2` when used within `ode45`.

ii. $\rho = 28$.

iii. $\rho = 90$.

iv. $\rho = 100$.

4. The **ABC**, or the Arnold–Beltrami–Childress, flow is defined by the following system of ODEs:

   $x' = A\sin z + C\cos y$, $y' = B\sin x + A\cos z$, $z' = C\sin y + B\cos x$.

   Let $A = 1$, $B = 0.1$, $C = -0.2$ and $(x_0, y_0, z_0) = (0.1, 0.2, 0.1)$. Plot the graph of the solution for $t \in (0, 50)$.

4.5 Asymptotic Behavior and Linearization

An important feature of analyzing fluid flows, especially in the context of the main theme of this text, the geophysical fluid flows, is understanding how the long-time behavior of currents depends on the various physical parameters in play. The qualitative theory of ODEs is particularly well-suited for this effort, and in the setting of nonlinear ordinary differential equations, linearization of the ODE system about solutions provides a critical tool. We introduce this concept for the system of two equations in two unknowns

$$\frac{dx}{dt} = f(t, x, y), \quad \frac{dy}{dt} = g(t, x, y), \tag{4.29}$$

although the ideas generalize readily to higher dimensional settings. We call the system in (4.4) *autonomous* if f and g do not explicitly depend on t. Most of the development below is dedicated to autonomous ODEs. Also, as is common in mathematics, we will often refer to (4.29) as a *dynamical system*, and to the xy-plane as its *phase plane*.

We say a point (a, b) is an *equilibrium point* of (4.29) if

$$f(t, a, b) = 0, \quad g(t, a, b) = 0, \tag{4.30}$$

for all t. Note that when (a, b) that satisfies (4.30), the functions

$$x(t) \equiv a, \quad y(t) \equiv b \tag{4.31}$$

form a solution to (4.29), which we will refer to as an *equilibrium solution*.

We say a solution $z(t) = \langle x(t), y(t) \rangle$ of (4.29) is *stable* (often called *Liapunov* stable) if solutions of (4.29) that start out close to $z(t)$ remain close to this solution for all time. To be precise, we say z is a stable solution of (4.29) if for every $\epsilon > 0$ there is a $\delta > 0$ such that

$$\text{if } ||\bar{z}(0) - z(0)|| < \delta \text{ then } ||\bar{z}(t) - z(t)|| < \epsilon \text{ for all } t > 0, \tag{4.32}$$

where $\bar{z}$ is a solution of (4.29). We say an equilibrium point (a, b) is *asymptotically stable* if in addition to being stable, it satisfies

$$\lim_{t \to \infty} ||\bar{z}(t) - z(t)|| = 0.$$

The concept of stability of solutions of ODEs has been studied extensively and several consequences of this working definition have been derived to aid a user to determine whether a solution is stable (see the references cited at the end of this section). We will elaborate on these ideas in the context of equilibrium points (4.30) and equilibrium solutions (4.31) for autonomous systems where it turns out that linearization and Taylor expansion about an equilibrium point are the main tools of analysis. Before developing these ideas we emphasize one important feature about the definition of stability in (4.32), that the bound on the distance between z and $\bar{z}$ is to hold for all $t > 0$ and not just for t in a finite interval. The latter is not much of a restriction on a dynamical system, since all that is required is continuity of the solution as a function of its initial condition, which is referred to as *continuous dependence* on initial data in the mathematical literature. Continuous dependence on initial data is a property of the dynamical system in (4.4) and typically holds for all solutions of such a system, stable or not, under rather mild assumptions on f and g in (4.4). Satisfying (4.32), however, is a property of individual solutions (equilibrium points) as we will see shortly.

Let (a, b) be an equilibrium point of the autonomous system

$$x' = f(x, y), \quad y' = g(x, y). \tag{4.33}$$

Let $z = \langle x, y \rangle$ be the equilibrium solution and let $\bar{z} = \langle a+\epsilon\bar{x}(t), b+\epsilon\bar{y}(t) \rangle$ be a perturbation of z, with the understanding that ϵ is a small number. Since $\bar{z}$ is a solution of (4.5) we have

$$\epsilon\bar{x}' = f(a + \epsilon\bar{x}(t), b + \epsilon\bar{y}(t)), \quad \epsilon\bar{y}' = g(a + \epsilon\bar{x}(t), b + \epsilon\bar{y}(t)). \tag{4.34}$$

We expand the right sides of (4.34) about $\epsilon = 0$ to get (recalling that $f(a, b) = g(a, b) = 0$)

$$\bar{x}' = \frac{\partial f}{\partial x}|_{(a,b)}\bar{x} + \frac{\partial f}{\partial y}|_{(a,b)}\bar{y} + \epsilon(\frac{1}{2}\frac{\partial^2 f}{\partial x^2}|_{(a,b)}\bar{x}^2 + \frac{\partial^2 f}{\partial x \partial y}|_{(a,b)}\bar{x}\,\bar{y} + \frac{1}{2}\frac{\partial^2 f}{\partial y^2}|_{(a,b)}\bar{y}^2) + \text{h.o.t.} \tag{4.35}$$

where h.o.t. stands for terms involving ϵ^2 and above. A similar expression holds for $\bar{y}'$:

$$\bar{y}' = \frac{\partial g}{\partial x}|_{(a,b)}\bar{x} + \frac{\partial g}{\partial y}|_{(a,b)}\bar{y} + \epsilon(\frac{1}{2}\frac{\partial^2 g}{\partial x^2}|_{(a,b)}\bar{x}^2 + \frac{\partial^2 g}{\partial x \partial y}|_{(a,b)}\bar{x}\bar{y} + \frac{1}{2}\frac{\partial^2 g}{\partial y^2}|_{(a,b)}\bar{y}^2) + \text{h.o.t.} \tag{4.36}$$

The linear part of the above equations

$$\bar{x}' = \frac{\partial f}{\partial x}|_{(a,b)}\bar{x} + \frac{\partial f}{\partial y}|_{(a,b)}\bar{y}, \qquad \bar{y}' = \frac{\partial g}{\partial x}|_{(a,b)}\bar{x} + \frac{\partial g}{\partial y}|_{(a,b)}\bar{y}, \tag{4.37}$$

contains the information of immediate interest, the argument being that the remaining terms, which depend on ϵ, remain small as time evolves. This observation ends up being the case for a large class of ODEs as long as the initial data associated with (4.5) is close enough to the equilibrium point. We now proceed to derive the precise conditions needed to arrive at this result.

The linearized system in (4.37) has the matrix form

$$\mathbf{x}' = A\mathbf{x} \quad \text{where} \quad \mathbf{x} = \begin{bmatrix} \bar{x} \\ \bar{y} \end{bmatrix} \quad \text{and} \quad A = \begin{bmatrix} \frac{\partial f}{\partial x} & \frac{\partial f}{\partial y} \\ \frac{\partial g}{\partial x} & \frac{\partial g}{\partial y} \end{bmatrix} \Big|_{(a,b)}. \tag{4.38}$$

The solutions of (4.38) are combinations of exponential functions. We seek them in the form

$$\mathbf{x}(t) = e^{\lambda t}\mathbf{v} \tag{4.39}$$

where λ and $\mathbf{v}$ are constants and yet to be determined. Substituting (4.38) into (4.38) yields the algebraic system $A\mathbf{v} = \lambda\mathbf{v}$, which states that the pair $(\lambda, \mathbf{v})$ is an eigenvalue-eigenvector pair associated with A. As discussed in Section 2.10, the eigenvalues of A are roots of the polynomial

$$\det\,(A - \lambda I)$$

and the associated eigenvector is found by applying Gaussian Elimination to

$$A\mathbf{x} = \lambda\mathbf{x}.$$

The general solution of (4.38) will be a linear combination of the special solutions in (4.39), analogous to our development of solutions of the second order ODE in (4.3). We are not considering all of the mathematical complications that could arise here, but as the reader can imagine, the same issues that arose for (4.3), such as real versus complex-valued solutions, and multiplicity of eigenvalues, also manifest themselves for (4.38). Suffice it to say that these mathematical complications can be addressed (see Refs. [1],[2] for details). The key feature to keep in mind, however, is that these issues do not alter the conclusion we will derive below regarding the stability of the equilibrium point (a, b).

Returning to (4.39) we note that the long-time behavior of this expression depends critically on the sign of the real part of λ: If all eigenvalues of A have the property that Re $\lambda < 0$ then we expect that the perturbation of the equilibrium point (a, b) will approach zero as time evolves, leading us to conclude that (a, b) is stable, while if one of these eigenvalues has its real part positive, then any perturbation of (a, b) will eventually grow and the equilibrium point (a, b) will be unstable. The case of Re $\lambda = 0$ remains ambiguous and arguments that draw upon the nonlinearities in (4.4) must be brought to bear in the analysis of such an equilibrium point. We summarize this discussion in the following theorem.

Theorem 4.1 (Asymptotic Stability of Equilibrium Points) *Consider the autonomous system of differential equations (4.5) with the point (a, b) as its equilibrium point. Let A, λ and $\mathbf{v}$ be defined as in*

(4.38). Then the equilibrium solution $\mathbf{x} = \langle a, b\rangle$ *of (4.5) is asymptotically stable in accordance with (4.32) if all eigenvalues of* A *have negative real parts. This equilibrium solution is unstable if any of the eigenvalues of* A *has a positive real part. If an eigenvalue of the linearization has zero real part, the state of stability of the equilibrium point must be determined by taking into account the nonlinear terms in the system.*

As an example of the utility of Theorem 4.1, consider the system of equations

$$x' = y, \quad y' = -\alpha \sin x - \beta y, \tag{4.40}$$

where α and β are non-negative. Note that all points of the form $(n\pi, 0)$, $n = 0, \pm 1, \pm 2, \ldots$ are equilibrium points of this system. Since $f(x, y) = y$ and $g(x, y) = -\alpha \sin x - \beta y$, the matrix A is

$$A = \begin{bmatrix} 0 & 1 \\ -\alpha \cos n\pi & -\beta \end{bmatrix} \tag{4.41}$$

where $a = n\pi$. The eigenvalues of this matrix are

$$\lambda_1(n) = \frac{1}{2}(-\beta - \sqrt{\beta^2 - 4\alpha \cos n\pi}), \quad \lambda_2(n) = \frac{1}{2}(-\beta + \sqrt{\beta^2 - 4\alpha \cos n\pi}). \tag{4.42}$$

We first consider the equilibrium solution $\langle 0, 0\rangle$. The two eigenvalues at $n = 0$ are $\lambda_1(0) = \frac{1}{2}(-\beta - \sqrt{\beta^2 - 4\alpha})$ and $\lambda_2(0) = \frac{1}{2}(-\beta + \sqrt{\beta^2 - 4\alpha})$. If β is small enough so that $\beta^2 - 4\alpha < 0$, then both λ_1 and λ_2 are both complex and Re $\lambda_i < 0$ so that $(0, 0)$ is stable in this case. If β is large enough so that $\beta^2 - 4\alpha > 0$ then both eigenvalues are real and negative. In this case also the equilibrium point $(0, 0)$ is stable, although as Figures 4.4 and 4.5 show, the behavior of the system near the origin differs in the two cases. This spiral in Figure 4.4 is due to the fact that eigenvalues of the linearization about $\langle 0, 0\rangle$ are complex-valued when $\alpha = 0.3$ and $\beta = 0.1$, while in Figure 4.5, with $\alpha = 0.3$ and $\beta = 1$, the orbits of (4.40) are attracted to the origin without undergoing any oscillations, under scoring the strength of the dissipation in this case. The equilibrium solution $\langle \pi, 0\rangle$ has eigenvalues $\frac{1}{2}(-\beta - \sqrt{\beta^2 + 4\alpha})$ and $\frac{1}{2}(-\beta + \sqrt{\beta^2 + 4\alpha})$, which are both real, one negative and the other positive. This equilibrium point is therefore unstable, as is also clear from Figures (4.4) and (4.5).

The two Figures 4.4 and 4.5 are the output of the following two MATLAB M-files:

```
%%% odedef.m %%%
function yprime=odedef(t,y);
global alpha beta
```

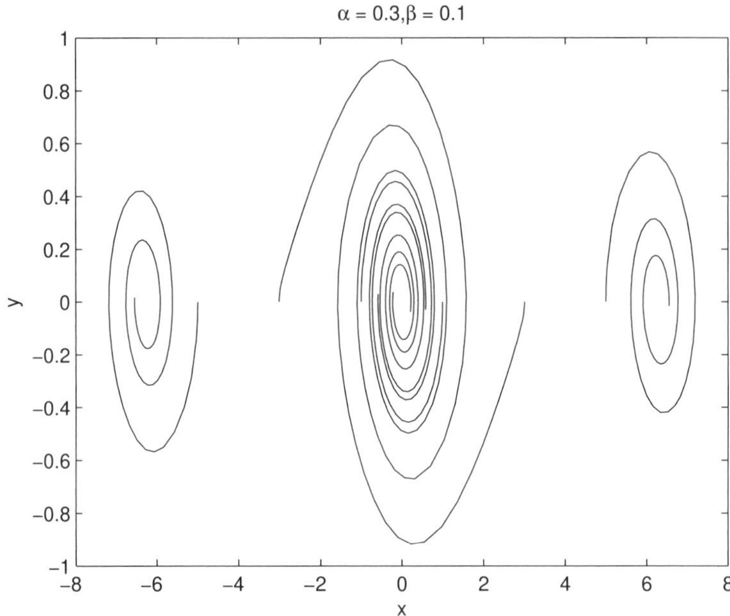

FIGURE 4.4: The phase plane for (4.40) with $\alpha = 0.3$, $\beta = 0.1$.

```
yprime=[y(2); -beta*y(2)-alpha*sin(y(1))];

%%% main.m  %%%
global alpha beta
alpha=0.3; beta=1;
for i=-5:2:5
[t,y]=ode45(@odedef,[0 30], [i 0]);
plot(y(:,1),y(:,2))
hold on
end
title(['\alpha = ',num2str(alpha),',', '\beta = ', ...
      num2str(beta)])
xlabel('x')
ylabel('y')
```

Problems 4.5

1. Find all equilibria of the following ODEs and determine their state of stability:

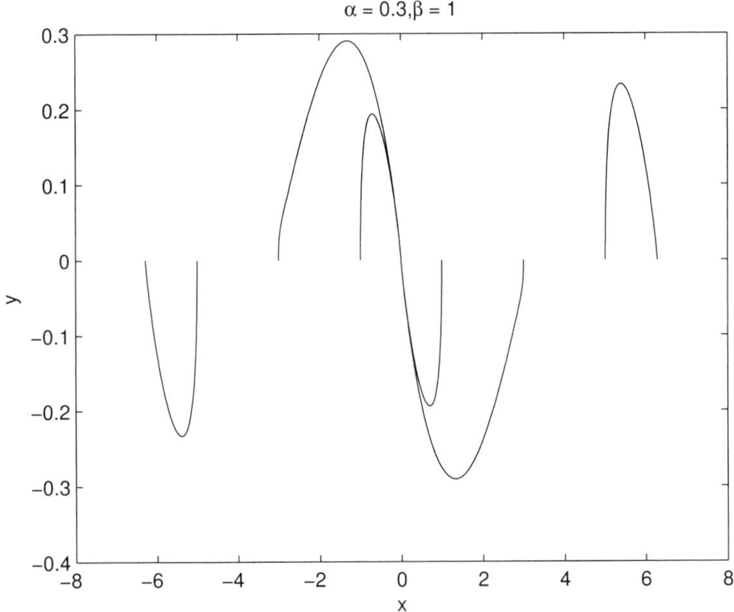

FIGURE 4.5: The phase plane for (4.40) with $\alpha = 0.3$, $\beta = 1$.

(a) $x' = y, y' = -y + 0.1x$.

(b) $x' = y, y' = -x + x^2$.

(c) $y'' + 0.1y' + y - y^3 = 0$.

2. Characterize the state of stability of the equilibria of the following ODEs in terms of the various parameters listed:

   (a) $x'' + ax' + bx = 0$.

   (b) $y'' + ay' + y + b^2y^3 = 0$.

4.6 Motion of Parcels of Fluid in MATLAB

In this section we present a MATLAB program that provides visual information of the behavior of solutions of a two- and three-dimensional systems of ordinary differential equations. The main idea is to take snapshots of a flow under the action of a system of ODEs such as

$$x' = f(t,x,y), \quad y' = g(t,x,y), \qquad x(0) = x_0, y(0) = y_0, \tag{4.43}$$

when we follow the development of a parcel of fluids whose initial state is typically a disk, when the problem is two dimensional as in (4.43), or a ball for the three-dimensional extension of this system. The program outlined below is written for a fundamental flow in fluid dynamics, flow past a cylinder, where the flow is induced by the stream function ψ

$$\psi(x,y) = y - \frac{y}{\sqrt{x^2+y^2}}. \tag{4.44}$$

Applying the formulas (3.32), where $u = \frac{\partial\psi}{\partial y}$ and $v = -\frac{\partial\psi}{\partial x}$, we find that f and g in (4.43) are

$$f(t,x,y) = \frac{1-x^2+y^2}{(x^2+y^2)^2}, \qquad g(t,x,y) = \frac{-2xy}{(x^2+y^2)^2}. \tag{4.45}$$

The program below leads to Figure 4.6. The main program, labeled `parcel.m`, calls on `fpc.m` for the definitions of the right sides of (4.45). It follows the evolution of seven parcels, which are located in a column at $x = -2$ and are initially in the shape of identical circles. The flow is from left to right, so as the ODEs influence the motion of each particle, these parcels react to the obstacle downstream, a circle of radius 1 centered at the origin — this idealized flow is intended to simulate the flow around an infinite cylinder and is assumed to be two-dimensional. This simulation shows the degree of stretching and rotation of each parcel — note that this behavior is by no means homogeneous and that parcels closer to the origin undergo considerably more severe deformations than the ones farther away.

The program below takes advantage of MATLAB's `ode45`, a differential equation solver that we have used already and whose properties we will discuss in detail throughout the text.

```
%%% fpc.m %%%
% This program defines the differential equations.
function yprime=fpc(t,y)
%
term=1/(y(1)^2+y(2)^2)^2;
%
yprime=[1-(y(1)^2-y(2)^2)*term; -2*y(1)*y(2)*term];
%
%%% parcel.m %%%
% This program plots snapshots of 7 parcels of fluids.
```

```
clf
noofparcels=7;
n=40; % number of points on the boundary of each parcel
tfinal=1; % time increment between snapshots
m=5;  % number of snapshots
pts=0:0.01:pi;
circle=[cos(pts);sin(pts)]';  % The obstacle, in this
%                               case a disk of radius one
plot(circle(:,1),circle(:,2));
set(gca,'DataAspectRatio',[1 1 1]);  % setting the aspect ratio
hold on
h=0;
t=0:1/n:1;
for i=1:noofparcels
    data1=-2+0.1*cos(2*pi*t);
    data2=0.2+h+0.1*sin(2*pi*t);
    plot(data1(:),data2(:));
    for k=1:m
        sol=[];
        for j=1:n+1
         [tt,y]=ode45('fpc',[0 tfinal],[data1(j) data2(j)]);
    sol=[sol;y(length(tt),:)];
        end
        plot(sol(:,1),sol(:,2))
        hold on
      data1 = sol(:,1); data2=sol(:,2);
    end
h=h+0.3;
end
title('Flow Past Cylinder')
xlabel('x')
ylabel('y')
```

Problems 4.6

1. Execute the MATLAB programs in this section to generate Figure 4.6.

2. Show that the flow past cylinder, defined by (4.45), is incompressible and irrotational.

3. Adapt the parcel deformation program of this section to apply to the following velocity fields. Select several initial circular parcels

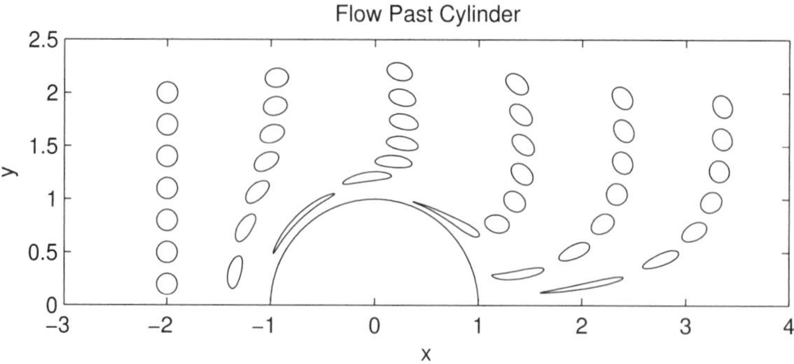

FIGURE 4.6: The deformation of parcels of fluid as they negotiate their way around a cylinder.

of your own choosing and track their deformation over time. Determine in each case if the flow is incompressible or irrotational.

(a) $\mathbf{v} = \langle y, -x \rangle$.

(b) $\mathbf{v} = \langle \frac{y}{\sqrt{x^2+y^2}}, -\frac{x}{\sqrt{x^2+y^2}} \rangle$.

(c) $\mathbf{v} = \langle \frac{y}{x^2+y^2}, -\frac{x}{x^2+y^2} \rangle$.

(d) $\mathbf{v} = \langle y, -0.1y - \sin x \rangle$.

4. For each stream function ψ defined below derive the associated velocity field and apply the parcel deformation program to it. Select several initial circular parcels of your own choosing and track their deformation over time.

(a) $\psi = \sin x \sin y$.

(b) $\psi = 2 \sin 3x \cos 4y - 0.5 \sin 4y \cos 3x - 1.3 \cos 3x \cos 4y$.

5. Generalize the parcel deformation program of this section to three-dimensional flows $\mathbf{v} = \langle f, g, h \rangle$, which lead to the system of three ODEs

$$\frac{dx}{dt} = f(t, x, y, z), \; \frac{dy}{dt} = g(t, x, y, z), \; \frac{dz}{dt} = h(t, x, y, z).$$

Apply the resulting program to the ABC Flow defines in Problem 4 of Section 4.4.

4.7 Project A: Ekman Layer

The Ekman layer, first described in the classic paper [3], is an important feature observed in rotating fluid flows. It turns out that the rotation of our planet induces a steady-state velocity field $\mathbf{v} = \langle u(z), v(z), 0\rangle$, having a distinct spiral shape, which is especially noticeable near boundaries such as the air-sea interface. The impact of this phenomenon is also observed in the motion of large pieces of ice floating in the oceans.

The development of equations of motion attributed to the Ekman layer is described in detail in many textbooks on Geophysical Fluid Dynamics, in particular in [4] (see pages 185–194). As it turns out, the motion in an Ekman layer can be described by a boundary-value problem for a fourth-order ODE, which is the subject of this project.

1. If you have access to [4], begin reading pages 185–194 and familiarize yourself with the concept of *geostrophic balance* and its connection with the concept of an Ekman layer.

2. The geometry we consider is the upper half-space $z > 0$. This region is occupied by a viscous fluid, which rotates about the z-axis at a fixed angular velocity Ω. The equations of geophysical fluid dynamics, after several simplifying assumptions (see [4] and [2] for details) result in the following BVP for $u(z)$:

$$\frac{d^4u}{dz^4} + \lambda^2 u = M, \tag{4.46}$$

where λ is a positive constant, whose value is related to the Coriolis parameter and the vertical contribution of the so-called Eddy viscosity. M, a constant, is a forcing term. This ODE is further supplemented by the following boundary conditions:

$$u(0) = 0, \quad u'(0) = 0, \quad \lim_{z\to\infty} u(z) < \infty. \tag{4.47}$$

The last condition states that we seek a solution to (4.46) with a finite asymptotic limit as z approaches infinity.

(a) Show that

$$u(z) = \frac{M}{\lambda^2} + c_1 e^{\mu z}\cos\mu z + c_2 e^{\mu z}\sin\mu z + \\ c_3 e^{-\mu z}\cos\mu z + c_4 e^{-\mu z}\sin\mu z. \tag{4.48}$$

is the general solution of (4.46) where $\mu = \sqrt{\frac{\lambda}{2}}$.

(b) Show that the boundary condition that u remains finite as z approaches infinity requires that $c_1 = c_2 = 0$.

(c) Apply the boundary conditions in (4.47) and determine c_3 and c_4.

(d) Plot the graph of u when $M = 1$ and $\lambda = 3$.

(e) Suppose that $\lim_{z\to\infty} = U$. Determine the value of the forcing term M. How would you obtain this answer without using the solution to the boundary value problem, but by appealing directly to the ODE in (4.46)?

(f) Let $v(z) = -u''(z)$. Plot the vector field $\langle u(z), v(z)\rangle$, for $z \in (0, 10)$, with $\lambda = 1$ and $M = 1$.

4.8 Project B: Lorenz 96 Model

Edward N. Lorenz, through a remarkable career, which spanned the entire second half of the twentieth century, has been responsible for calling our attention to several systems of ordinary differential equations, each having reached the status of a classic example, each rich enough to illustrate a variety of fundamental concepts and challenges in mathematics and physics. Lorenz either derived these systems from the basic principles of Geophysical Fluid Dynamics, or constructed them based on his unique intuitive understanding of which scientific issue he wished to highlight. We have already encountered, in Problem 3 in Section 4.4 of this chapter, the system that is universally referred to as the *Lorenz Equations*. In this project, we will study a second system, see (4.49) below, which is fast gaining grounds in occupying a special place in the annals of applied mathematics because of its simplicity, and because of its implication in modeling how instabilities may propagate in the atmosphere.

The system of ODEs (4.49) first appeared in 1996 in [6], and for that reason it is often referred to as the *Lorenz 96* or the *L96* system. In a second paper, listed in [7], a more detailed study of this system appeared in 1998. Subsequently, dozens of papers have been dedicated to the various mathematical and statistical properties of this system. Most notable among these studies is the extensive discussion in [8] (see pages 239–255). The reader is strongly encouraged to consult these developments to gain some insight into why L96 is considered such an important platform for computational experimentation.

The L96 system of equations is

$$\frac{dX_j}{dt} = (X_{j+1} - X_{j-2})X_{j-1} - X_j + F \tag{4.49}$$

where $j = 1, ..., J$. The variables X_j are periodic, so that

$$X_{-1} = X_{J-1},\ X_0 = X_j,\ X_{J+1} = X_1.$$

Here F is a positive constant. The discrete system size is $J = 40$, so that (4.49) constitutes a system of 40 nonlinear ordinary differential equations.

1. Show that $X_j = F, j = 1, ..., J$, is an equilibrium solution of (4.49).

2. Linearize (4.49) about this equilibrium solution. To that end, let $X_j = F + Y_j$, substitute in (4.49), ignore higher order terms in Y_j, to arrive at the linearized system

$$\frac{dY_j}{dt} = F(Y_{j+1} - Y_{j-2}) - Y_j, \tag{4.50}$$

   $j = 1, 2, ..., J$.

3. Write a MATLAB program to compute the eigenvalues of the linearized problem (4.50) as a function F. Begin with $F = 0.1$ and compute the associated 40 eigenvalues and report if the equilibrium solution is stable.

4. Increase the values of F with 0.01 until you arrive at the first value of F at which the equilibrium solution is unstable. Let F_0 denote this value.

5. Write a MATLAB program, using `ode45`, to solve (4.49) with initial values

$$X_j = F,\ j \neq 20,\ X_{20} = F + 0.01,$$

   where $F = F_0 + 1$ and F_0 is the value discovered in the previous problem. Plot the graph of $\{X_j(t)\}$. Experiment with various increments of t to obtain a graph similar to FIG. 1, on page 401 of [7].

6. Repeat the above simulation for $F = F_0 + i$, i an integer between 2 and 10, and report on the qualitative change in the evolution of $\{X_j(t)\}$. In particular, discuss if you observe the phenomenon that the initial perturbation at $J = 20$ moves westward.

4.9 References

1. Hale, J., *Ordinary Differential Equations*, Wiley Interscience, 1972.

2. Hartman, P., *Ordinary Differential Equations*, Cambridge University Press, 2002.

3. Ekman, V. W., "On the influence of the Earth's rotation on ocean currents," *Arch. Math. Astro. Phys.*, Vol 2, 1905, pp. 1–52.

4. Pedlosky, J., *Geophysical Fluid Dynamics*, Springer-Verlag, 1987.

5. Malek-Madani, R., *Advanced Engineering Mathematics with Mathematica® and MATLAB®*, Addison-Wesley, 1998.

6. Lorenz, E. N., "Predictability: A problem partly solved," *Proceeding of Seminar on Predictability*, Vol 1., ECMWF, Reading, Berkshire, UK, pp. 1–8, 1996.

7. Lorenz, E., Emanuel, K., "Optimal sites for supplementary weather observations: Simulation with a small model," *J. Atmos. Sci.*, Vol 55, pp. 399–414, 1998.

8. Majda, A. J., Wang, X., *Nonlinear Dynamics and Statistical Theories for Basic Geophysical Flows*, Cambridge University Press, 2006.

Chapter 5

Numerical Methods for ODEs

In the previous chapter we introduced several methods commonly used to obtain solutions of initial and boundary value problems to systems of ordinary differential equations. One of the tools we introduced was MATLAB's `ode45`, a powerful numerical solver that is based on applying a finite difference approach to first discretize an ordinary differential equation and converting the underlying continuous problem to an algebraic problem, which we can then treat with matrix theory methods. In this chapter we will look at the details of finite difference methods to understand their scope of applicability. Although we will introduce finite difference methodology in the context of ordinary differential equations, one of our main goals is to prepare for applying these techniques to partial differential equations.

5.1 Finite Difference Methods

The MATLAB function `ode45` is one of several numerical schemes designed for solving systems of differential equations. To appreciate how accurate and powerful `ode45` is, we now review some of the basic ideas that have been developed for solving differential equations, and display simple MATLAB code that implements these ideas. We carry out this discussion for the simplest of initial-value problems (IVP) where the differential equation is

$$y' = \lambda y, \tag{5.1}$$

complemented with the initial data

$$y(t_0) = y_0. \tag{5.2}$$

The exact solution to this IVP is

$$y(t) = y_0 e^{\lambda(t-t_0)}, \tag{5.3}$$

which grows unboundedly when Re $\lambda > 0$ as t approaches infinity, while it remains bounded when Re $\lambda \leq 0$. The motivation behind a numerical approximation of (5.1)–(5.2) is to capture the basic features of this system as accurately as possible.

We begin by recalling the various definitions of derivatives presented in (3.1)–(3.4). These formulas offer several choices to replace $y'(t)$ in (5.1) by a *finite difference* formula. Three commonly used formulas and their corresponding titles are

$$\frac{y(t+h)-y(t)}{h}, \quad \text{forward difference,} \tag{5.4}$$

$$\frac{y(t)-y(t-h)}{h}, \quad \text{backward difference,} \tag{5.5}$$

$$\frac{y(t+h)-y(t-h)}{2h}, \quad \text{centered difference.} \tag{5.6}$$

Each of these formulas, when substituted into (5.1), results in a *Finite Difference Equation* (FDE). For example, the forward difference approximation of $y' = \lambda y$ is

$$\frac{y_p(t+h)-y_p(t)}{h} = \lambda y_p(t)$$

where the subscript p is introduced to emphasize that y_p is only an approximation to the exact solution $y(t)$. The above expression is equivalent to

$$y_p(t+h) = (1+\lambda h)y_p(t) \tag{5.7}$$

which suggests that if we have a value for $y_p(t)$, we may then compute $y_p(t+h)$ from (5.7), this way obtaining an approximation to the exact value $y(t+h)$. In a typical initial-value problem such as (5.1)–(5.2) we know the value of the solution at $t = t_0$, hence a formula such as (5.7) will then enable us compute approximate values at t_0+h, t_0+2h, t_0+3h, ..., recursively. To give a concrete example, consider the IVP with $\lambda = -0.1$, $h = 0.2$, $t_0 = 0$ and $y_0 = 2$. With these parameter values we are considering the IVP

$$y' = -0.1y, \qquad y(0) = 3, \tag{5.8}$$

whose solution is $y(t) = 3e^{-0.1t}$. The Forward Difference formula for this IVP is (see (5.7) and set $h = 0.2$)

$$y_p(t+0.2) = 0.98\,y_p(t), \qquad \text{with } y_p(0) = y(0) = 3. \tag{5.9}$$

Setting $t = 0$ in (5.9) yields

$$y_p(0.2) = 0.98\,y_p(0) = 2.94.$$

This value is an approximation to $y(0.2) = 3e^{-0.02} = 2.9406$. The absolute error incurred is $|y(0.2) - y_p(0.2)| = 0.0006$ with the corresponding relative error

$$\left|\frac{y(0.2) - y_p(0.2)}{y(0.2)}\right| = 0.0002.$$

Now that we have $y_p(0.2)$ in hand, we can continue with using the expression in (5.9) to arrive at approximate values at $t = 0.2$, $t = 0.4$, etc. For example, after setting $t = 0.2$ in (5.9) we have

$$y_p(0.4) = 0.98\, y_p(0.2).$$

Since $y_p(0.2) = 2.94$, we end up with

$$y_p(0.4) = 2.8812,$$

which is an approximation to the solution $y(t)$ at $t = 0.4$ or $3e^{-0.04} = 2.8824$. We note that the absolute and relative errors are 0.0012 and 0.0004, respectively, both of which are larger than the corresponding errors when $t = 0.2$.

We will refer to the above approach, where the forward difference formula (5.4) is used to approximate y' in the IVP, as the *Forward Euler Method* (FEM). The following table summarizes the results of the FEM approach for the IVP in (5.8).

Forward Euler Method (FEM)				
$y' = -0.1y$, $y(0) = 3$, with $t_0 = 0, h = 0.2$				
t	y_p	y	Abs Err	Rel Err
0	3	3	0	0
0.2	2.9400	2.9406	0.0006	0.0002
0.4	2.8812	2.8824	0.0012	0.0004

It is instructive to see how `ode45` handles the IVP given by (5.8). If needed, refer back to Section 4.4 to refresh your memory on how one implements `ode45` in MATLAB. Because of the simplicity of the right side of (5.8), we enter this function into MATLAB using `inline`:

```
rhs = inline('-0.1*y','t','y');
```

Next we invoke `ode45`:

```
[t,y]=ode45(rhs, [0 0.2],3);
```

`ode45` applies its adaptive method and finds an approximate solution. The expression `y(length(t))` results in

```
ans =

    2.9406
```

providing an excellent approximation to the exact value.

Although we have demonstrated the Forward Euler Method for a specific example, this method is equally effective with any first-order IVP of the form

$$y' = f(t, y), \qquad y(t_0) = y_0. \tag{5.10}$$

Substitution of the expression (5.4) in (5.10) leads to the FDE

$$y_p(t + h) = y_p(t) + hf(t, y_p(t)), \qquad y_p(t_0) = y_0. \tag{5.11}$$

Once f, h, t_0 and y_0 are known explicitly, the above expression leads to values for $y_p(t_0 + h)$, $y_p(t_0 + 2h)$, etc., as shown in the case of the above example where $f(t, y) = -0.1y$. As seen in the concrete example (5.8), the implementation of FEM leads to a set of discrete evaluations of the FDE at t_i, where t_i is given by

$$\begin{array}{rcl} t_0 &=& t_0, \\ t_1 &=& t_0 + h, \\ t_2 &=& t_0 + 2h, \\ \ldots && \ldots \\ t_i &=& t_0 + ih, \\ \ldots && \ldots \end{array} \tag{5.12}$$

at which we determine the approximate values y_i given by

$$\begin{array}{rcl} y_0 &=& y_0, \\ y_1 &=& y_0 + hf(t_0, y_0) \\ y_2 &=& y_1 + hf(t_1, y_1), \\ \ldots && \ldots \\ y_{i+1} &=& y_i + hf(t_i, y_i), \\ \ldots && \ldots \end{array} \tag{5.13}$$

With this notation the FDE in (5.11) takes the form

$$y_{i+1} = y_i + hf(t_i, y_i), \quad y_0 = \text{given, and} \quad i = 1, 2, 3, \ldots \tag{5.14}$$

Because of the recursive nature of the dependence of y_{i+1} on y_i, it is relatively easy to compute all of the necessary y_i's using an application of the `for ... end` loop in MATLAB, as shown below:

```
....
....
```

```
yold=y0;
output=[yold];
for i=1:n
     ynew=yold+h*f(t0+i*h,yold);
     output=[output ynew];
     yold=ynew;
end
...
...
```

For example, to apply FEM to the IVP given by

$$y' = t/(1+t) + e^{\sin y}, \quad y(0) = 1, \tag{5.15}$$

which is a nonlinear IVP and hence does not lend itself to analytical methods, we first construct the FDE from (5.14):

$$y_{i+1} = y_i + h(t_i/(1+t_i) + e^{\sin y_i}), \quad y_0 = 1. \tag{5.16}$$

The following table shows the first two iterations of the FEM and its comparison with the "exact" solution, which in this case is obtained by an application of `ode45`:

Forward Euler Method (FEM)				
$y' = t/(1+t) + e^{\sin y}, \quad y(0) = 1$, with $t_0 = 0, h = 0.1$				
t	y_p	y (using `ode45`)	Abs Err	Rel Err
0	1	1	0	0
0.1	1.2411	1.2559	0.00004	0.00003
0.2	1.5153	1.5530	0.0377	0.0243

Notice how dramatically the absolute and relative errors have increased at the second iteration.

We have introduced the Forward Euler Method for scalar (single) differential equations. The generalization of this method to systems of ordinary differential equations is straightforward. Consider the IVP

$$\mathbf{y}' = \mathbf{f}(t, \mathbf{y}), \qquad \mathbf{y}(t_0) = \mathbf{y}_0, \tag{5.17}$$

where

$$\mathbf{y} = \begin{bmatrix} y_1 \\ y_2 \\ \cdots \\ \cdots \\ y_n \end{bmatrix}$$

is an n-dimensional column vector of unknowns, and

$$\mathbf{f} = \begin{bmatrix} f_1(t,\mathbf{y}) \\ f_2(t,\mathbf{y}) \\ \cdots \\ \cdots \\ f_n(t,\mathbf{y}) \end{bmatrix}$$

defines the right side of (5.17). Let

$$\mathbf{y}_i = \begin{bmatrix} y_{i,1} \\ y_{i,2} \\ \cdots \\ \cdots \\ y_{i,n} \end{bmatrix}$$

denote the values of $\mathbf{y}$ at the i-th discretization point t_i. Then $\mathbf{y}_{i+1}$ is determined from

$$\mathbf{y}_{i+1} = \mathbf{y}_i + h\mathbf{f}(t_i, \mathbf{y}_i)$$

or in component form

$$\begin{bmatrix} y_{i+1,1} \\ y_{i+1,2} \\ \cdots \\ \cdots \\ y_{i+1,n} \end{bmatrix} = \begin{bmatrix} y_{i,1} \\ y_{i,2} \\ \cdots \\ \cdots \\ y_{i,n} \end{bmatrix} + h \begin{bmatrix} f_1(t,\mathbf{y}) \\ f_2(t,\mathbf{y}) \\ \cdots \\ \cdots \\ f_n(t,\mathbf{y}) \end{bmatrix}. \tag{5.18}$$

To see FEM in action on a system of ODEs, consider the example of the Rotating Duffing System given by the system of the two equations

$$\begin{cases} x' &= x\sin 2\beta t + y(\beta + \cos 2\beta t) + N(x,y)\sin\beta t, \\ y' &= x(-\beta + \cos 2\beta t) - y\sin 2\beta t + N(x,y)\cos\beta t, \end{cases} \tag{5.19}$$

where the nonlinearity N is

$$N(x,y) = -(x\cos\beta t - y\sin\beta t)^3 + \epsilon\sin\omega t, \tag{5.20}$$

with β, ω, and ϵ are various physical parameters. This important system of equations is the subject a detailed study in [4] by S. Wiggins and co-authors. If the right side of (5.19)–(5.20) represents a fluid flow, then $(x(t), y(t))$ represents the position of the fluid particle at time t that occupies position (x_0, y_0) at time zero. We apply FEM to this system of ODES to find the path of a particle that is located at $(0.1, 0.2)$ at time zero. We consider the following values for the parameters:

$$\beta = 0.1, \qquad \omega = 1, \qquad \epsilon = 0.3.$$

We track the trajectory of this particle for 3 units of time and compare FEM's result with that of `ode45`. It would be a bit easier to carry out all of the necessary MATLAB computations if we define the right side of (5.19) in an M-File, which we name `RotDuff.m`; the variable `y` in the M-file is a vector and its components contain both of the physical variables x and y, so that `y(1)` stands for x and `y(2)` stands for y:

```
function yprime = RotDuff(t,y)
global beta epsilon omega

N = -y(1)*cos(beta*t)-y(2)*sin(beta*t)^3 + ...
     epsilon*sin(omega*t);

yprime=[y(1)*sin(2*beta*t)+y(2)*(beta+cos(2*beta*t))+N;...
  y(1)*(-beta+cos(2*beta*t))-y(2)*sin(2*beta*t)+ N];
```

With the parameter values defined earlier, and letting $x_0 = 0.1$ and $y_0 = 0.2$, first we combine `ode45` with this M-file to obtain an excellent approximation to the solution of this IVP:

```
global beta epsilon omega
%
beta=0.1;
epsilon=0.3;
omega=1;
%
[t,y]=ode45(@RotDuff,[0 3],[0.1;0.2]);
%
plot(y(:,1),y(:,2),'o')
hold on
```

Next we obtain the FEM approximation as follows (the M-file could be labeled `RotDuffFEM.m`):

```
beta=0.1;
epsilon=0.3;
omega=1;
%
n=100;
h=3/n;
y0=[0.1;0.2];
yold=y0;
output=[yold];
for i=1:n
    t=(i-1)*h;
```

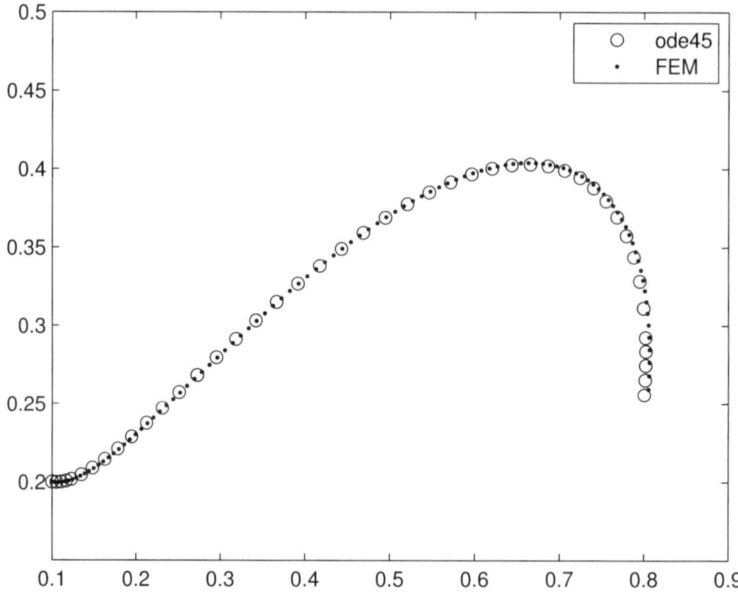

FIGURE 5.1: The Forward Euler Method and `ode45` applied to (5.19)–(5.20).

```
    ynew=yold+h*RotDuff(t,yold);
    output=[output ynew];
    yold=ynew;
end
plot(output(1,:),output(2,:), '.')
legend('ode45','FEM','Location','NorthEast')
```

Figure 5.1 shows the output of FEM and `ODE45` on the Rotating Duffing system. Referring to this figure, it is worth taking notice of how well the Forward Euler Method has succeeded in approximating the solution of a system of ODEs as nonlinear and complex as the ones expressed by (5.19)–(5.20).

Problems 5.1

1. Consider the IVP

$$y' = 0.2y, \qquad y(0) = 3.$$

   (a) Determine the exact solution of this IVP.

   (b) With $h = 0.1$, compute the values of y_1 and y_2 using FEM.

Compare the value of y_2 to the exact value y at the appropriate x and compute the absolute and relative errors at this x.

(c) With $h = 0.05$, compute the values of y_1 through y_4. Write down the values of the x_i's to which these y_i's correspond. Which y_i when $h = 0.05$ corresponds to y_2 when $h = 0.1$?

(d) Compare the two approximate values of the exact solution $y(x)$ with $x = 0.2$ when $h = 0.1$ and $h = 0.05$ and determine which approximation is more accurate.

(e) Apply MATLAB to this IVP and

i. Plot the graph of the exact solution on the interval $(0, 3)$.

ii. With $h = 0.1$, compute the necessary y_i's from the Forward Euler Method and plot the graph of the approximation together with the exact solution.

iii. Let h now be 0.05 and repeat the previous problem.

2. Consider the IVP

$$y' = 0.2y - 1, \qquad y(0) = 2.$$

Repeat the contents of Problem 1 for this IVP.

3. Consider the IVP

$$y' = -y + \sin t, \qquad y(0) = 2.$$

(a) Show that

$$y(t) = \frac{1}{2}\left(5e^{-t} + \sin t - \cos t\right)$$

is the exact solution of the IVP. Plot its graph on the interval $(0, 15)$.

(b) Let $h = 0.1$. Use FEM and determine y_1 and y_2. What is the relative error in each case?

(c) Let $h = 0.05$. Determine y_1 through y_4. What are the relative errors in each case? By comparing these values to the associated values when $h = 0.1$, describe which approximations are more accurate.

(d) Let $h = 0.1$ and apply FEM in MATLAB to this IVP. Plot the graph of the approximate y together with the exact solution on the interval $(0, 15)$. Next let $h = 0.1$ and compute the approximation of y and graph it together with the two graphs already obtained.

4. Consider the IVP

$$y' = \sin(ty), \qquad y(0) = 1.$$

   (a) Apply MATLAB's `ode45` to this problem and plot the graph of the solution in the interval $(0, 3)$.

   (b) With $h = 0.1$, apply FEM and compute y_1 and y_2.

   (c) With $h = 0.1$, apply FEM in MATLAB and plot the graph of the approximation against the graph of the solution obtained from `ode45`.

5. Apply FEM to the IVP

$$x' = y, \quad y' = -0.1y + \sin x, \qquad x(0) = 0.1, \qquad y(0) = 0.$$

   and compute the first two iterations (x_1, y_1) and (x_2, y_2) when $h = 0.1$. Next, Use MATLAB and apply both `ode45` and FEM to this IVP and plot the graph of the two approximate solutions on the interval $(0, 3)$; for FEM use a step-size h of your own choosing.

6. Generate the graph in Figure 5.1.

7. Consider the Rotating Duffing system of ODEs described in (5.19)–(5.20). Use the parameter values

$$\beta = 1, \ \epsilon 0.1, \ \omega = 3.$$

   Apply MATLAB's `ode45` and plot the graphs of the trajectories of the particles that at time zero are located on the perimeter of a circle of radius 0.1 centered at $(1, 0)$. Select at least 30 particles on the perimeter of the initial circle and track their evolution for 3 units of time. A good visual representation of the data you will obtain from `ode45` is to use MATLAB's capabilities to display in different colors the location of the particles at a set increment of time. See Figure 5.2 for how the original parcel of fluid deforms under the action of the Rotating Duffing system. This figure is obtained by appropriate modification of the MATLAB code `fpc.m` in Section 4.6. The following is one way of modifying `fpc.m` to obtain Figure 5.2.

```
t=0:1/n:1;
data1=1+0.1*cos(2*pi*t); data2=0.1*sin(2*pi*t);
plot(data1(:),data2(:));
for k=1:snapshots
    sol=[];
```

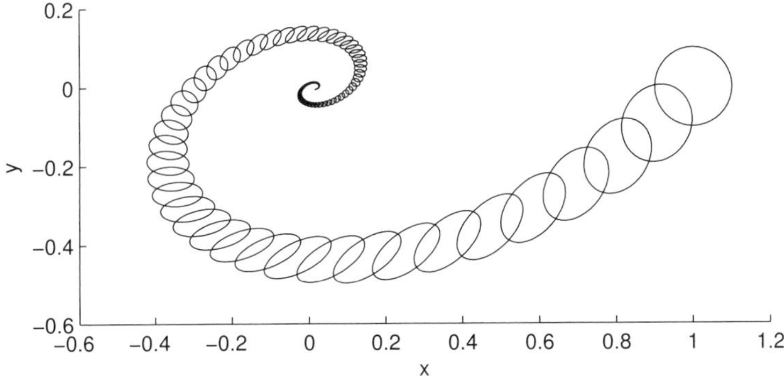

FIGURE 5.2: The deformation of a parcel of fluid under the action of the system of ODEs in (5.19)–(5.20) with parameter values $\beta = 1$, $\epsilon = 0.1$ and $\omega = 3$.

```
        for j=1:n+1
        [tt,y]=ode45('RotDuff',[0 tfinal],[data1(j) data2(j)]);
        sol=[sol;y(length(tt),:)];
        end
        plot(sol(:,1),sol(:,2))
        hold on
        data1 = sol(:,1); data2=sol(:,2);
    end
```

5.2 Backward Euler Method (BEM)

In the previous section we discussed how the formula $(f(a+h) - f(a))/h$, with $h > 0$ (see (5.4)), gives rise to a natural approximation of $f'(a)$ and leads to the Forward Euler Method for solving IVPs. This expression, which is just one of an infinitely many ways of approximating $f'(a)$, is suited well for initial-value problems when a single side condition, such as $y(0) = y_0$, is just the right amount of information to begin the process of obtaining y_i for $i > 0$. We now discuss an alternative approximation to $f'(a)$, based on the backward formula

$$\frac{f(a) - f(a-h)}{h}$$

(see (5.5)), again with $h > 0$, which is equally as effective as FEM, but with some significant advantages and disadvantages.

As in the case of FEM, we begin the Backward Euler Method approximation of the IVP $y' = f(t, y)$, $y(0) = y_0$, by evaluating the ODE at a typical point t_i and obtaining the usual expression

$$y'(x_i) = f(t_i, y(x_i)), \quad y(x_0) = y_0.$$

In BEM the term $y'(x_i)$ is replaced by (as before y_i stands for $y(x_i)$)

$$\frac{y_i - y_{i-1}}{h}$$

leading to the FDE

$$y_i = y_{i-1} + hf(t_i, y_i), \qquad y_0 = y_0. \tag{5.21}$$

Our task, as in the case of FEM, is to obtain y_i in terms of y_{i-1}. In the case of FEM, where the equivalent FDE is $y_i = y_{i-1} + hf(t_{i-1}, y_{i-1})$, this task is trivial, since every term on the right side is known at the i-th iteration. But for BEM, when f is a nonlinear function of y, the task of writing out an explicit formula for y_i from (5.21) is somewhat more complicated since it involves solving an equation. The fact that y_i is expressed implicitly in (5.21) is perhaps the main disadvantage of this method, although we can certainly think of relatively simple techniques to obtain a good approximation of y_i from (5.21), or modify (5.21) slightly to end up dealing with an explicit expression; see Project A, the Modified Euler Method, at the end of this chapter for a more detailed discussion of this point. The main advantage of BEM ends up being its stability properties, which will be discussed in the next section, which are substantial enough that we continue further with the development of this method.

When the function f on the right side of (5.21) is a linear function of y, the expression in (5.21) reduces to a simple algebraic equation in y_i that can easily be solved for y_i, and the rest of the implementation of BEM follows the pattern of FEM. When the function f depends nonlinearly on y_i, we may view y_i in (5.21) as a fixed-point of the expression in (5.21) and use simple iterative methods for obtaining accurate and robust approximations of y_i in terms of y_{i-1}. We will elaborate on this point later, after presenting a simple example to illustrate the case when f depends linearly on y_i. Consider the IVP

$$y' = ay + \sin bt, \qquad y(0) = c, \tag{5.22}$$

whose exact solution is

$$y(t) = (c + \frac{b}{a^2 + b^2})e^{at} - \frac{1}{a^2 + b^2}(a \sin bt + b \cos bt). \tag{5.23}$$

The backward Euler Method applied to this IVP leads to the FDE

$$y_i = y_{i-1} + h(ay_i + \sin bt_i), \quad y_0 = c, \quad i = 1, 2, 3, ...$$

While this FDE is implicit in y_i, because the equation is linear, we can solve for y_i to obtain the explicit FDE

$$y_i = \frac{1}{1 - ah}(y_{i-1} + h \sin bt_i), \quad y_0 = c, \quad i = 1, 2, 3, ... \tag{5.24}$$

Similar to all of the Finite Difference Equations we have encountered when applying the Forward Euler Method, in the above FDE y_i is explicitly expressed in terms of y_{i-1} and hence is easily coded in MATLAB. The following is an example of BEM applied to (5.22) when $a = -0.1$, $b = 6\pi$ and $c = 3$, and its comparison with `ode45`. See Figure 5.3.

```
%
% Backward Euler Method applied to
%      y'= -0.1 y + sin (6 pi t)
%
a=-0.1; b = 6*pi; c = 3;
n=100; h= 0.01;
y0=c;
yold=y0;
output = [yold];
factor = 1/(1-a*h);
for i=1:n
    t=i*h;
    ynew=factor*(yold+h*sin(b*t));
    output=[output ynew];
    yold=ynew;
end
plot(0:h:n*h,output,'.')
hold on
f=inline('-0.1*y+sin(6*pi*t)','t','y');
[t,y]=ode45(f,[0 n*h], y0);
plot(t,y,'o')
title('BEM, with n = 100 and h = 0.01, versus ode45')
legend('BEM','ode45','Location','Northeast')
```

In the next example we attempt to solve a nonlinear IVP by the Backward Euler Method and comment on the difficulties we normally face when we wish to arrive at the equivalent of the explicit formula we obtained in (5.24). Consider the IVP

$$y' = -0.1y^2 + \sin t, \quad y(0) = 3. \tag{5.25}$$

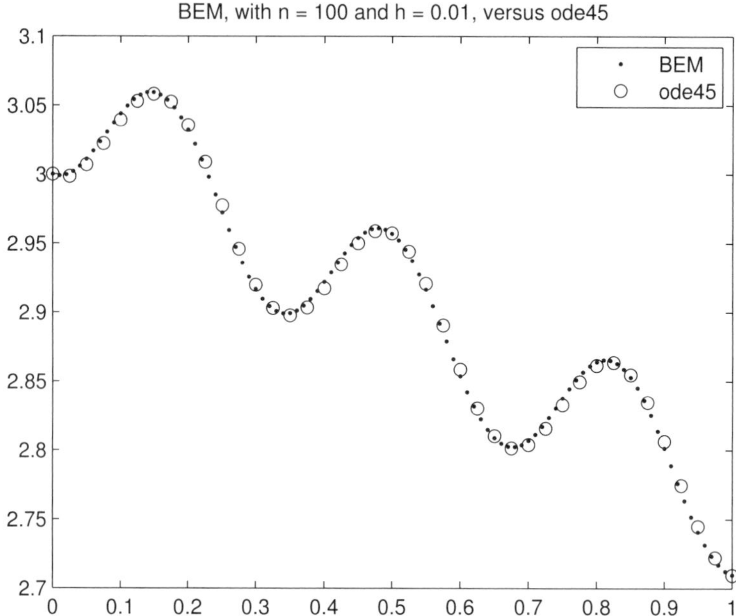

FIGURE 5.3: BEM and ode45 applied to (5.22) with $a = -0.1$, $b = 6\pi$, and $c = 3$.

An application of BEM to this ODE leads to the Finite Difference Equation

$$y_i = y_{i-1} - 0.1hy_i^2 + h\sin t_i, \quad y_0 = 3. \tag{5.26}$$

In order to obtain an explicit formula for y_i from (5.26) we need to appeal to the quadratic formula. We will then obtain two solutions, from which we select the correct branch by noting that $y_0 = 3$. However, instead of taking this path, we present a different approach that applies to a much larger class of implicit/nonlinear Finite Difference Equations similar to the one in (5.26).

We view the right side of (5.26) as a function of y_i and denote it, temporarily, by $g(y_i)$. Hence g is give by

$$g(z) = -0.1hz^2 + y_{i-1} + h\sin t_i. \tag{5.27}$$

With this definition of g in mind, we note that determining y_i in (5.26) is equivalent to obtaining the *fixed-point* of g, i.e., y_i satisfies

$$g(y_i) = y_i. \tag{5.28}$$

We recall that a general and natural approach to finding a fixed-point of a function g is by applying the iterative method

$$z_i = g(z_{i-1}) \tag{5.29}$$

for $i = 1, 2, 3, ...$, with z_0 a given starting point for this iteration. If this sequence z_i converges to a value z, and if g is a sufficiently smooth function, then it is clear that $g(z) = z$.

What is often important in obtaining a convergent sequence when applying the fixed-point algorithm is a judicious choice of the initial guess z_0; if the starting guess is sufficiently close to the fixed-point, the probability of obtaining a convergent sequence is quite high. Fortunately, when applying this technique in the context of BEM we always have a good starting point in hand, namely y_{i-1}, because when h is small we expect that y_i is relatively close to y_{i-1}. This is the approach we adopt in the following MATLAB code in obtaining a relatively accurate solution to the IVP in (5.26). The following code does the job; note that the structure of the code is that of FEM, not BEM, in that the formula $y_i = y_{i-1} + hf(t_i, y_i)$ is the expression to which `yoldInterm` and `ynewInterim` apply. The key new structure is the inner loop, indexed by `k`, where the fixed-point idea is implemented:

```
clf;
clear all;
n=200; h= 0.01;
y0=3;
yold=y0;
output = [yold];
f=inline('-0.1*y.^2+sin(t)','t','y');
for i=1:n
t=i*h;
    yoldInterim=yold;
for k=1:50
        ynewInterim=yold+h*f(t,yoldInterim);
        yoldInterim=ynewInterim;
    end
    ynew=ynewInterim;
output=[output ynew];
yold=ynew;
end
plot(0:h:n*h,output,'.')
hold on
[t,y]=ode45(f,[0 n*h], y0);
plot(t,y,'o')
string=['BEM, Fixed-Point Method',', n = ',num2str(n),
```

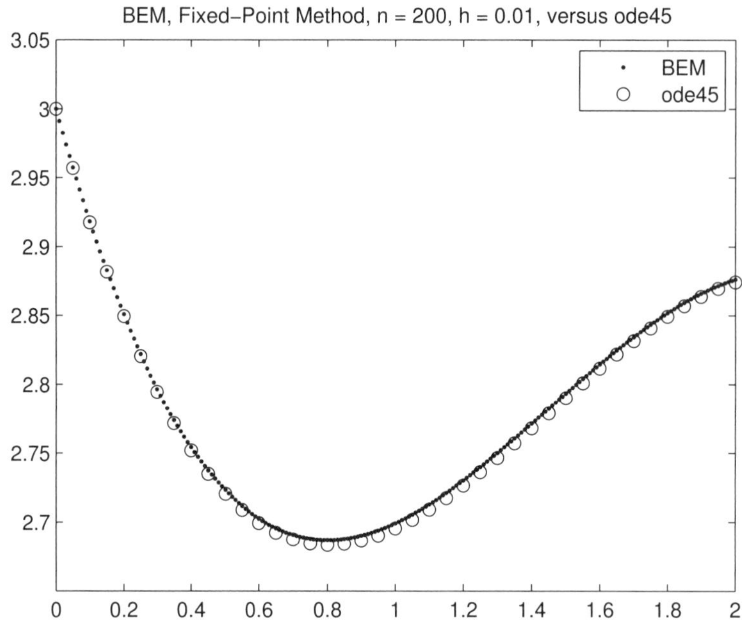

FIGURE 5.4: BEM, implemented with the fixed-point method, and `ode45` applied to (5.26) with $a = -0.1$, $b = 6\pi$ and $c = 3$.

```
',h = ', num2str(h),', versus ode45']
title(string)
legend('BEM','ode45','Location','Northeast')
```

Figure 5.4 contains the output.

5.3 Stability of Numerical Methods

We are able to solve the FDEs that result from discretizing an ODE in a similar fashion that we obtained the analytic solution (5.3) of the ODE in (5.1). Comparison of the exact solution to each of the three FDEs will shed considerable light on the properties of the three finite difference formulas in (5.4)–(5.6). Before proceeding further, we call the reader's attention to [3], which is a an excellent resource for studying numerical methods of ordinary differential equations.

We begin with the analysis of the Forward Difference Method we obtain by replacing the y' term in (5.1) by (5.4). Without loss of generality, let $t_0 = 0$ and consider a finite time interval $[0, T)$ as the domain for the independent variable t. Discretize this domain into n subintervals by choosing the $n + 1$ equidistant points

$$t_0 = 0,\ t_1 = h,\ t_2 = 2h,\ ...,\ t_n = nh = T, \tag{5.30}$$

so that $h = \frac{T}{n}$. We denote by y_i the approximate value for $y(t_i)$, replace $y'(t_i)$ in (5.1) by $(y_{i+1} - y_i)/h$, and solve for y_{i+1} and get the FDE

$$y_{i+1} = (1 + \lambda h)y_i, \tag{5.31}$$

where i ranges from 0 to $n - 1$, and y_0 is given. We seek the general solution to (5.31) in the form

$$y_i = \gamma^i, \tag{5.32}$$

which we substitute into (5.31) and arrive at $\gamma = (1 + \lambda h)$. Hence the general solution to (5.31) is $y_i = c(1 + \lambda h)^i$ for any constant c. The initial condition (5.2) determines the constant c as y_0. Hence, the unique solution to the initial-value problem (5.31) is

$$y_i = y_0(1 + \lambda h)^i, \quad i = 0, 1, ..., n. \tag{5.33}$$

We observe that the sequence y_i in (5.33) actually converges to $y(t)$ for some t if n approaches infinity. To see this, recall that the limit definition of e^a (that $\lim_{m\to\infty}(1 + \frac{a}{m})^m = e^a$) suggests that the limit of (5.33) as n approaches infinity is $y_0 e^{\lambda t}$: let t^* be a fixed point in $[0, T)$ and consider the index i so that $\lim_{n\to\infty} ih = t^*$. Note that both i and h in the latter expression depend on n, and although the existence of this limit may not be obvious, a little experimentation with the definitions of $h = \frac{T}{n}$ and i should convince the reader that there is a sequence $t(n) = ih$ such that $t(n) \to t^*$. With $t(n)$ in hand, we consider the following sequence of equalities:

$$\lim_{n\to\infty} (1 + \lambda h)^i = \lim_{n\to\infty} (1 + \frac{\lambda T}{n})^{\frac{nt(n)}{T}} = \lim_{m\to\infty} (1 + \frac{\lambda t(n)}{m})^m = e^{\lambda t^*},$$

where we have made the substitution $m = \frac{nt(n)}{n}$. Hence

$$\lim_{n\to\infty} y_i = y(t^*), \tag{5.34}$$

which is reassuring in that we can be confident that the forward finite difference scheme will, at least theoretically, converge to the exact solution if we are allowed to discretize the interval $(0, T)$ with as fine a mesh as we wish. In this sense we say that the forward finite difference scheme is *consistent* with the initial value problem (5.1)–(5.2).

Unfortunately we don't have the luxury of taking the time-step h as small as we wish and at some point we must confront the reality of implementing this scheme on a computing platform with hardware limitations. We thus need to analyze the finite difference scheme further in terms of its practicality. One of the practical attributes of any numerical scheme is the requirement of the *stability* of that scheme.

Definition 5.3.1 *A scheme is said to be stable if the approximate sequence* $\{y_i\}$ *is bounded, that is, there is a constant* M *such that* $|y_i| < M$ *for all* $0 \leq i \leq n$ *and* n *approaching infinity.*

In the next section we will apply this definition to understand the restrictions on the forward difference scheme.

Problems 5.3

1. Review and complete the analysis that led to the proof of (5.34).

2. Consider the FDE

$$y_{i+1} + 3y_i = 0, \quad y_0 = -2, \quad i = 0, 1, ...$$

   Find the solution y_i. (Hint: Start with the template $y_i = \gamma^i$ and find γ.)

3. Consider the FDE

$$y_{i+1} + 0.1y_i + y_{i-1} = 0, \quad y_0 = 1, y_1 = 2, \quad i = 1, 2, ...$$

   Find the solution y_i. (Hint: Start with the template $y_i = \gamma^i$ and find γ by solving a quadratic equation.)

5.4 Stability Analysis of Numerical Schemes

Returning to (5.33), the Forward Euler Method, we see that

$$|y_i| = |y_0|\,|(1 + \lambda h)|^i.$$

The above sequence is bounded if and only if $|1+\lambda h| \leq 1$ (recall that a^i is unbounded in i if $|a| > 1$). Mindful that λ may be complex, let $\lambda = a+bi$, where now $i = \sqrt{-1}$. The inequality $|1 + \lambda h| \leq 1$ is equivalent to (a, b)

satisfying $(a+\frac{1}{h})^2+b^2 \leq \frac{1}{h^2}$. Geometrically, this expression is equivalent to the point (a, b), i.e., λ, being located inside a circle of radius $\frac{1}{h}$ and centered at $(-\frac{1}{h}, 0)$ in the complex plane. Putting it a little differently, *for a fixed h*, the Forward Finite Difference scheme is stable for (5.1)–(5.2) if and only if the physical parameter λ is within the circle of $\frac{1}{h}$ and centered at $(-\frac{1}{h}, 0)$.

The above result is somewhat surprising in how restrictive it is. It states that the range of physical parameters for which we should trust the Forward Finite Difference scheme is limited to the interior of the above disk in the left-half complex plane. In particular, this region excludes all $\lambda = i\omega$ where ω is any nonzero real number, which are quite important parameter values in many physical problems.

As the experiments below show, the forward Euler scheme performs well when the stability property holds but could behave poorly when it is violated.

Figure 5.5a shows the graphs of the approximate and exact solutions of $y' = -1.2y$ with $y(0) = 1$ in the interval $[0, 2)$ with $n = 20$. Here $\lambda = -1.2$ is within the disk of radius $\frac{1}{h} = 10$, centered at $(-\frac{1}{h}, 0) = (-10, 0)$. The approximate solution, with circles denoting the computed values, provides a good estimate of the solution, although it consistently underestimates the exact solution. Note that the absolute error is tolerable (see Figure 5.5b), considering the size of h, and that the relative error grows steadily at a linear rate (see Figure 5.5c). Every aspect of this computation improves by adopting a smaller h (or equivalently a larger value of n), although the relative error will continue to grow as a function of t even when we choose smaller and smaller h.

Figure 5.6 shows the exact solution and its approximation for the initial value problem $y' = -19y$, $y(0) = 1$. As before, $h = 0.1$. Now while $\lambda = -19$ is still inside the region of stability of the Euler method, it is close to the boundary of this region and we note the worsening behavior in the computed solution, in particular that the approximate solution begins to oscillate (see Figure 5.6a). The absolute error remains relatively small, while the relative error begins to grow, at a considerably faster rate that the case $\lambda = -1.2$.

Figure 5.7 shows the output for the case where $\lambda = -21$, where now the stability of the scheme is violated. Note the deterioration in the approximate solution, that the oscillations are growing now and will not be bounded, a fact that can also be observed in the absolute error graph. The relative error continues to grow exponentially.

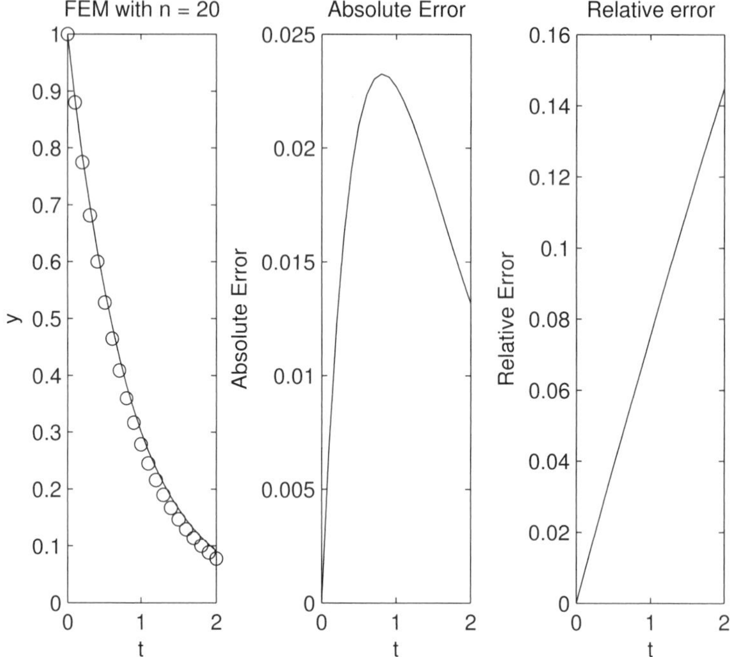

FIGURE 5.5: The Forward Euler Method applied to $y' = -1.2y$. Here $h = 0.1$ and $\lambda = -1.2$. Note that FEM underestimates the exact solution in this example, and while the absolute error may be tolerable, that the relative error grows steadily. Adopting a smaller h improves the approximate solution and reduces both the absolute and relative error. The relative error, however, will continue to grow as a function of t.

5.5 MATLAB Programs for the Forward Finite Difference Method

We digress momentarily to discuss how Figures 5.6–5.7 were obtained in MATLAB. The following lines, when applied to formula (5.33), result in Figure 5.5:

```
clf;
% Parameter Definitions
%
lambda=-1.2; y0=1; T=2;
%
```

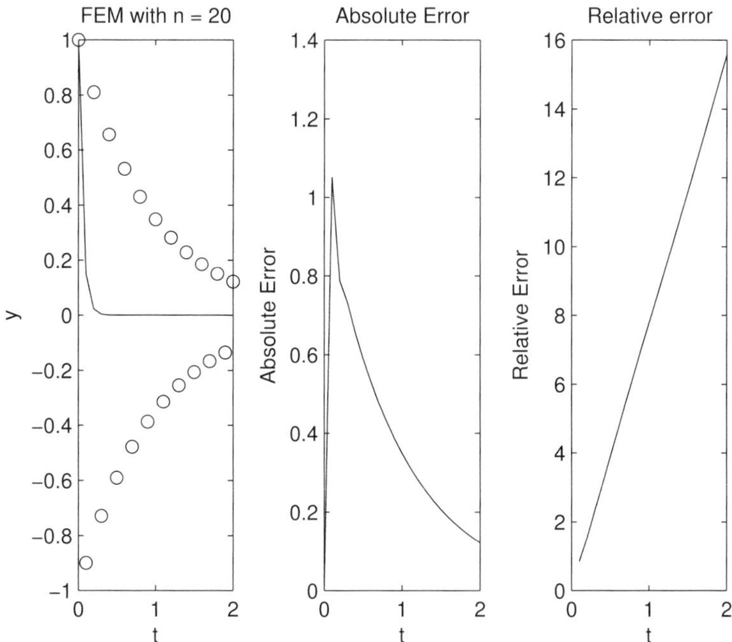

FIGURE 5.6: FEM applied to $y' = -19y$. Here h is still 0.1. Note the bounded oscillation of the approximate solution. The absolute error continues to remain tolerable, but the relative error is now growing at a faster rate than in the case $\lambda = -1.2$.

```
n=20; h=T/n;
%
i=0:n;  % defines the index
t=i*h;  % defines the t domain
%
% Analytic solution of the FDE
%
y=y0*(1+lambda*h).^i;
%
% Plotting output
%
subplot(1,3,1)
plot(t,y,'o')
hold on
%
% Exact solution and its plot
```

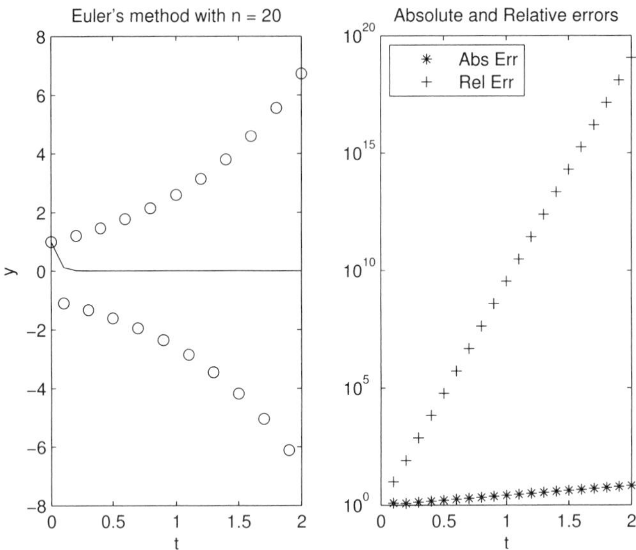

FIGURE 5.7: FEM applied to $y' = -21y$, with $h = 0.1$. Note that the values of the approximate solution, the absolute error, and the relative error are all growing. Here MATLAB's `semilogy` is used to display the absolute and relative errors.

```
%
exact=y0*exp(lambda*t);
plot(t,exact);
xlabel('t')
ylabel('y')
title(['FEM with n = ',num2str(n)])
%
% Computing the various measures of error
%
error=abs(exact-y);
subplot(1,3,2)
plot(t,error)
title('Absolute Error')
xlabel('t')
ylabel('Absolute Error')
relerr=abs((exact-y)./exact);
subplot(1,3,3)
plot(t,relerr)
```

```
title('Relative error')
xlabel('t')
ylabel('Relative Error')
```

The above program uses the analytic solution of the FDE (5.31). Unfortunately, in most of the problems we encounter we won't have the luxury of determining the analytic solution of the underlying FDE. It turns out that with a little care, we can actually obtain the same outcome by implementing the FDE (5.31) directly into MATLAB. To that end we replace the line

```
y=(1+lambda*h).^i;
```

with the following lines

```
y=[y0];
oldy=y0;
factor=1+lambda*h;
for j=1:n-1
    newy=factor*oldy;
    y=[y;newy];
    oldy=newy;
end
```

The above code uses two important features of MATLAB. One, the `for ... end` loop capability, which allows us to repeat the lines in between `for` and `end` as often as the index `j` runs through its counter (in this case `j` begins with 1 and ends at `n-1` with the default increment of 1). The second feature is employed in the line `y=[y;newy]`—notice that the vector `y` is first initialized outside of the `for ... end` loop, at that stage having a single entry, namely, `y0`, and then its value is updated each time the loop is executed. The line `y=[y;newy]` allows us to enlarge the size of the vector `y` by appending the newly computed `newy` to it each time the loop is executed.

Given a general initial value problem

$$y' = f(t, y) \quad y(0) = y_0,$$

the associated forward difference approximation for $y'(t_i)$ leads to

$$y_{i+1} = y_i + hf(t_i, y_i), \qquad i = 0, 1, ...$$

which can be implemented in MATLAB as follows:

```
% Define parameter values n, T, h, y0, i,
%    as before. The function f
```

```
%    needs to defined either using the inline
%    command or in an M-file
%
y=[y0];
oldt=0;
for j=1:n-1
    newy=oldy + h*f(oldt,oldy);
    y=[y;newy];
    oldt=oldt+h;
    oldy=newy;
end;
```

Problems 5.5

1. Implement the MATLAB code presented in this section and generate Figure 5.5.

2. Generate Figures 5.6 and 5.7.

3. Modify the code that led to Figure 5.5 and replace the exact solution formula with a code that computes the solution of the FDE recursively. How should the concepts of absolute and relative errors be modified if one does not have the exact solution?

4. Apply the Forward Euler scheme to the following initial value problems. In each case compare the approximate solution to the exact solution, whether the exact solution is obtained analytically or by using `ode45` to obtain a very good approximation to the exact solution and using that as proxy for the exact solution.

   (a) $y' = 0.1y$, $y(0) = 2$, $T = 2$, $n = 10$, $n = 50$ and $n = 100$.

   (b) $y' = -0.1y + 1$, $y(0) = 2$, $T = 2$, $n = 10$, $n = 50$ and $n = 100$.

5. Apply the Forward Euler method to (5.45) to obtain the formula

$$y_{i+1} = y_i + hf(t_i, y_i),\ y_0 = \text{given}. \tag{5.35}$$

   Write a MATLAB program to implement this scheme to the following initial value problems. Compare the approximate solution to the analytic solution or the one obtained from `ode45`.

   (a) $y' = \cos t \sin y$, $y(0) = \frac{\pi}{2}$, $T = 4\pi$, $n = 100$.

   (b) $y' = \cos(ty)$, $y(0) = 0$, $T = 4\pi$, $n = 100$.

5.6 Stability Analysis of Numerical Schemes (continued)

Returning now to the discussion of stability analysis, we will see shortly that the analysis presented for the forward finite difference method leads to a different result when applied to the backward difference method, which is derived when the formula (see (5.5))

$$\frac{y(t) - y(t-h)}{h},$$

is used to replace $y'(t)$. Applying this formula to the differential equation $y' = \lambda y$ results in the FDE

$$y_i = \frac{1}{1-\lambda h} y_{i-1}, \quad i = 1, 2, ..., n, \text{ with } \quad y_0 = \text{given}, \tag{5.36}$$

whose solution is

$$y_i = \frac{y_0}{(1-\lambda h)^i}, \quad i = 1, 2, ... \tag{5.37}$$

This scheme is stable if $|y_i|$ is bounded, which is guaranteed if $|\frac{1}{1-\lambda h}| \leq 1$ or

$$|1 - \lambda h| \geq 1.$$

As before we let $\lambda = a + bi$ and observe that the above inequality holds if

$$(1 - ah)^2 + b^2h^2 \geq 1 \tag{5.38}$$

which always holds if $a = \text{Re } \lambda \leq 0$. In general the inequality in (5.38) is satisfied for any (a, b) located outside of a circle of radius $\frac{1}{h}$ and centered at $(\frac{1}{h}, 0)$. See Problem 1. Thus the region of stability of the backward Euler scheme is considerably larger than the forward Euler scheme. In particular this region includes any $\lambda = i\omega$, $\omega \in R$.

Figure 5.8 shows the exact and approximate solutions to $y' = -1.2y$, $y(0) = 1$ using the Backward Euler Method, where we have used the same parameter values for T and n as before. Figures 5.9 and 5.10 show the same circumstances for the ODEs $y' = -19y$ and $y' = -21y$, respectively. Notice that in all three cases the approximate solution remains bounded, the absolute error is relatively small, while the relative error in the latter two cases are substantially larger. In particular, notice that the oscillatory behavior of FEM in the case $\lambda = -19$ (see Figure 5.6), which corresponds to a parameter value in the stable region of FEM, is not present in the BEM simulation.

Problems 5.6

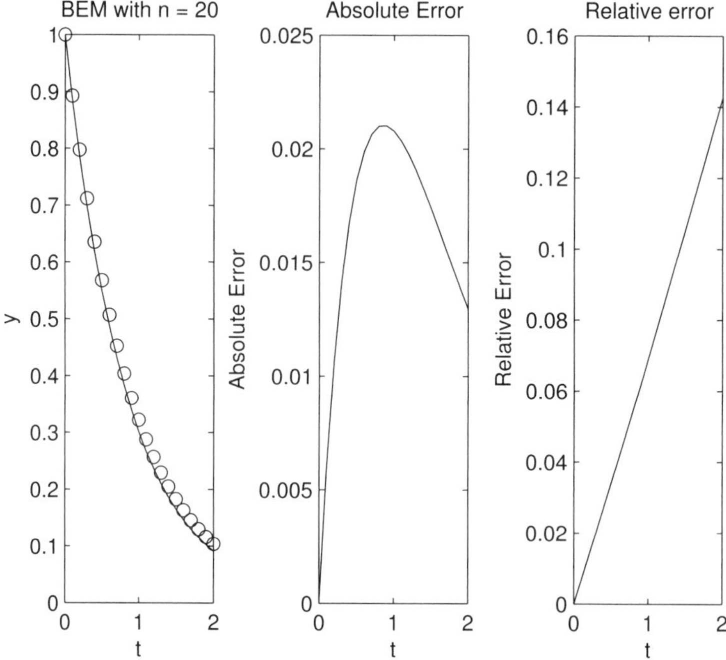

FIGURE 5.8: BEM applied to $y' = -1.2y$. Compare with Figure 5.5 and note how BEM seems to overestimate the exact solution, as opposed to FEM that has a tendency to underestimate the solution in this example.

1. Show that the inequality in (5.38) holds if λ is outside of a circle of radius $\frac{1}{h}$ and centered at $(\frac{1}{h}, 0)$.

2. Alter the MATLAB code presented in this section and obtain Figures 5.9 and 5.10.

3. Apply the Backward Euler scheme to the problems listed in Problem 4 from the previous section.

4. Apply the Centered difference formula in (5.6) to (5.1)–(5.2). Find the region of stability of this scheme.

5. Repeat Problem 5 of the previous section for the Backward Euler scheme.

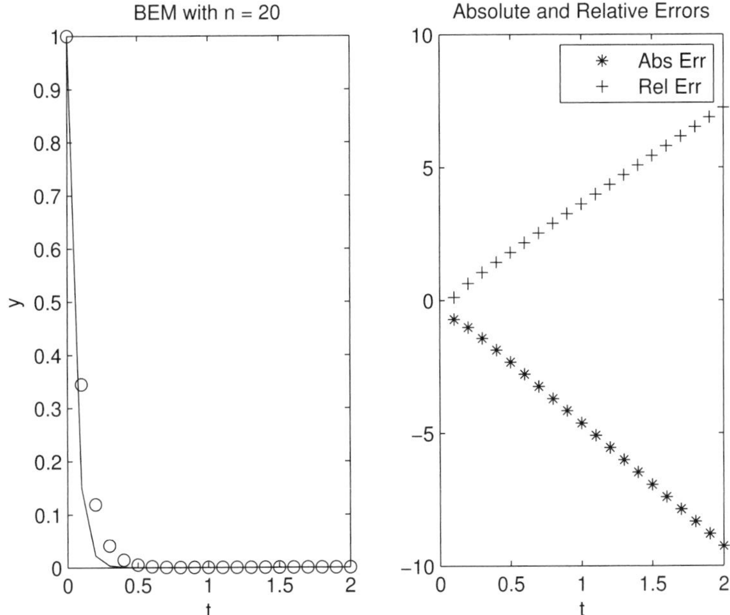

FIGURE 5.9: BEM applied to $y' = -19y$. Here we have plotted the graphs of the absolute and relative errors in log coordinates. Compare with Figure 5.6. Note that BEM does not have the bounded oscillation we see in the case of FEM.

6. Consider the system of equations

$$y' = f(t,x,y), \quad y' = g(t,x,y), \qquad x(t_0) = x_0,\ y(t_0) = y_0. \tag{5.39}$$

Show the Forward Euler method for this system is given by

$$x_{i+1} = x_i + hf(t_i, x_i, y_i), y_{i+1} = y_i + hg(t_i, x_i, y_i), \tag{5.40}$$

with x_0 and y_0 given in (5.39). Apply this result to the following systems of equations. Compare the graph of each approximate solution with the one obtained by using `ode45`.

(a) $x' = y$, $y' = -\sin x$, $x_0 = 0$, $y_0 = 1$, $T = 2\pi$, $n = 100$.

(b) $x' = \frac{y}{\sqrt{x^2+y^2}}$, $y' = -\frac{x}{\sqrt{x^2+y^2}}$, $x(0) = -2$, $y(0) = 0$, $T = 3$, $n = 100$.

7. Repeat Problem 6 for the Backward Euler scheme.

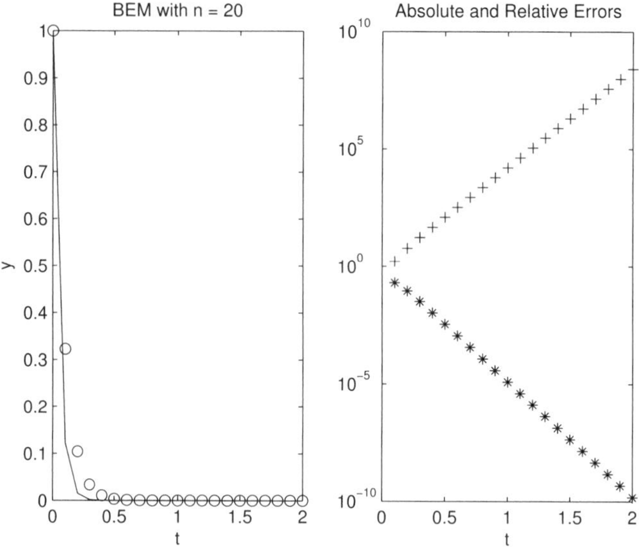

FIGURE 5.10: BEM applied to $y' = -21y$. The graphs of the absolute and relative errors are plotted in log coordinates. Compare with Figure 5.6. Note that BEM does not have the unbounded oscillation we see in the case of FEM for this value of λ, which is due to the fact that $\lambda = -21$ is in the stable region of BEM.

5.7 Truncation Error

FEM and BEM are first order schemes in that the residual between the exact solution $y(t_i)$ and y_i is proportional to h, which we usually denote by $O(h)$. To see this point in the case of FEM, let $L(y)$ denote the finite difference operator in this scheme, that is,

$$L(y(t_i)) = \frac{y(t_{i+1}) - y(t_i)}{h} - \lambda y(t_i). \tag{5.41}$$

Of interest to us is the impact of this operator on the exact solution of (5.1), that is, when $y(t) = y_0 e^{\lambda t}$. The expression $L(y_0 e^{\lambda t})$ will not vanish in general, so we are interested in estimating how far this expression is from zero. This residual is called the *Truncation Error* of the numerical scheme. In general, for any finite difference scheme $L(y)$, the truncation error of that scheme is the value of $L(y)$ when the analytic solution y of the differential equation is used in the evaluation of $L(y)$. For the Euler scheme and (5.41) we have (recall $t_0 = 0$)

$$L(y(ih)) = \frac{y_{i+1} - y_i}{h} - \lambda y_i = \frac{y((i+1)h) - y(ih)}{h} - \lambda y(ih). \quad (5.42)$$

Applying Taylor's formula to $y(ih + h)$ we have

$$y(ih + h) = y(ih) + hy'(ih) + \frac{h^2}{2}y''(ih) + ...,$$

so that

$$\begin{aligned} L(y(ih)) &= \frac{y(ih + h) - y(ih)}{h} - \lambda y(ih) \\ &= y'(ih) + \frac{h}{2}y''(ih) + \text{h.o.t} - \lambda y(ih), \end{aligned} \quad (5.43)$$

where h.o.t. stands for higher order terms in h. Since y is the analytic solution of (5.1), we have $y'(ih) = \lambda y(ih)$. The expression in (5.43) now reduces to

$$L(y(ih)) = \frac{h}{2}y''(ih) + \text{h.o.t.} \quad (5.44)$$

Because the leading term of the truncation error depends on the first power of h we call FEM a *first order* method. It can be shown similarly that BEM is also first order, while the scheme we get from the centered difference formula (5.6) is second order.

The methods we have described in this section are just two methods for obtaining approximate solutions to (5.1)–(5.2). As we have already seen, they are easily extended to general initial value problems

$$y' = f(t, y), \ y(t_0) = y_0 \quad (5.45)$$

and to systems of equations

$$\mathbf{y}' = f(t, \mathbf{y}), \ \mathbf{y}(t_0) = \mathbf{y}_0. \quad (5.46)$$

It is not difficult to show that FEM and BEM are first order schemes for the general scalar IVP in (5.45) or the general system in (5.46). A more important point, however, is the generalization of these methods to numerical schemes that have larger regions of stability and are higher order, as well as methods that are **adaptive**, i.e., methods that take advantage of variations in f and the solution of the initial value problem in (5.46) to adapt the discretization of the domain $[0, T)$ to variable step-size h. We will take up several of these schemes in the projects at the end of this chapter, but conclude this section by pointing out that MATLAB's `ode45`, in conjunction with several other ODE solvers available in MATLAB, already incorporates state-of-the-art advances made in this field and provides us with one of the most accurate, powerful and versatile numerical schemes for solving initial value problems involving ODEs.

Problems 5.7

1. Verify the calculations in (5.43) and 5.44) to complete the proof that FEM is a first order scheme.

2. Show that BEM is a first order scheme.

3. Show that the scheme we obtain from the Centered Euler Method is second order.

5.8 Boundary Value Problems and the Shooting Method

As evidence of the versatility of MATLAB and its `ode45` function, in this section we present a MATLAB code that combines two numerical techniques designed to solve an important problem in applied mathematics, namely that of obtaining a solution to a **boundary Value Problem** (BVP) of the form

$$x' = f(t, x, y), \quad y' = g(t, x, y), \qquad x(0) = a, \, x(T) = b. \tag{5.47}$$

Note that (5.47) is a boundary value problem because we have specified the value of x at $t = 0$ and at $t = T$, as opposed to specifying the values of x and y at $t = 0$. The function `ode45` is designed to solve initial value problems, as are FEM and BEM described in the previous sections, and our main task in this section is to convert (5.47) to an appropriate initial value problem to which we can apply `ode45`. The technique we will employ is called the *Shooting Method*, whereby in place of solving (5.47) we solve the initial value problem (IVP)

$$x' = f(t, x, y), \quad y' = g(t, x, y), \qquad x(0) = a, \, y(0) = y_0. \tag{5.48}$$

The solution $\langle x(t), y(t)\rangle$ we obtain in this way will probably not satisfy the boundary condition

$$x(T) = b, \tag{5.49}$$

unless we are very lucky. Our objective will be to experiment with y_0 in (5.48) and compute $x(T)$ for each new guess of y_0 and try to minimize the residual $x(T) - b$ as a function of y_0.

The heart of the shooting method is in what was just described: we think of the quantity $x(T) - b$ as a function of y_0 and seek a zero of this function. The program listed below accomplishes this task. It involves three MATLAB M-files, `RightSide.m`, ShootFirst.m and `Bisection.m`. The M-file `RightSide.m` makes the ODEs in (5.47) accessible to MATLAB. The second M-file `ShootFirst.m` uses `RightSide.m` with `ode45` and solves the IVP (5.48) and returns the value $x(T) - b$. The third M-file, `Bisection.m` implements the **bisection Method** and computes a root of $x(T) - b$ as y_0 varies.

The bisection method is one of the simplest algorithms for finding zeros of a function $y = f(x)$. Its implementation is based on the notion that if f is continuous on an interval (a, b) with f having different signs at a and b, then f must have a zero in (a, b). To find such a point, we begin by evaluating f at the midpoint $m = \frac{a+b}{2}$ and comparing the sign of this value with those of $f(a)$ and $f(b)$. The next step of the algorithm is to replace the interval (a, b) by either (a, m) or by (m, b) depending on whether $f(a)f(m) < 0$ or $f(b)f(m) < 0$. We then repeat this process, that is, consider the midpoint of the new interval and proceed to compare the evaluation of f at this point relative to the endpoints. It is easy to see that the algorithm always converges and that at each step of the algorithm, the length of the interval containing the zero is cut in half. The program below lists the three M-files whose execution in MATLAB leads to Figure 5.11. This code is written for the system

$$x' = \frac{y}{\sqrt{x^2 + y^2}}, \quad y' = -\frac{x}{\sqrt{x^2 + y^2}}, \qquad x(0) = -3,\ x(10) = 2. \quad (5.50)$$

```
%%%  RightSide.m %%%

function yprime=RightSide(t,y);
term=1./(sqrt(y(1).^2+y(2).^2));
yprime=term.*[y(2); -y(1)];

%%%  ShootFirst.m %%%

function target=ShootFirst(a,boundaryvalue);
[t y]=ode45('RightSide',[0 10],[-3 a]);
target=y(length(t),1)-boundaryvalue;

%%%  Bisection.m %%%

function root=bisection(a0,b0,boundaryvalue,n)
a=a0:0.1:b0;
l=length(a);
b=[];
for i=1:l
    b=[b ShootFirst(a(i),boundaryvalue)];
```

```
end
b1=b(1:l-1);
b2=b(2:l);
y=b1.*b2;
[z,j]=min(y)
z
left=a(j);right=a(j+1);
for i=1:n
    mid=(left+right)/2;
    term1=ShootFirst(left,boundaryvalue);
    term2=ShootFirst(mid,boundaryvalue);
    if term1*term2 < 0
        right = mid;
    else left=mid;
        eval(['left = ',num2str(left),', mid = ', ...
          num2str(mid), ', right = ', num2str(right)])
    end
end
root=mid;
```

The starting point of this algorithm requires a guess for a and b, which we arrive at by running `ShootFirst.m` at various values of a until we obtain a negative `target` value and a positive one. The trials `ShootFirst(-3, 2)` and `ShootFirst(1,2)` give us the appropriate a and b. With this information in hand, we next apply `Bisection.m`:

```
Bisection(-3,1,2,10)
```

which gives us ten iteration of the bisection algorithm, leading to the `target` value of `-2.0483`. Having found the right shooting value for y_0 we run the following lines in MATLAB to get Figure 5.11:

```
[t,y]=ode45('RightSide',[0 10], [-3 -2.0483]);
plot(y(:,1),y(:,2))
set(gca,'DataAspectRatio',[1 1 1])
xlabel('x')
ylabel('y')
```

Problems 5.8

1. Apply the shooting method to find the solution to the following boundary value problems:

   (a) $x' = y$, $y' = -x$, $x(0) = 0$, $x(\frac{\pi}{2}) = 1$. Compare the approximate solution to the exact solution of this problem—note that

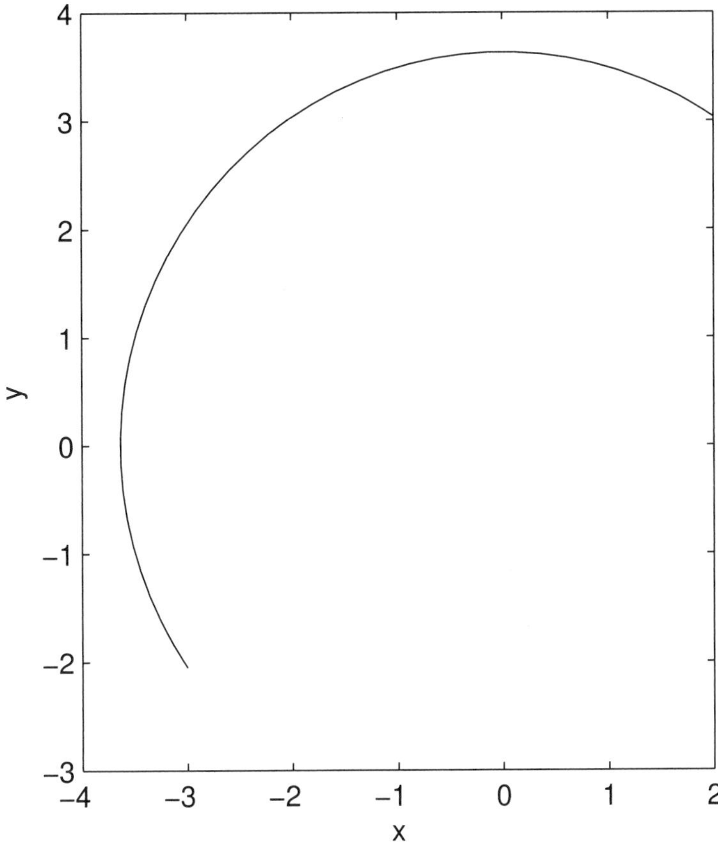

FIGURE 5.11: The output of the shooting algorithm.

this system of ODEs is equivalent to the second order ODE $x'' + x = 0$.

(b) $x' = y$, $y' = -\sin x$, $x(0) = 0$, $x(\pi) = 1$.

(c) $x' = y$, $y' = -0.1y - \sin x + 0.2\cos 3t$, $x(1) = 0$, $x(2\pi) = 2$.

(d) $x'' + 0.1x' + (1 + t^2)x = 0$, $x(-1) = 0$, $x(1) = a$. Experiment with a to discover if there are any $a \geq 0$ for which this problem has a solution. Recall that any second order ODE $x'' = f(t, x, x')$ can be converted to a system of first order equations by defining a new variable y with $x' = y$ and noting that $y' = f(t, x, y)$.

2. Modify the shooting method to apply to the following BVPs:

(a) $y'' = f(t, y, y')$, $y(a) = y_1$, $y'(b) = y_2$. Write the program with a and b as input parameters.

(b) $y'' + 4y = 0$, $y(0) = 1$, $y'(1) = 0$.

(c) $y'' + 0.1y' + \sin y = 0$, $y'(1) = 1$, $y(2) = 3$.

(d) $x'' + x' + (1 - x^2) = 0$, $x'(0) = 0$, $x'(1) = 1$.

(e) $yy'' + y^2 = 0$, $y(0) = 1.1$, $y(3) = 4$.

3. A well-known problem in the flow past a flat plate, called the **Blasius Boundary Layer** problem, is modeled by the BVP

$$f''' + ff'' = 0, \quad f(0) = f'(0) = 0, \quad f'(\infty) = 1. \tag{5.51}$$

Modify the shooting method to apply to (5.51). See Figure 5.12 for the expected output. (Hint: First convert (5.51) to a system of three ODEs by defining $y_1 = f$, $y_2 = f'$, $y_3 = f''$, and noting that $y_1' = y_2$, $y_2' = y_3$ and $y_3' = -y_1 y_3$. Next modify the `Bisection.m` file to apply the shooting method on y_3 with the range $(0.3, 0.5)$, which is the range along which $y_2 - 1$ changes sign. It is sufficient to replace the domain $t \in (0, \infty)$ with $(0, 10)$ because the system converges to its equilibrium very quickly. Use MATLAB's `legend` command to get the legend shown in Figure 5.12.)

5.9 Project A: Modified Euler Method

The Backward Euler Method is based on the difference formula (5.5). When this formula is applied to the nonlinear equation

$$y' = f(t, y), \quad y(0) = y_0, \tag{5.52}$$

it leads to the nonlinear difference equation

$$y_i = y_{i-1} + hf(t_i, y_i), \quad i = 1, 2, 3... \tag{5.53}$$

Because this equation depends implicitly on y_i, we need to take an extra step of inverting (5.53) before proceeding to find its general solution. A fixed-point approach that leads successfully to this inversion was proposed and implemented in MATLAB in Section 5.2 of this chapter.

Instead of taking that route, an alternative approach is to replace the y_i on the right side of (5.53) by a formula involving lower-indexed y_i's. One way to accomplish this is to replace the y_i on the right side

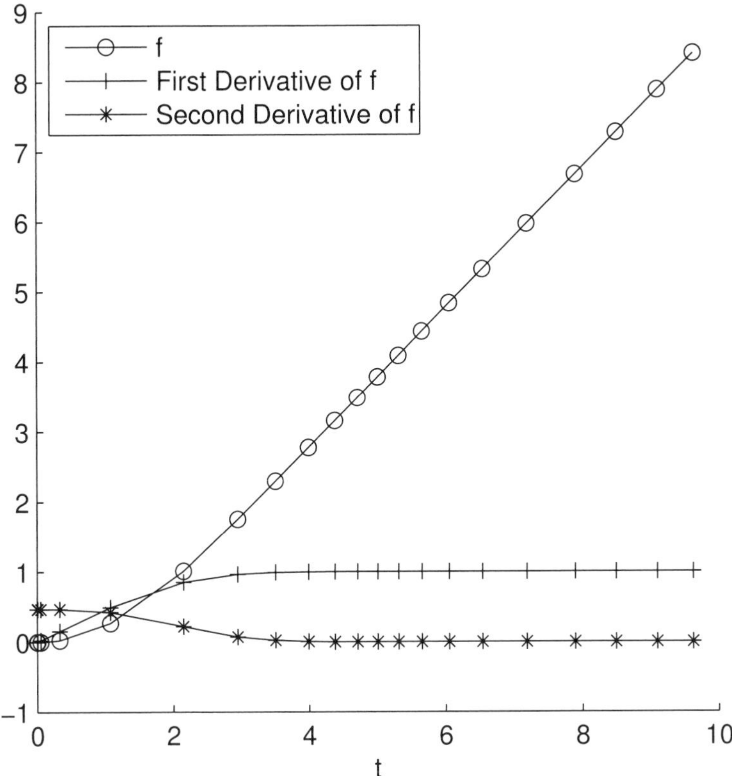

FIGURE 5.12: The output of the shooting algorithm for the Blasius equation (5.51).

with its approximation using FEM, namely with $y_{i-1} + hf(t_{i-1}, y_{i-1})$, so the modified formula for computing y_i is

$$y_i = y_{i-1} + hf(t_i, y_{i-1} + hf(t_{i-1}, y_{i-1})).$$

Note that now the difference equation is explicit in y_i. Another approach is to first write the slope $f(t_i, y_i)$ as the average of the slopes at (t_{i-1}, y_{i-1}) and (t_i, y_i)

$$y_i = y_{i-1} + \frac{h}{2}(f(t_{i-1}, y_{i-1}) + f(t_i, y_i))$$

and then introduce the FEM formula on the right side:

$$y_i = y_{i-1} + \frac{h}{2}(f(t_{i-1}, y_{i-1}) + f(t_i, y_{i-1} + hf(t_{i-1}, y_{i-1} + hf(t_{i-1}, y_{i-1})))). \tag{5.54}$$

The motivation behind the averaging approach actually comes from resetting (5.52) in its integral form:

$$y(t+h) - y(t) = \int_t^{t+h} f(s, y(s))\, ds. \tag{5.55}$$

The integral on the right side of (5.55) can be approximated in several ways. The simplest one would be to replace the integrand $f(s, y(s))$ by the constant value $hf(t, x(t))$, thus obtaining FEM, or by the constant value $hf(t+h, y(t+h))$ to get BEM. Alternatively, we could approximate the integral by the area of the trapezoid with vertices $(t, 0)$, $(t, f(t, y(t)))$, $(t+h, f(t+h, y(t+h)))$ and $(t+h, 0)$, which is

$$\frac{h}{2}(f(t, y(t)) + f(t+h, y(t+h))),$$

which is the basis of (5.54).

The formula in (5.54) is known as the *Modified Euler Method,* which we now state slightly differently. This method is an example of a *predictor-corrector* algorithm, where one typically applies a known method (in this case FEM) to get a first approximation to the output, and then uses other means (in this case the averaging of the slopes) to correct this value further.

(Modified Euler Method)
The difference equation for the Modified Euler Method is

$$y_0 = given, \quad y_i = y_{i-1} + \frac{h}{2}(s_1 + s_2) \tag{5.56}$$

where s_1 *and* s_2 *are the approximate slopes at* (t_{i-1}, y_{i-1}) *and* $(t_i, y_i + hs_1)$, *or*

$$s_1 = f(t_{i-1}, y_{i-1}), \quad s_2 = f(t_i, y_i + hs_1). \tag{5.57}$$

The following MATLAB code will implement this scheme:

```
%
% Initialize t, T, y0, n,
% f is defined by an M-file or by the inline command
%
h = T/n;
h2=h/2;
```

```
out = [t y0];
y=y0;
for i=1:n
        s1=f(t,y);
        t=t+h;
        pred=y+h*s1;
        s2=f(t,pred);
        y=y+h2*(s1+s2);
        out=[out; t y];
end
```

1. Consider the differential equation

$$y' = y^2 + \sin t, \quad y(0) = 1. \tag{5.58}$$

   Write a MATLAB program to compute y_i with $n = 10$ and $T = 0.8$.

   The first five entries of the output of this program are

0	1.0000
0.0800	1.0899
0.1600	1.2037
0.2400	1.3485
0.3200	1.5349

   Figure 5.13 shows the output of this program versus `ode45`'s.

2. Modify this program to allow for $n = 100$ and compare the result with the `ode45` output.

3. Analyze this problem as T approaches 1. What happens to the `ode45` output? And what happens to the output of the Modified Euler Method?

4. Apply the Taylor expansion formula to show that the Modified Euler Method is second order and determine its truncation error.

5.10 Project B: Runge–Kutta Methods

The modified Euler method is a second-order finite difference scheme whose implementation requires function evaluations of f but does not

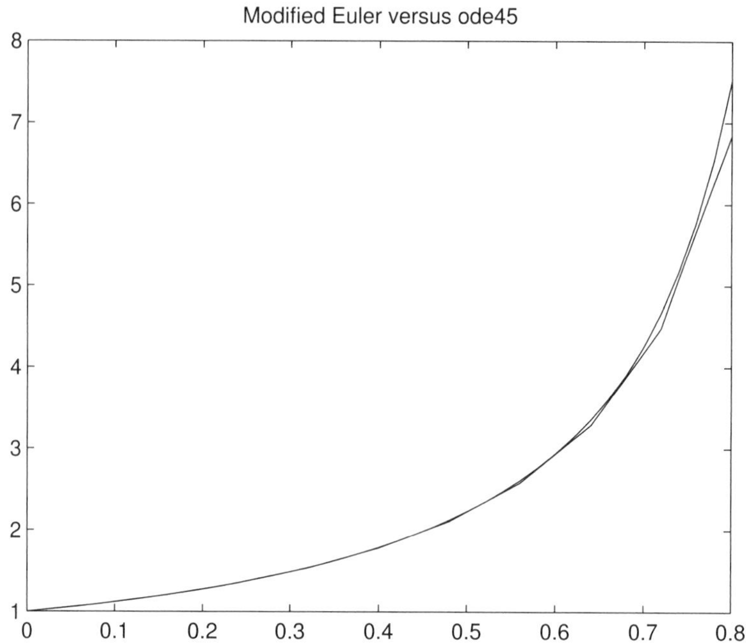

FIGURE 5.13: The Modified Euler Method applied to $y' = y^2 + \sin t$, y(0) = 1 and its comparison with the output from `ode45`.

rely on evaluating any of the derivatives of f. Methods having this property are quite desirable and attempts have been made to construct higher order methods which rely only the evaluation of the known right side of the system of ODEs $\mathbf{y}' = f(t, \mathbf{y})$. These methods are collectively referred to as *Runge–Kutta* schemes. The fourth–order Runge–Kutta scheme is very similar to the modified Euler method in that its implementation requires computing predictors and corrector. For the scalar equation $y' = f(t, y)$, this scheme's finite difference formula is

$$y_{n+1} = y_n + \frac{h}{6}(s_1 + 2s_2 + 2s_3 + s_4), \tag{5.59}$$

where the term that multiplies h is essentially a weighted average of the slopes of the tangent lines to the solution at t, at $t + \frac{h}{2}$ and at $t + h$. Here we give the formulas for s_i's and refer the reader the references at the end of this chapter for details. The term s_1 is the slope y' of the solution at t_n, where we use the ODE to determine y':

$$s_1 = f(t_n, x_n). \tag{5.60}$$

The term s_2 estimates the slope at the midpoint $t_n + \frac{h}{2}$; Euler's method is used to compute $y_{n+\frac{1}{2}}$:

$$s_2 = f\left(t_n + \frac{h}{2}, x_n + \frac{h}{2}s_1\right). \tag{5.61}$$

The term s_3 is a correction of this predicted value where s_2 replaces s_1:

$$s_3 = f\left(y_n + \frac{h}{2}, y_n + \frac{h}{2}s_2\right). \tag{5.62}$$

the term s_4 is a correction of the slope at t_{n+1} using s_3:

$$s_4 = f(t_{n+1}, y_n + hs_3). \tag{5.63}$$

We will not justify why this scheme works as well as it does, although the reader is encouraged to apply Taylor's formula to the difference equation (5.59) that this method does lead to a fourth-order scheme.

Algorithm 5.10.1 (Runge–Kutta Method)
The difference equation for the 4th-order Runge–Kutta method is given by (5.59), where the slopes s_i's are defined by (5.60)–(5.63).

The modified Euler and the 4th-order Runge–Kutta methods generalize in a straightforward manner to systems of ODEs. The finite difference formulas (5.56) and (5.59) are simply applied to each differential equation in a given system.

In addition to 4th-order Runge–Kutta method just described, MATLAB's `ode45` incorporates *adaptivity* in selecting its step size at each step n. Stepsize adaptivity results in an algorithm that is considerably more efficient in solving systems of ODEs especially when the solution behaves quite differently in various parts of the domain. For example, the solution $y(t)$ of the initial value problem $y' = y^2$ with $y(0) = 1$ is $y(t) = \frac{1}{1-t}$, which becomes undefined as t approaches 1. Hence, we expect that we need to use more grid points near $t = 1$ to represent the solution accurately relative to the neighborhood of $t = 0$, where the function $\frac{1}{1-t}$ is well-behaved.The goal of adaptivity is to devise an algorithm to anticipate the change in behavior of the solution to adjust the step size h to achieve desired accuracy while keeping the function evaluations to a minimum.

Adaptive schemes keep track of two error tolerances, the minimum and maximum errors of the method, while computing the solution at the n-th iteration step. Regardless of which of the several methods we discussed earlier are being implemented, FEM, BEM, or Modified Euler, we can compute two approximate values at T, one with step size h, which we denote by y_h, and the other with step size $\frac{h}{2}$ and denoted by $y_{\frac{h}{2}}$. We typically suspect that $y_{\frac{h}{2}}$ is more accurate than y_h. The difference between y_n and $y_{\frac{h}{2}}$ is the quantity that one tests against the minimum and maximum tolerance errors. If this difference falls between the tolerance errors, y_h is assumed to be tolerable and a good approximation to the true solution and one stays with the step size h. If the difference is below the minimum tolerance error, then the step size is doubled for next iteration. If the difference is above the maximum tolerance error, one retreats to the step prior to T and computes again but this time with step size $\frac{h}{2}$. This type of adaptivity is an elegant way of treating sharp transitions that may appear in a solution.

Problems

1. Apply the modified Euler method and obtain a table of values for the following differential equations. Use a step size h and a number of iterations n of your own choosing. In each case compare the output with the solution from `ode45`.

   (a) $y' = -y + 1, \quad y(0) = 0.$

   (b) $y' = -y^2 + t(1 - t), \quad y(0) = 1.$

   (c) $y' = t \sin y, \quad y(\pi) = -1.$

2. Write a MATLAB program to implement the fourth-order Runge–Kutta scheme for the initial value problem $y' = f(t, y)$, $y(t_0) = y_0$. The program should be structured to access f through an M-file. As always, the first few lines of the program should introduce the parameter values n, h, etc. It should output a table of values for t_n and y_n. apply this program to the following IVPs while using a step size h and a number of iterations n of your own choosing. In each case, compare the output with the solution one obtains from MATLAB.

   (a) $y' = -2y + t, \quad y(0) = 0.$

   (b) $y' = -\sin y + \sin t, \quad y(1) = 0.$

   (c) $z' = z^2, \quad z(0) = 0.1$. Recall that the analytic solution of this problems blows up in finite time. How does this effect appear in the approximate solution?

   (d) $y' = \tan y, \quad y(0) = \frac{\pi}{4}.$

3. Develop a MATLAB program for the 4th-order Runge–Kutta scheme for the 2×2 system

$$x' = f(t, x, y), \quad y' = g(t, x, y).$$

Apply this program to the following systems. Use h, n, and initial conditions of your own choosing. In each case, graphically compare the output with that of `ode45`.

   (a) $x' = y, \quad y' = -x.$
   (b) $x' = x - 2y, \quad y' = x + 2y.$
   (c) $x' = y, \quad y' = -\sin x.$
   (d) $x' = y, \quad y' = -0.1y - \sin x.$
   (e) $x' = 1 - \frac{x^2-y^2}{(x^2+y^2)^2}, \quad y' = -\frac{2xy}{(x^2+y^2)^2}.$

5.11 Project C: Finite Difference Methods and BVPs

Consider the BVP

$$u'' = f(x), \qquad u(0) = u(1) = 0. \tag{5.64}$$

In this project, a finite difference scheme is applied to (5.64) to obtain its approximate solution.

1. Show that

$$u(x) = -x \int_0^1 \int_0^s f(\tau)\, d\tau\, ds + \int_0^x \int_0^s f(\tau)\, d\tau\, ds) \tag{5.65}$$

   is the exact solution of (5.64).

2. Let $(0, x_1, x_2, ..., x_{n-1}, x_n, 1)$ be a discretization of the interval $(0, 1)$, define the step-size $h = x_i - x_{i-1}$ and let u_i stand for $u(x_i)$. Use Taylor's formula to show that

$$u''(x_i) = \frac{1}{h^2}(u_{i+1} - 2u_i + u_{i-1}) + \frac{h^2}{12}u''''(x_i) + ... \tag{5.66}$$

   The expression

$$\frac{1}{h^2}(u_{i+1} - 2u_i + u_{i-1}) \tag{5.67}$$

   is a three-point approximation of $u''(x_i)$. The formula in (5.66) shows that this approximation has an h^2 *local truncation error* (denoted by $O(h^2)$), assuming u'''' is a well-behaved function.

3. Approximate the original BVP in (5.64) with the finite-difference approximation

$$u_{i+1} - 2u_i + u_{i-1} = h^2 f_i, i = 1, 2, ..., n, \tag{5.68}$$

where $f_i = f(x_i)$. Show that the system of n simultaneous equations in (5.68) is equivalent to $A\mathbf{u} = f$ where

$$A = \begin{bmatrix} -2 & 1 & 0 & 0 & \cdots & \cdots & 0 \\ 1 & -2 & 1 & 0 & \cdots & \cdots & 0 \\ 0 & 1 & -2 & 1 & 0 & \cdots & 0 \\ \cdots & \cdots & \cdots & \cdots & \cdots & \cdots & \cdots \\ 0 & \cdots & 0 & 1 & -2 & 1 & 0 \\ 0 & \cdots & \cdots & 0 & 1 & -2 & 1 \\ 0 & \cdots & \cdots & \cdots & 0 & 1 & -2 \end{bmatrix}, \qquad \mathbf{u} = \begin{bmatrix} u_1 \\ u_2 \\ u_3 \\ \cdots \\ u_{n-2} \\ u_{n-1} \\ u_n \end{bmatrix},$$

$$\mathbf{f} = \begin{bmatrix} h^2 f_1 & h^2 f_2 & h^2 f_3 & ... & h^2 f_{n-2} & h^2 f_{n-1} & h^2 f_n \end{bmatrix}^{\mathrm{T}}. \tag{5.69}$$

Note that $u_0 = u_{n+1} = 0$ from the boundary conditions.

4. Write a MATLAB program to solve (5.69) when $f(x) = 100 \sin \pi x$. Plot the graphs of the approximate and the exact solution on the same screen and compare with Figure 5.14. Hint: Look up the syntax for `diag` to come up with the following convenient way of entering A into MATLAB:

```
vec1=ones(n,1);
vec2=ones(n-1,1);
A = -2*diag(vec1)+diag(vec2,1)+diag(vec2,-1);
```

5. Consider the BVP $u'' = f(x)$ subject to the boundary conditions $u(0) = a$ and $u(1) = b$.

   (a) Find the exact solution of this problem.

   (b) Discretize this BVP to obtain the equivalent of (5.69).

   (c) Apply the results to the BVP $u'' = x \sin x$ with $u(-1) = 1$ and $u(2) = -3$.

6. Consider the BVP $u'' = f(x)$ subject to the boundary conditions $u(0) = a$ and $u_x(1) = b$.

   (a) Find the exact solution of this problem.

   (b) Discretize this BVP to obtain the equivalent of (5.69).

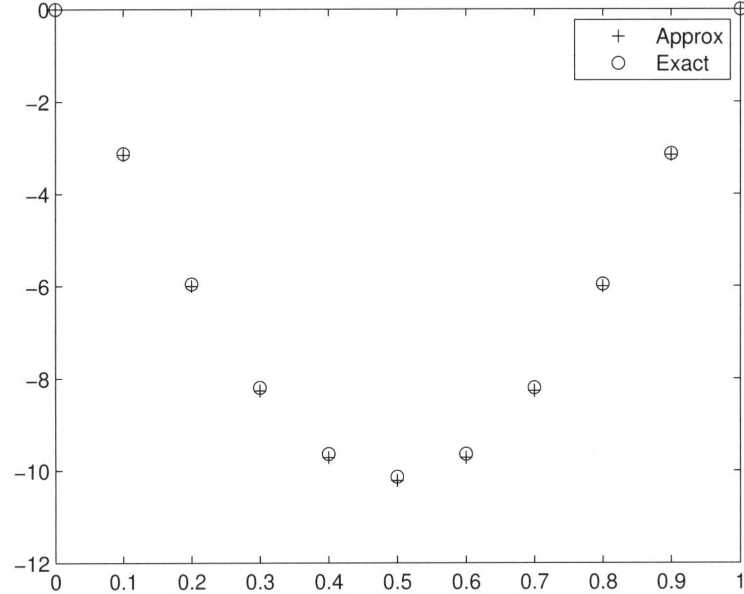

FIGURE 5.14: The approximate solution to $u'' = 100 \sin \pi x, u(0) = u(1) = 0$ and its exact solution when $n = 10$. The absolute and relative errors are 0.0684 and 0.0068, respectively.

(c) Apply the results to the BVP $u'' = \frac{x}{1+x^2}$ with $u(1) = -1$ and $u_x(3) = 2$.

7. Apply the above method to the BVP

$$u'' + \lambda u = f(x), \quad u(0) = 0, u(1) = 0. \tag{5.70}$$

Show that the matrix A in (5.69) must be replaced by

$$A = \begin{bmatrix} \gamma & 1 & 0 & \dots & 0 & & \\ 1 & \gamma & 1 & 0 & \dots & \dots & 0 \\ 0 & 1 & \gamma & 1 & 0 & \dots & 0 \\ \dots & \dots & \dots & \dots & \dots & \dots & \dots \\ 0 & \dots & 0 & 1 & \gamma & 1 & 0 \\ 0 & \dots & \dots & 0 & 1 & \gamma & 1 \\ 0 & \dots & \dots & \dots & 0 & 1 & \gamma \end{bmatrix},$$

where $\gamma = -2 + \lambda h^2$. Apply this method to the BVP $u'' + 4u = (1-4x^2) \sin \pi x$, $u(0) = u(1) = 0$ to obtain an approximate solution similar to the one in Figure 5.15. This solution is remarkably close to the exact solution of the problem, which can readily be obtained using any symbolic manipulator such as *Mathematica*. If the reader has access to this software, then

```
DSolve[{u''[x] + 4 u[x] ==
(1 - 4 x^2)*Sin[Pi*x], u[0] == 0,
  u[1] == 0}, u[x], x]
```

leads to the expression

$$\frac{1}{(-4+\pi^2)^3}(16\pi\left(-4+\pi^2\right)x\cos(\pi x)+$$

$$16\pi\left(-4+\pi^2\right)\csc(2)\sin(2x)+$$

$$\left(64x^2-16\pi^2\left(2x^2+1\right)+\pi^4\left(4x^2-1\right)-48\right)\sin(\pi x))$$

for the exact solution. The plot of this function and the approximate solution by our method is shown in Figure (5.15). This figure is obtained by executing the following lines in MATLAB:

```
clear all
clf
n=10;
lambda=4;
h=1/(n+1);
x=h:h:1-h;
vec1=ones(n,1);
vec2=ones(n-1,1);
A=(-2+h^2*lambda)*diag(vec1)+diag(vec2,1)+...
    diag(vec2,-1);
f=h^2*(1-4*x'.^2).*sin(pi*x');
u=A\f;
plot(x,u)
exact=(1/(-4 + pi^2)^3)*(16*pi*(-4 + ...
   pi^2)*x.*cos(pi*x) + ....
  16*pi*(-4 + pi^2)*csc(2)*sin(2*x) + ...
     (-48 + 64*x.^2 - 16*pi^2*(1 + 2*x.^2) + ...
        pi^4*(-1 + 4*x.^2)).*sin(pi*x));
plot(x,exact,'*', x, u, '+r')
legend('exact','approximate','location','NorthWest')
error=max(abs(exact-u'));
relerr=error/max(abs(exact));
```

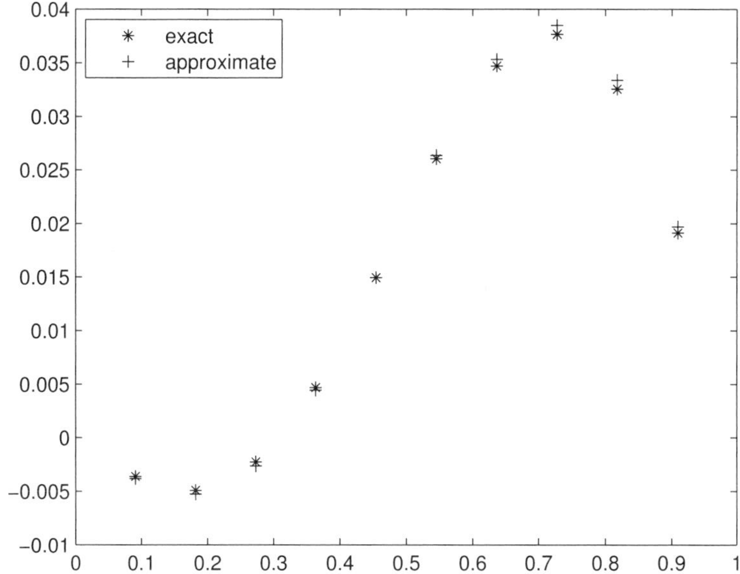

FIGURE 5.15: The approximate solution to $u'' + 4u = (1 - 4x^2)\sin \pi x, u(0) = u(1) = 0$ and its exact solution when $n = 10$. The absolute and relative errors are 0.00083 and 0.0221, respectively.

5.12 Project D: Method of Lines

This project was first developed in [2], Vol 1, pp. 511–515, for the linear heat equation, where the PDE

$$u_t = \lambda u_{xx} \tag{5.71}$$

with the boundary conditions

$$u(0, t) = u(L, t) = 0, \tag{5.72}$$

and initial data

$$u(x, 0) = u_0(x) \tag{5.73}$$

is solved by converting the problem to a set of infinitely many coupled ODEs and the resulting system is solved using an ODE solver such as `ode45`. The method we describe below applied equally well to nonlinear PDEs of the form

$$u_t = \lambda u_{xx} + f(t, x, u)$$

or wave equations of the form

$$u_{tt} = c^2 u_{xx} + f(t, x, u).$$

The Method of Lines seeks approximate solutions of a PDE by only replacing the spatial derivatives in the PDE by finite differences. This results in a large system of simultaneous ODEs, which can then be solved in MATLAB. For example, referring to (5.16), we first discretize the domain $(0, L)$ into equal subintervals, with a typical subinterval of the form (x_{i-1}, x_i), and then approximate $u_{xx}(t, x_i)$ by the centered difference scheme

$$\frac{1}{h^2}\left(u(x_{i+1}, t) - 2u(x_i, t) + 2u(x_{i-1}, t)\right). \tag{5.74}$$

With this approximation the heat equation now takes the form

$$u_t(x_i, t) = \frac{1}{h^2}\left(u(x_{i+1}, t) - 2u(x_i, t) + u(x_{i-1}, t)\right), \tag{5.75}$$

with i ranging from 0 to n. Thus the PDE in (5.71) is replaced by a set of ODEs. The following program in MATLAB, called `mol1D.m`, is written with the following set of parameters $l = 0.79$, $n = 64$, $u_0(x) = \sin\frac{\pi x}{L}$, and $\lambda = 0.1$. It solves a system of odes, consisting of 64 coupled differential equations, and compares the result to the exact solution of this problem, which is

$$u(x, t) = e^{-\frac{\lambda\pi^2}{L^2}t}\sin\frac{\pi x}{L}. \tag{5.76}$$

```
global n h lambda;
clf;
nographs=5; lambda = 0.12; L=0.79; n=64; h=L/n;
x=h:h:L-h;
u0=sin(pi*x/L);
x=[0 x L];
exact=inline('exp(-lambda*pi^2*t/L^2)*sin(pi*x/L)','x',...
          't','lambda','L');
[t,u]=ode45(@OneDheat,[0 0.5],u0,10^(-7));
deltat=floor(length(t)/nographs);
for i=1:nographs
      approximate=[0 u(i*deltat,:) 0];
      subplot(211)
      plot(x,approximate)
      title(['1D Heat Equation, Method of Line, n=',...
           num2str(n)]);
      hold on
      subplot(212)
```

```
        Exact=exact(x,t(i*deltat),lambda,L);
        Error=abs(Exact-approximate)
        plot(x,Error)
        ylabel('Error')
        hold on
end
```

The M-file `mol1D.m` calls the M-file `oneDheat.m` which is

```
function uprime=oneDheat(t,u);
%
global n h lambda;
y=length(u);
uleft=[u(2:y); 0];
uright=[0; u(1:y-1)];
uprime=(lambda/h^2)*(uleft-2*u+uright);
```

1. Create the two M-files `mol1d.m` and `oneDheat.m`. Note the small differences in syntax between these two M-files and their counterparts in [2]. Compare the output of `mol1D.m` with Figure 5.16.

2. (a) Experiment with different values of n and report on the qualitative differences in the `Error`.

   (b) Let $n = 2^j$, j ranging from 2 to 10. Write a program that computes

   $$err(j) = max_{0 \le x \le L} |u(x,1) - u_{app}(x,1)|, \quad j = 2, ..., 10,$$

   where u_{app} is the output of `mol1d.m`. Plot the graph of err.

3. Modify the above program to apply it to the initial condition

   $$u(x,0) = \sum_{n=1}^{N} \frac{1}{n} \sin \frac{n\pi x}{L}. \tag{5.77}$$

   Report on the absolute error (i.e., `Error` in `mol1D.m`) as N varies from 2 to 16.

4. Consider the initial-boundary value problem

   $$u_t = u_{xx} + \sin u, \quad u(0,t) = u(1,t) = 0$$

   and

   $$u(x,0) = x(1-x).$$

   Modify `mol1d.m` appropriately to solve this nonlinear initial-boundary value problem.

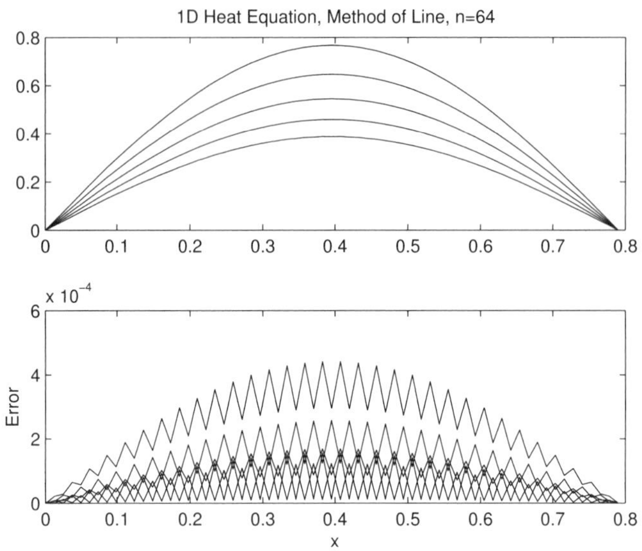

FIGURE 5.16: The graph of $u(x, t)$ with $u(x, 0) = \sin \frac{\pi x}{L}$.

5. Consider the nonlinear initial-boundary value problem

$$u_t = u_{xx} + u(1 - u), \quad u(0, t) = u(1, t) = 0$$

and

$$u(x, 0) = \sin \pi x.$$

report on the asymptotic state of u, i.e., on the behavior of u as t approaches infinity.

5.13 Project E: Burgers Equation (Method of Characteristics)

The partial differential equation

$$u_t + (f(u))_x = 0, \tag{5.78}$$

is a fundamental PDE in mathematical physics serving as the prototype for many phenomena of interest in physics where turbulence and formation of discontinuities play significant roles. Typically this equation governs the behavior of a quantity represented by u whose value changes in t but is balanced by a *flux* $f(u)$, representing the advection of the quantity u in x.

In this project we study the IVP

$$u(x,0) = u_0(x) \tag{5.79}$$

together with the PDE (5.78). The method we address first leads to determining a set of special curves called characteristics, along which the solution u happens to take on a simple structure, which we are then able to take advantage of and on occasion write down the exact solution. When obtaining an exact and analytical solution is too difficult, the characteristics method will still lead to accurate approximate solutions.

The *characteristic curves* of the Burgers equation are defined as curves $\hat{x}(t)$ such that the initial-value problem

$$\frac{d\hat{x}}{dt} = f'(u(\hat{x}(t),t)), \quad \hat{x}(0) = x_0 \tag{5.80}$$

holds. Here f' is the derivative of $f(u)$ in (5.78). As we will see shortly, these curves characterize the behavior of solutions of the PDE (5.78) by reducing this equation to a system of ODEs which we are able to solve, often analytically, and always numerically.

1. Let $f(u) = au$ in (5.78), where a is a constant. Show by direct differentiation that the solution to the reduced IVP

   $$u_t + au_x = 0, \quad u(x,0) = u_0(x) \tag{5.81}$$

   is

   $$u(x,t) = u_0(x - at). \tag{5.82}$$

   The form of u in (5.82) suggests that u has exactly the same shape as u_0 and travels to the right with speed a, if a is positive. Write a MATLAB program that generates a graph similar to the one in Figure 5.17 to demonstrate the wave propagation character of (5.81).

   We next develop the solution $u(x,t) = u_0(x - at)$ by a different method, the method of characteristics, and generalize this method to functions f in (5.78) which are considerably more general than the linear function $f(u) = au$.

2. Returning to the definition of characteristics (5.80), observe that in the special case when $f(u) = au$ this definition reduces to the ODE

   $$\frac{d\hat{x}}{dt} = a, \quad \hat{x}(0) = x_0. \tag{5.83}$$

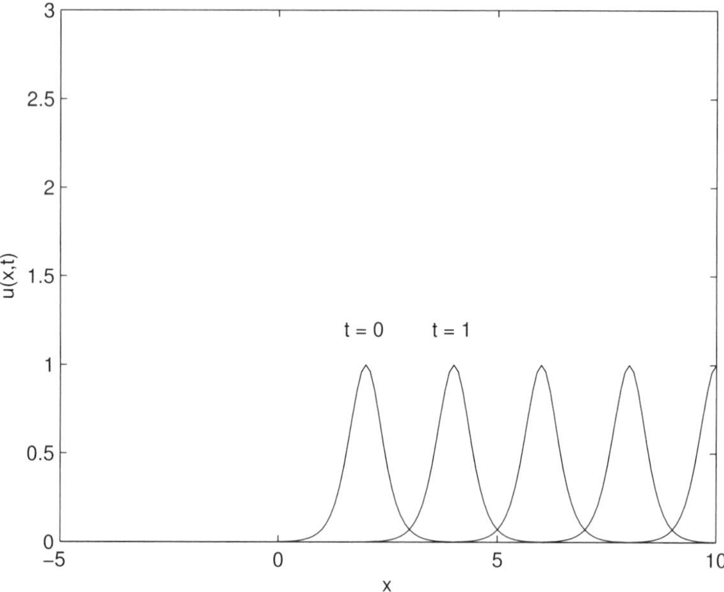

FIGURE 5.17: The graph of a typical solution to (5.81) showing the wave character of this PDE. The initial disturbance $u_0(x) = \operatorname{sech}^2(x)$ travels to the right with speed $a = 2$.

Show that the solution to the IVP in (5.83) is

$$\hat{x}(t) = x_0 + at. \tag{5.84}$$

The expression in (5.84) states that the characteristics of the PDE $u_t + au_x = 0$ are all straight lines, all having the same slope a, independent of the initial condition u_0.

3. Let U be the solution u confined to the characteristic line $\hat{x} = x_0 + at$, i.e.,

$$U(t) = u(x_0 + at, t). \tag{5.85}$$

Show by direct differentiation that $U'(t) = 0$ for all t. Hence $U(t) = U(0)$. Since $U(0) = u(x_0, 0)$ and $u(x, 0) = u_0(x)$ by (5.79), show that

$$U(t) = u_0(x_0). \tag{5.86}$$

Expression (5.86) states that $U(t)$ is in fact constant and its value is determined by the initial condition u_0 and the intercept x_0 of the characteristic that passes through (x, t). See Figure 5.18.

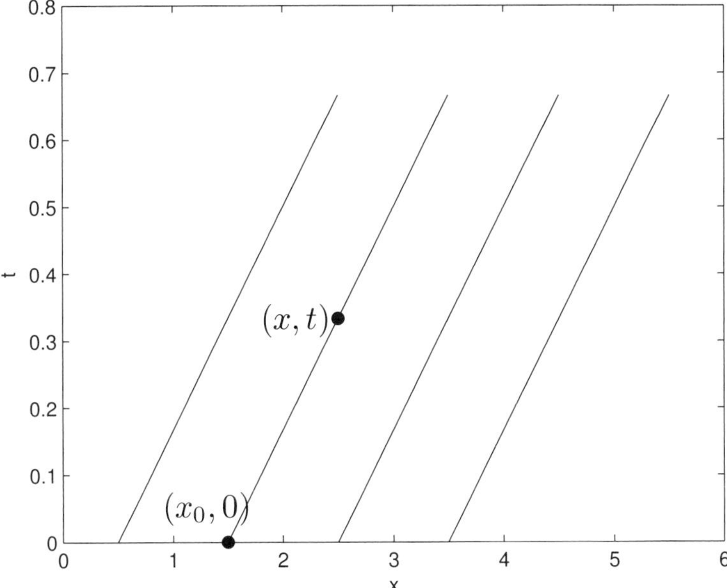

FIGURE 5.18: The characteristic curves for the IVP in (5.81) are straight lines $\hat{x} = x_0 + at$ in the (x, t) plane. The solution $u(x, t)$ remains constant along each characteristic, the constant value being $u_0(x_0)$. Thus the value of u at any (x, t) shown in the graph is determined the value of the initial function u_0 at x_0.

4. Next, observe that since $x_0 = x - at$, the function $U(t)$ is then

$$U(t) = u_0(x - at)$$

which is precisely the expression we see in (5.82), that is, the method of characteristics has brought us full circle back to the exact solution of (5.81). The advantage of this method, however, is that it removes the guess work, albeit a very good guess work, that led to (5.82).

As stated earlier, many of the features of the characteristics method go over to the case when f in (5.78) is nonlinear, which we will pursue in the next project.

5.14 Project F: Burgers Equation (Method of Characteristics – Nonlinear Case)

In the previous project we observed that the solution to Burgers IVP (5.78)–(5.79) can be obtained by studying the solution to a set of ODEs. We developed that analysis in the context of a linear flux, when $f(u) = au$. Here we continue with developing the Characteristics Method for the Burgers equation when the flux term f depends nonlinearly on u.

1. Consider the characteristics method for (5.78) by defining the curve $\hat{x}$ as in (5.80). Let us denote by $\hat{x}(t, x_0)$ the characteristic curve that satisfies the initial condition $\hat{x}(0, x_0) = x_0$. Let U be the solution of the Burgers IVP when confined to this curve, that is, define U by
$$U(t) = u(\hat{x}(t, x_0), t). \tag{5.87}$$
Show that $U'(t) \equiv 0$. Use this fact to show, as in the case of the linear flux f, that
$$U(t) = U(0) = u_0(x_0). \tag{5.88}$$

2. Returning to the definition of a characteristic curve, and recalling that $\hat{x}'(t, x_0) = f'(u(\hat{x}(t, x_0), t)$ and using the definition of U, arrive at the relation
$$\hat{x}' = f'(U(t))$$
which in turn, using (5.88), leads to the remarkable result
$$\hat{x}' = f'(u_0(x_0)),$$
i.e., $\hat{x}'$ is again *constant*, much like the case when f is linear! The main difference now is that the slope of $\hat{x}$ is dependent on f', and through it, it depends on u_0, unlike the linear case. To see the impact of this dependence, write a MATLAB program to compute the shape of the characteristic curves when
$$f(u) = \frac{1}{2}u^2$$
for two sets of initial conditions: First $u_0(x) = x$, an increasing function of x, and next for $u_0(x) = e^{-(x-1)^2}$, which is a decreasing function of x near $x = 1$. Your graphs should resemble the ones in Figure (5.19).

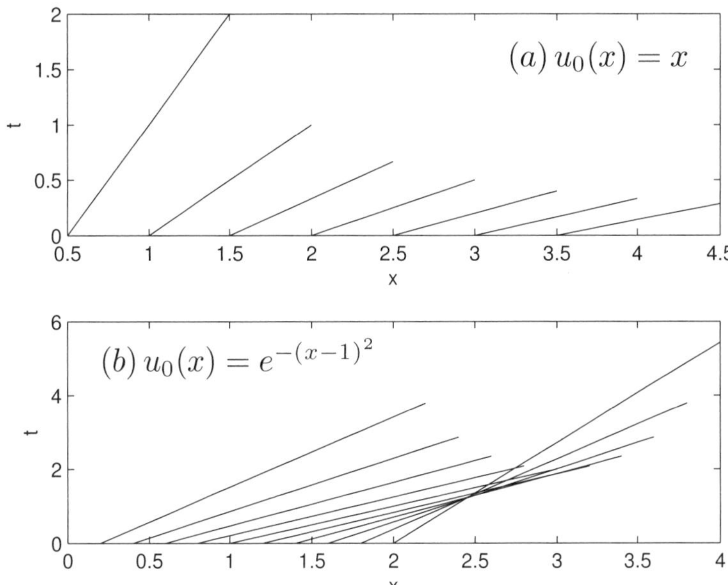

FIGURE 5.19: Characteristic curves for the Burgers equation with $f(u) = \frac{1}{2}u^2$ and (a) $u(x,0) = x$, an increasing function, and (b) $u(x,0) = e^{-(x-1)^2}$, a decreasing function near $x = 1$. Note the curves in (a) do not intersect in forward time, while those in (b) do, causing solutions of the IVP with $u_0(x) = e^{-(x-1)^2}$ to become multi-valued in finite time.

3. Returning to the case $u_0(x) = x$ and Figure 5.19a, note that the characteristics fan out into the (x,t)-space and do not intersect. Since the solution u is constant along each characteristic, show that we have

$$U(t) = x_0, \quad x = x_0 + x_0 t. \tag{5.89}$$

Eliminate x_0 between the two expressions in (5.89) to arrive at the analytical description of the exact solution:

$$u(x,t) = \frac{x}{1+t}. \tag{5.90}$$

Plot the graph of the exact solution to obtain a graph similar to the first graph in Figure 5.20.

4. Consider the Burgers equations with initial data

$$u(x,0) = e^{-10(x-1)^2} \tag{5.91}$$

(a) Apply the Method of Characteristics to show that the exact solution of this IVP can be written in a parametric form as:

$$\langle x, t, u(x,t)\rangle = \langle x_0 + u_0(x_0)t, t, u_0(x_0)\rangle, \qquad x_0 \in R. \quad (5.92)$$

(b) Execute the following MATLAB program to get the second graph in Figure 5.20:

```
u0=inline('exp(-10*(x-1).^2)','x');
x0=0:0.01:5;
dt=0.1;
Y = dt*repmat(1:20,size(x0,2),1);
X=zeros(length(x0),20);
for i=1:20
    X(:,i)=x0+u0(x0)*i*dt;
end
Z = zeros(size(X));
for i=1:20
    Z(:,i)=u0(x0);
end
subplot(2,1,2)
plot3(X, Y, Z, 'k');
xlabel x;ylabel t; zlabel u;
view(10,30)
```

As for the program, note the use of `repmat` and `plot3`. As for the graph, note how the graph of $u(., t)$ becomes multi-valued in finite time, signaling the formation of singularity that was predicted by plotting the characteristic curves as in Figure (5.20).

5.15 Project G: Burgers Equation (Formation of Singularities)

In the previous project we saw that the solution to the Burgers IVP (5.78)–(5.79) may develop singularities if the initial data u_0 decreases at any point $xi \in R$. In this project we will purse an alternative approach that leads to the same observation. This new approach requires following the evolution of u_x of (5.78)–(5.79) along characteristics.

Confining our analysis to the special case of $f(u) = \frac{1}{2}u^2$, let $\hat{x}(t, x_0)$

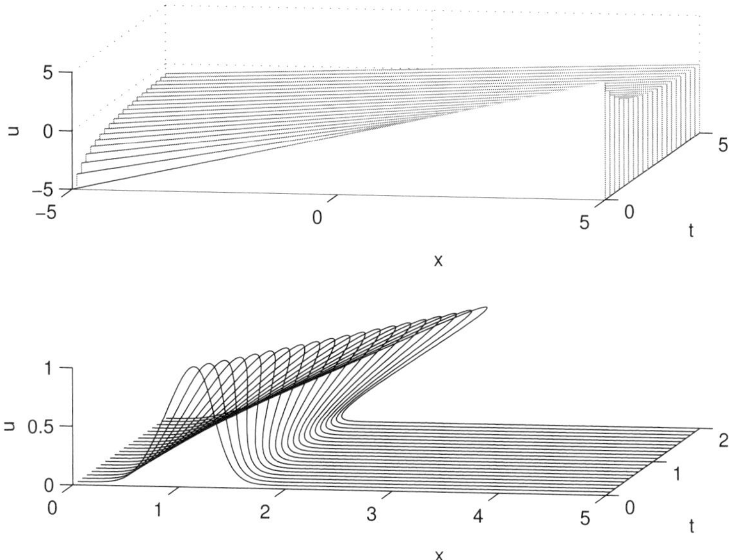

FIGURE 5.20: Graphs of two solutions to the Burgers equation, one with $u(x,0) = x$, which leads to the exact solution $u(x,t) = \frac{x}{1+t}$. The figure shows a smooth solution that decays to zero for large t as predicted by the analytical solution: $\lim_{t\to\infty} \frac{x}{1+t} = 0$ for all x. The second is the graph of the exact solution corresponding to the initial data $u(x,0) = e^{-10(x-1)^2}$. Note that the graph of this solution becomes multi-valued in finite time, as predicted by the characteristics method.

denote a typical characteristic curve with $\hat{x}(0, x_0) = x_0$. Define $W(t, x_0)$ by

$$W(t, x_0) = u_x(\hat{x}(t, x_0), t). \tag{5.93}$$

The goal now is to understand the evolution of W as a function of time t. To that end,

1. differentiate W with respect to t and observe

$$\frac{dW}{dt} = u_{xx}\frac{d\hat{x}}{dt} + u_{xt} = u_{xt} + uu_{xx}. \tag{5.94}$$

Returning to the Burgers equation (5.78) and differentiating it with the respect to x, we arrive at the following expression:

$$u_{tx} + u_x^2 + uu_{xx} = 0. \tag{5.95}$$

After evaluating the latter along the characteristic curve $x =$

$\hat{x}(t, x_0)$ and comparing the result to (5.94), obtain the following ODE initial-value problem for W:

$$\frac{dW}{dt} + W^2 = 0, \qquad W(0, x_0) = u_0'(x_0). \tag{5.96}$$

2. Show that the above IVP, which is an example of a *Riccati* equation, can be solved to yield:

$$W(t, x_0) = \frac{u_0'(x_0)}{1 + tu_0'(x_0)}. \tag{5.97}$$

3. Show that the expression W, which is u_x, becomes unbounded in finite time if $u_0'(x_0) < 0$. How does this result reflect on the behavior of the characteristics in Figure 5.18?

4. Let $T(x_0)$ be the time at which $W(T, x_0)$ becomes unbounded. Find the value x_0 at which T is minimized. This special T value is the time beyond which the analytic solution described in (5.92) is no longer valid and new ideas are needed to construct an appropriate solution of (5.78)–(5.79). Reference [5] is an excellent book on this subject.

5. Generalize the analysis that led to (5.96) to the general case of $f(u)$ in (5.78.

5.16 Project H: Burgers Equation and the Method of Lines

The Method of Lines (MOL) was developed in Project D. According to this method, an IVP such as the one we have been studying in the previous three projects can be reduced to a large system of ODEs, which in turn can be solved efficiently in MATLAB. The key idea in implementing MOL is to replace all of the spatial derivatives with approximate expressions that only involve function evaluations. In the case of (5.78) where u satisfies the PDE $u_t + (f(u))_x = 0$, the single x-derivative may be replaced by any of the various finite-difference approximations listed in (5.4) (forward-difference) or (5.5) (backward-difference) or (5.6) (centered-difference), just to name a few such approximations. One of the new features we encounter now is that the domain must be finite in order to implement MOL, which suggests that we should be considering

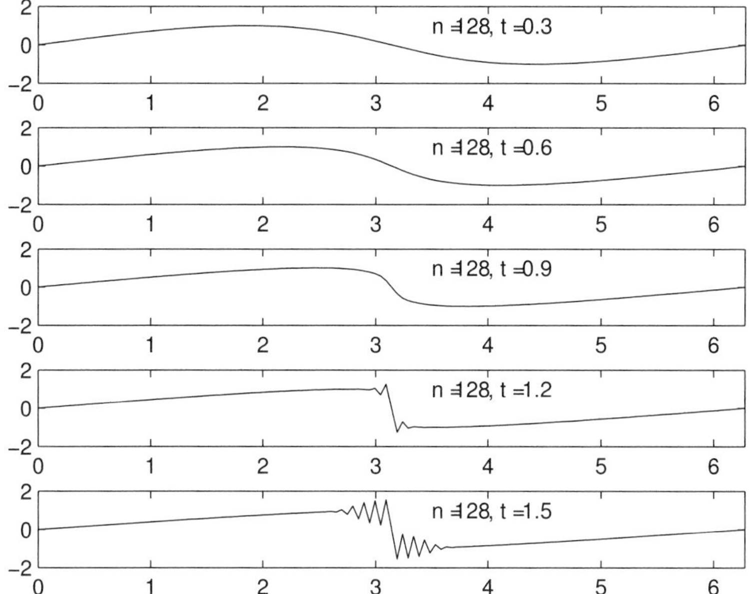

FIGURE 5.21: The output of the Method of Line algorithm when applied to IBVP (5.98) with $N = 128$. Note that the solution begins to show numerical instability around $t = 1$, which is the critical value at which the original PDE develops a singularity.

an initial-boundary value problem, rather than an initial-value problem over the initial real line. Motivated by the special paper in [6], we consider the following IBVP:

$$u_t + \frac{1}{2}(u^2)_x = 0, \quad u(x,0) = \sin x, \qquad u(0,t) = u(2\pi, t) \tag{5.98}$$

that is, we seek a solution to a 2π-periodic solution of the Burgers equation with initial data $\sin x$.

1. Show by the Method of Characteristics that we should expect that the smooth solution to this problem develops a singularity in finite time (in fact, at $t = \frac{1}{2}$).

2. Let $N = 16$ and dicretize the domain $[0, 2\pi)$ into N equal intervals. With $h = \frac{2\pi}{N}$ let $x_i = ih$, with $i = 0, ..., N-1$. Let $U_i(t) = u(x_i, t)$. Evaluate (5.98) at x_i to get

$$U_i' + \frac{1}{2}(u^2)_x)|_{x_i} = 0. \tag{5.99}$$

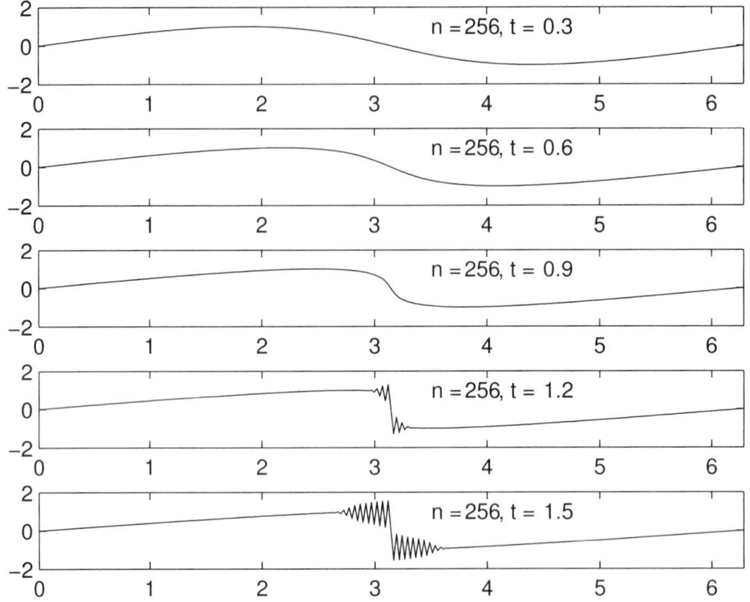

FIGURE 5.22: The output of the Method of Line algorithm when applied to IBVP (5.98) with $N = 256$.

Replace the second term in (5.99) using the centered difference approximation of the first derivative i.e.,

$$\frac{1}{2}(u^2)_x)|_{x_i} \approx \frac{1}{4h}(U_{i+1}^2 - U_{i-1}^2). \tag{5.100}$$

Hence, the original PDE in (5.98) is replaced by the following system of ODEs:

$$U_i' + \frac{1}{4h}(U_{i+1}^2 - U_{i-1}^2) = 0, \qquad i = 0, 1, ..., N-1, \tag{5.101}$$

where by periodicity,

$$U_{-1}(t) = U_{N-1}(t) \qquad \text{and} \qquad U_{N+1}(t) = U_1(t).$$

Show that (5.101) is equivalent to the following system of ODEs:

$$\mathbf{U}' = \mathbf{F}(\mathbf{U}), \qquad \mathbf{U}(0) = \mathbf{U}_0, \tag{5.102}$$

where

$$\mathbf{U} = \begin{bmatrix} U_0 \\ U_1 \\ U_2 \\ \cdots \\ \cdots \\ U_{N-1} \end{bmatrix}, \quad \mathbf{F} = -\frac{1}{4h} \begin{bmatrix} U_1^2 - U_{N-1}^2 \\ U_2^2 - U_0^2 \\ U_3^2 - U_1^2 \\ \cdots \\ \cdots \\ U_0^2 - U_{N-2}^2 \end{bmatrix},$$

$$\mathbf{U}_0 = \begin{bmatrix} \sin x_0 \\ \sin x_1 \\ \sin x_2 \\ \cdots \\ \cdots \\ \sin x_{N-1} \end{bmatrix}. \tag{5.103}$$

3. Write a MATLAB program that uses `ode45` to solve the above system of ODEs. Plot the graph of the solution $u(x, t)$ at $t = 0, 0.1$ through $t = 1.2$ and compare your graphs with the ones in Figure 5.21) and (5.22), which are the output with $N = 128$ and $N = 256$.

5.17 References

1. Vallis, G., *Atmospheric and Ocean Dynamics*, Cambridge University Press, 2006.

2. Malek-Madani, R., *Advanced Engineering Mathematics with Mathematica® and MATLAB®*, Addison-Wesley, 1998.

3. Moin, P., *Engineering Numerical Analysis*, Cambridge University Press, 2001.

4. Ide, K., Small, D., Wiggins, S., "Distinguished hyperbolic trajectories in time-dependent fluid flows: analytical and computational approach for velocity fields defined as data sets," *Nonlinear Processes in Geophysics*, 2002, Vol. 9, pp. 237–263.

5. Leveque, R. J., *Numerical Methods for Conservation Laws*, Birkhauser, 1992.

6. Majda, A. J., Timofeyev, I., "Remarkable statistical behavior for truncated Burgers-Hopf dynamics," *Proceedings of the National Academy of Sciences*, Vol 97, 2000, pp. 12413 – 12417.

Chapter 6

Equations of Fluid Dynamics

In this chapter we develop the equations of motion for fluid flows in an inertial coordinate system and present examples of exact solutions of this fundamental set of equations. In the subsequent chapter we develop the same equations in a rotating coordinate system, the equations of *Geophysical Fluid Dynamics*, and concentrate on the properties that distinguish motions of fluids in an inertial frame from the ones in a rotating frame.

The equations we develop in this section are general and describe flows with temporal and spatial scales that typically arise in laboratory experiments. We consider fluid flows that are smooth, i.e., do not have sharp discontinuities and shock waves, and are often referred to as *laminar flows*. We do not develop the description of turbulent flows in this text, since the proper description of these flows requires a different set of mathematical tools, including statistical tools, which deserve their own dedicated treatment. One way to view our approach here and in the next chapter, where we consider large-scale oceanographic flows, is to note that the mathematical models we study are intended to exhibit the averaged behavior of the phenomena of interest. We refer the reader to Chapter 8 of [1] for a thorough treatment and development of the methodology that is commonly adopted when dealing with turbulent flows.

We will now develop the traditional mathematical description of a fluid flow, its *Eulerian* description, where the attributes of a fluid element are described by the *position* occupied by the fluid element. That is, the velocity of the fluid element, or its temperature and salinity, or the pressure induced on it, are quantified by position and time, and not by the distinguished fluid element. This viewpoint agrees with our daily experience with fluid flows, whether we make measurements in a wind tunnel or in the Chesapeake Bay—we often fix our attention at a particular position in the flow and measure the attributes of the fluid element that passes through that position at any instance. We hardly care which fluid element is passing the measuring instrument at that moment; we care considerably more whether the characteristics of the flow vary from position to position. This approach has seen enormous success for several centuries. As we will see, however, there is a significant price to be paid for this approach, in that the acceleration of the fluid element ends up depending *nonlinearly* on the velocity of fluid element. As a consequence of this fact, the governing equations of fluid dynamics, the *Euler* and the *Navier–Stokes* equations, end up being very difficult and formidable PDEs to deal with. On the other hand, because of the nonlinear nature of these PDEs, we are able to capture rich structures that have distinct counterparts in nature, features that linear PDEs simply are not capable of capturing.

An interested reader may wish to consult Chapter 12 of Reference [1] for a similar development, especially if the reader is familiar with *Mathematica* and its capabilities.

6.1 Flow Representations — Eulerian and Lagrangian

As is common in the mathematical formulation of fluid dynamics, we distinguish between two ways of viewing motion, one where the measurements of the various physical quantities are made while keeping the position fixed, and the other when the same physical quantities are measured as functions of a fixed particle or fluid element. We refer to the first representation as *Eulerian* and the second as *Lagrangian.*

To make these ideas precise, let $\mathbf{X}$ denote a (fluid) particle or element with $\mathbf{x}$ the position it occupies at time t. Let $\mathbf{p}$ be the vector-valued function that maps $\mathbf{X}$ into $\mathbf{x}$:

$$\mathbf{x} = \mathbf{p}(t, \mathbf{X}). \tag{6.1}$$

It is common practice to identify the original position of a particle with the particle itself, that is, we assume that $\mathbf{p}$ satisfies the following condition

$$\mathbf{p}(0, \mathbf{X}) = \mathbf{X}. \tag{6.2}$$

All subsequent positions of $\mathbf{X}$ are labeled $\mathbf{p}(t, \mathbf{X})$. Function $\mathbf{p}$ is called the *motion* or the *deformation* of the flow. The velocity and acceleration of this motion, denoted by $\mathbf{V}$ and $\mathbf{A}$, are defined by

$$\mathbf{V}(t, \mathbf{X}) = \frac{\partial \mathbf{p}}{\partial t} \quad \text{and} \quad \mathbf{A}(t, \mathbf{X}) = \frac{\partial^2 \mathbf{p}}{\partial t^2}. \tag{6.3}$$

The functions $\mathbf{V}$ and $\mathbf{A}$ defined in (6.3) are part of the Lagrangian representation of the motion; these quantities are computed in reference to the particle itself, or putting it slightly differently, by the position this particle occupied in its *reference configuration* (see (6.2) where the particle is identified with its position at time zero). This representation is particle-centric in the sense that the instruments that actually perform the measurements are moving with the particle.

The Eulerian description of the flow, by contrast, is defined by two new functions, $\mathbf{v}$ and $\mathbf{a}$, which are related to $\mathbf{V}$ and $\mathbf{A}$ by

$$\mathbf{v}(t,\mathbf{x}) = \mathbf{V}(t,\mathbf{X}) \quad \text{and} \quad \mathbf{a}(t,\mathbf{x}) = \mathbf{A}(t,\mathbf{X}), \tag{6.4}$$

where $\mathbf{x}$ and $\mathbf{X}$ are related through the motion (6.1). Thus the relations in (6.4) can be rewritten as

$$\mathbf{V}(t,\mathbf{X}) = \mathbf{v}(t,\mathbf{p}(t,\mathbf{X})) \quad \text{and} \quad \mathbf{A}(t,\mathbf{X}) = \mathbf{a}(t,\mathbf{p}(t,\mathbf{X})). \tag{6.5}$$

The Eulerian representation clearly emphasizes positions over particles. In the description of the function $\mathbf{v}(t,\mathbf{x})$ in (6.4) we lose all information about the earlier velocities the particle $\mathbf{X}$, which is currently occupying position $\mathbf{x}$, may have attained—by contrast, the function $\mathbf{V}(t,\mathbf{X})$ retains the history of the velocities $\mathbf{X}$ has experienced.

The following example illustrates the relationship between $\mathbf{V}$ and $\mathbf{v}$. Consider the two-dimensional motion $\mathbf{p} = \langle p_1, p_2\rangle$ given by

$$x_1 = p_1(t,\mathbf{X}) = X_1 + t^2 X_2, \quad x_2 = p_2(t,\mathbf{X}) = X_2 - tX_1, \tag{6.6}$$

where each particle $\mathbf{X}$ is identified by its position at time 0, as in (6.2). In Figure 6.1 we start with a collection of particles located on a circle of radius one and plot their subsequent positions according to (6.6) in the time interval $t \in (0, 1.5)$, every 0.3 units of time. This figure is obtained by executing the following program in MATLAB:

```
clf;
s=0:2*pi/100:2*pi;
X=[cos(s);sin(s)];
plot(X(1,:),X(2,:),'*');
hold on
for i=1:5
    x=chi(0.3*i,X);
    plot(x(1,:),x(2,:));
    hold on
end
set(gca,'DataAspectRatio',[1 1 1]);
```

where `chi.m` is (see (6.6))

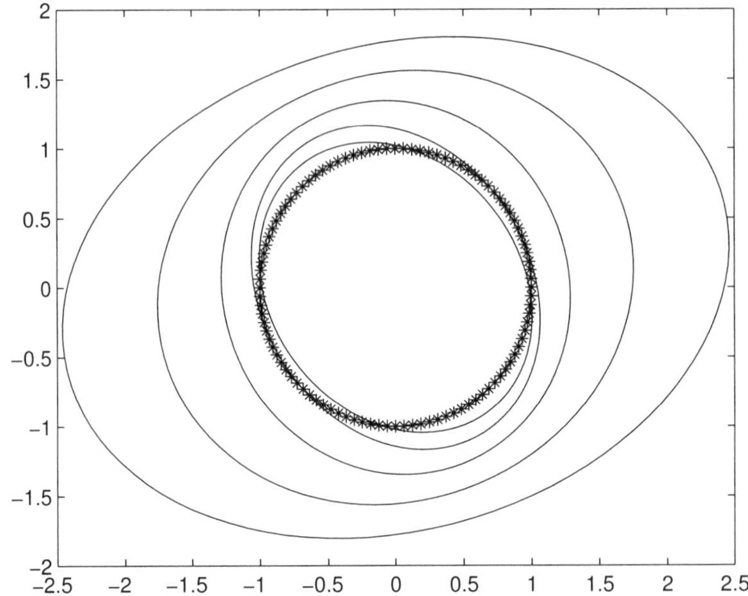

FIGURE 6.1: The motion defined by $p_1 = X_1 + t^2X_2, p_2 = X_2 - tX_1$. The figure shows the position of particles originally located on a circle of radius one and their subsequent positions as time evolves. Under the action of the fluid flow, this circle deforms to an ellipse at time t, and this ellipse enlarges as t increases.

```
function x=chi(t,X)

A=[1 t^2; -t 1];
x=A*X;
```

The velocity field of this motion in its Lagrangian representation is determined by simply differentiating (6.6) with respect to t, that is,

$$V_1(t, \mathbf{X}) = \frac{\partial p_1}{\partial t} = 2tX_2, \quad V_2(t, \mathbf{X}) = \frac{\partial p_2}{\partial t} = -X_1. \tag{6.7}$$

The same velocity field in Eulerian representation is determined as follows: Begin by computing $\mathbf{p}^{-1}$, the inverse of the function $\mathbf{p}$, by solving the two relations in (6.6) for X_1 and X_2:

$$X_1 = \frac{x_1 - t^2x_2}{1+t^3}, \quad X_2 = \frac{tx_1 + x_2}{1+t^3}. \tag{6.8}$$

By definition $\mathbf{v}(t,\mathbf{x}) = \mathbf{V}(t,\mathbf{X})$, so we substitute (6.8) in (6.7) to get

$$v_1(t,\mathbf{x}) = \frac{2t^2x_1 + 2tx_2}{1+t^3}, \quad v_2(t,\mathbf{x}) = -\frac{x_1 - t^2x_2}{1+t^3}. \tag{6.9}$$

It is instructive to see that if we use (6.7) or (6.9), we arrive at the same velocity for a typical particle. For example, consider the particle $\mathbf{P}$ that is located at $\mathbf{x} = \mathbf{X} = \langle 1, 1\rangle$ at time $t = 0$. We now determine its velocity at $t = 2$: First, from (6.7), we have

$$V_1(2, \langle 1,1\rangle) = 4, \quad V_2(2, \langle 1,1\rangle) = -1. \tag{6.10}$$

Similarly, we find from (6.6) that $\mathbf{X}$ occupies the position $\langle 5, -1\rangle$ at time 2. Thus, using (6.9), we have

$$v_1(2, \langle 5,-1\rangle) = 4, \quad v_2(2, \langle 5,-1\rangle) = -1, \tag{6.11}$$

which agrees with (6.10).

Continuing with this example, the determination of the acceleration, whether we compute $\mathbf{A}$ or $\mathbf{a}$, will result in the same description. To see this, note that, by (6.3) and (6.6)

$$A_1 = \frac{\partial^2 p_1}{\partial t^2} = 2X_2, \quad A_2 = \frac{\partial^2 p_2}{\partial t^2} = 0. \tag{6.12}$$

Since X_1 and X_2 are related to x_1 and x_2 by the relations in (6.6), we have

$$a_1 = A_1 = 2X_2 = \frac{2(tx_1 + x_2)}{1+t^3}, \quad a_2 = A_2 = 0. \tag{6.13}$$

Let us now return to the relations in (6.9) to see that we are able to obtain the above expressions for a_1 and a_2 by differentiating (6.9) directly. Starting with v_1 in (6.9)a, we have

$$a_1 = \frac{dv_1}{dt} = 2\frac{d}{dt}(\frac{t^2x_1 + tx_2}{1+t^3}) =$$

$$2\frac{-3t^2(t^2x_1 + tx_2)}{(1+t^3)^2} + 2\frac{t^2x_1 + t^2\frac{dx_1}{dt} + x_2 + t\frac{dx_2}{dt}}{1+t^3}. \tag{6.14}$$

Since $\frac{dx_1}{dt} = v_1$ and $\frac{dx_2}{dt} = v_2$, we can apply (6.9) again to simply (6.14):

$$a_1 = 2\frac{-3t^2(t^2x_1 + tx_2)}{(1+t^3)^2} + 2\frac{t^2x_1 + 4t^2\frac{t^2x_1+tx_2}{1+t^3} + x_2 - 2t\frac{x_1-t^2x_2}{1+t^3}}{1+t^3},$$

which simplifies to the expression for a_1 in (6.13). Similarly, starting with v_2 in (6.9) and differentiating it with respect to t and again noting that $v_1 = \frac{dx_1}{dt}$ and $v_2 = \frac{dx_2}{dt}$, we arrive at the Eulerian expression a_2, which agrees with the one in (6.13). See also Problem 1 below.

The calculations we just went through to arrive at the expression for the Eulerian acceleration $\mathbf{a}$, relying only on Eulerian relations, may be formalized in the formula

$$\mathbf{a} = \frac{\partial \mathbf{v}}{\partial t} + \mathbf{v} \cdot \nabla \mathbf{v}, \tag{6.15}$$

where the notation $\mathbf{v} \cdot \nabla$ stands for the scalar operation

$$\mathbf{v} \cdot \nabla = v_1 \frac{\partial}{\partial x} + v_2 \frac{\partial}{\partial y} + v_3 \frac{\partial}{\partial z}.$$

We will derive this formula later in this chapter, the derivation relying on a simple application of the chain rule. It is interesting to note that the somewhat inelegant algebra that we had to employ (for example, in (6.14)), has a nice, simple and compact representation in (6.15). As a byproduct of this formula, we will encounter the difference between $\frac{d}{dt}$, as in $\mathbf{a} = \frac{d\mathbf{v}}{dt}$, and $\frac{\partial}{\partial t}$, as in $\frac{\partial \mathbf{v}}{\partial t}$, used in (6.15).

Problems 6.1

1. Complete the calculation that leads to the expressions that follow (6.13), when we use the Eulerian representation of the flow defined by (6.9).

2. Find $\mathbf{V}$ and $\mathbf{v}$ for each deformation.

   (a) $x_1 = X_1 + te^{-t}X_2, x_2 = X_2 + te^{-2t}X_1$.

   (b) $x_1 = X_1 + tX_2$, $x_2 = X_2 - tX_1 + tX_2$.

   (c) $x_1 = X_1 + tX_1 - tX_2$, $x_2 = X_2 + tX_1 - t^2X_2$.

   (d) $x_1 = X_1 + t^2X_2$, $x_2 = X_2 - t^2X_1$.

   (e) $x_1 = X_1 + (\sin t)X_2$, $x_2 = X_2 - (1 - \cos t)X_1$.

   (f) $x_1 = X_1 + f(t)X_2$, $x_2 = X_2 + g(t)X_1$.

3. Let $\mathbf{v} = \langle v_1(t, x_1, x_2), v_2(t, x_1, x_2), 0\rangle$ be the velocity field of a flow. Show that $\mathbf{a}$, its Eulerian acceleration, is the vector

$$\langle \frac{\partial v_1}{\partial t} + v_1 \frac{\partial v_1}{\partial x} + v_2 \frac{\partial v_2}{\partial y}, \frac{\partial v_2}{\partial t} + v_1 \frac{\partial v_2}{\partial x} + v_2 \frac{\partial v_2}{\partial y}, 0\rangle.$$

4. Consider the following velocity vector fields. In each case, compute the associated acceleration field.

(a) $\mathbf{v} = \langle y, -x, 0 \rangle$.

(b) $\mathbf{v} = \langle y, z, -y \rangle$.

(c) $\mathbf{v} = \langle \sin y \cos z, \sin x \cos z, \cos x \sin y \rangle$.

5. Consider the deformation

$$x_1 = X_1 + 2t^2 X_2, \quad x_2 = X_2 + \frac{1}{1+t} X_1.$$

(a) Let D consist of the set of particles that occupy the disk of radius one centered at the origin at time 0. Draw the graphs of the image at times $t = 1$ and $t = 2$.

(b) Determine the Lagrangian and Eulerian formulations of the acceleration of this motion.

6.2 Deformation Gradient and Conservation of Mass

As Figure 6.1 shows, motions described by (6.1) deform parcels of fluid as time evolves. It is of particular interest to us to develop analytical and computational tools that measure deformations and eventually relate them to the forces that act on the parcels of fluid.

We denote by F the gradient of a deformation given by (6.1), that is,

$$F = \frac{\partial \mathbf{p}}{\partial \mathbf{X}}. \tag{6.16}$$

F is a matrix with components $F_{ij} = \partial p_i / \partial X_j$. We will refer to F as the *deformation gradient* of the motion. For example, the deformation gradient of the motion $\mathbf{p}$ given by

$$\mathbf{p} = \langle X_1 + te^{-t} X_2, X_2 + te^{-2t} X_1 \rangle$$

is

$$F = \begin{bmatrix} \partial p_1/\partial X_1 & \partial p_1/\partial X_2 \\ \partial p_2/\partial X_1 & \partial p_2/\partial X_2 \end{bmatrix} = \begin{bmatrix} 1 & te^{-t} \\ te^{-2t} & 1 \end{bmatrix}. \tag{6.17}$$

The notion of deformation gradient first appears in elementary calculus when formulas are derived that describe how a change of variables will affect the evaluation of double and triple integrals. Given a region

D, in R^2 or R^3, and an integrand f, computing the integral of f over the region D,

$$\int_D f\,dV,$$

is sometimes simplified if we are able to find a change of variables that converts the region D into a new region D^* where the integration over D^* is reduced to an evaluation of iterated integrals. The price for this change of variables appears as $f|J|$ in the new integrand, where the additional term J is the *Jacobian* of the change of variables. The original integral and the transformed one are related as follows:

$$\int_D f\,dV = \int_{D^*} f|J|\,dV^*.$$

For example, we recall that when using polar coordinates a double integral $\int\int_D f(x,y)\,dxdy$ is transformed into

$$\int\int_{D^*} f(r\cos\theta, r\sin\theta)\,rdrd\theta,$$

where the factor r is the determinant of the gradient of the map that relates rectangular coordinates to polar coordinates, that is, with $x = r\cos\theta$ and $y = r\sin\theta$, with $r \geq 0$, then

$$|J| = \left|\det \begin{bmatrix} \partial x/\partial r & \partial x/\partial\theta \\ \partial y/\partial r & \partial y/\partial\theta \end{bmatrix}\right| = r.$$

The next theorem shows that this notion plays a critical role in fluid flows as well, since we are able to quantify how the time rate of change of any deformation gradient is related to the deformation gradient itself.

Theorem 6.2.1
Let $\mathbf{p}$ *be a deformation with* F *and* $\mathbf{v}$ *as its deformation gradient and velocity, respectively. Then*

$$\frac{d}{dt}(\det F) = (\text{div}\,\mathbf{v})\det F. \tag{6.18}$$

Proof: We present here the proof for the two-dimensional case and leave the proof in the three-dimensional case, which is similar, as an exercise.

Let $\mathbf{p} \in E^2$. The determinant of F is

$$\det F = F_{11}F_{22} - F_{12}F_{21}. \tag{6.19}$$

Differentiate (6.19) with respect to t:

$$\frac{\partial}{\partial t}(\det F) = \frac{\partial F_{11}}{\partial t}F_{22} + F_{11}\frac{\partial F_{22}}{\partial t} - \frac{\partial F_{12}}{\partial t}F_{21} - F_{12}\frac{\partial F_{21}}{\partial t}. \tag{6.20}$$

The four derivatives $\partial F_{ij}/\partial t$ are related to the velocity field because $F_{ij} = \partial p_i/\partial X_j$. Since $V_i = \partial p_i/\partial t$, we can rewrite

$$\frac{\partial F_{ij}}{\partial t}$$

as

$$\frac{\partial V_i}{\partial X_j},$$

the gradient of velocity, where the gradient is computed with respect to a moving particle's positions X_j, and not a fixed position $\mathbf{x}$. We can relate this gradient to the gradient of $\mathbf{v}$ with respect to $\mathbf{x}$ as follows: We recall that $V_1 = v_1(t, \mathbf{x}) = v_1(t, \mathbf{p}(t, \mathbf{X}))$. Hence, by differentiating the latter expression with respect to X_1, say, we get

$$\frac{\partial V_1}{\partial X_1} = \frac{\partial v_1}{\partial x_1}\frac{\partial p_1}{\partial X_1} + \frac{\partial v_1}{\partial x_2}\frac{\partial p_2}{\partial X_1}. \tag{6.21}$$

We can therefore express $\frac{\partial F_{11}}{\partial t}$ in (6.20) in terms of $\mathbf{v}$:

$$\frac{\partial F_{11}}{\partial t} = \frac{\partial V_1}{\partial X_1} = \frac{\partial v_1}{\partial x_1}F_{11} + \frac{\partial v_1}{\partial x_2}F_{21}. \tag{6.22}$$

All other terms on the right side of (6.20) have a similar form in terms of $\mathbf{v}$ (for example, $\frac{\partial F_{22}}{\partial t} = \frac{\partial v_2}{\partial x_1}F_{12} + \frac{\partial v_2}{\partial x_2}F_{22}$). Thus, (6.20) reduces to

$$\frac{\partial}{\partial t}(\det F) = \frac{\partial v_1}{\partial x_1}(F_{11}F_{22} - F_{12}F_{21}) + \frac{\partial v_2}{\partial x_2}(F_{11}F_{22} - F_{12}F_{21}) \tag{6.23}$$

which equals (div $\mathbf{v}$)det F. This completes the proof.

An important consequence of Theorem 6.2.1 is that the determinant of F is independent of t if and only if div $\mathbf{v} = 0$, which we state as corollary to Theorem 6.2.1.

Corollary 6.2.1 (Incompressible Flows)
The velocity field $\mathbf{v}$ *is divergence free if and only if* det F *is time independent. In particular, when the velocity field* $\mathbf{v}$ *is divergence free, then the flow is* **incompressible**, *i.e.,* det $F = 1$ *for all time.*

Proof: It follows directly from the expression in (6.18) that det F is time independent if and only if div $\mathbf{v} = 0$. Note that det $F(0, \mathbf{X}) = 1$ since $F(0, \mathbf{X})$ is the identity matrix, as seen from the identity in (6.2). Hence, if det F is time independent, det $F \equiv 1$ for all t.

We will see shortly that this definition of incompressibility, that det $F = 1$ for all time, is equivalent to the more common definition of incompressibility in physics that the density of a particle remains unchanged under any deformation the material undergoes.

To recap, a fluid flow is incompressible if div $\mathbf{v} = 0$ for all time. This relation in turn implies that if a flow is incompressible, then det F, which for all flows, whether incompressible or not, starts out being one, remains one for all time. A consequence of this observation is that the volume of a parcel of fluid remains constant for an incompressible fluid as it undergoes a deformation. To see this recall that

$$|J| = |\det F|$$

is the factor that enters into the computation of an integral under a change of variable. With that in mind, let D be a parcel of fluid with D_0 denoting the volume of the region it occupies at time 0. Let D_t denote the volume of the region occupied by the same parcel of fluid at time t, after it has been deformed by the deformation $\mathbf{p}$. Note that

$$D_0 = \int\int\int_D dxdydz \qquad D_t = \int\int\int_{\mathbf{p}(t,D)} dxdydz.$$

But $\int\int\int_{\mathbf{p}(t,D)} dxdydz = \int\int\int_D |J|dxdydz$, which equals D_0 since $|J| = 1$. We have therefore arrived at the important geometrical property of incompressible flows that, as complicated as these flows may be, including exhibiting chaotic behavior, the volume of parcels of fluids remain invariant under the action of fluid deformations. We have proved this result for three-dimensional incompressible flows, which has a straightforward restriction to two-dimensional flows when "volume" is replaced by "area."

A second consequence of Theorem 6.2.1 is the *Transport Formula* stated below. This formula, which can be thought of as an analog of *Leibniz*'s formula for functions of several variables, will allow us to differentiate integrals involving integrands and integration domains that vary with a variable, say t. To prepare for stating this result, consider a deformation (motion) $\mathbf{p}$ and an arbitrary region Ω in R^n, which is mapped to the region $\mathbf{p}(t, \Omega)$ by the motion. Let $f(t, \mathbf{x})$ be a quantity of interest (such as density or pressure), and now consider the quantity

$$\int_{\mathbf{p}(t,\Omega)} f(t, \mathbf{x})\, dv,$$

where by dv, as distinguished from dV, we mean that the volume integration is carried out in the deformed domain. The theorem below shows how one computes the time rate of change of this quantity.

Theorem 6.2.2 (Transport Formula)
Let f be a sufficiently smooth function of its variables. Let $\mathbf{x}$ and $\mathbf{X}$ be related by $\mathbf{x} = \mathbf{p}(t, \mathbf{X})$. Then

$$\frac{d}{dt}\left(\int_{\mathbf{p}(t,\Omega)} f(t, \mathbf{x})\, dv\right) = \int_{\mathbf{p}(t,\Omega)} [f_t + \text{div } (f\mathbf{v})]\, dv, \tag{6.24}$$

where $\mathbf{v}$ is the velocity field associated with $\mathbf{p}$.

Proof: We consider $\mathbf{x} = \mathbf{p}(t, \mathbf{X})$ as a change of variables that relates the original domain Ω to $\mathbf{p}(t, \Omega)$. With this interpretation in mind, we have

$$\int_{\mathbf{p}(t,\Omega)} f(t, \mathbf{x})\, dv = \int_{\Omega} f(t, \mathbf{p}(t, \mathbf{X})) \det F\, dV,$$

where dV stands for the volume element in the undeformed domain, and where we assume, without loss of generality that $\det F$ is positive. Since Ω is time independent, the derivative of the above expression with respect to t can be passed directly to the integrand:

$$\frac{\partial}{\partial t}\left[\int_{\Omega} f(t, \mathbf{p}(t, \mathbf{x})) \det F\, dV\right] = \int_{\Omega} \frac{\partial}{\partial t}[f(t, \mathbf{p}(t, \mathbf{X})) \det F]\, dV$$

$$= \int_{\Omega} \left[(f_t + \text{grad} f \cdot \mathbf{v}) \det F + f\frac{\partial}{\partial t}(\det F)\right] dV$$

$$= \int_{\Omega} [f_t + \text{grad} \cdot \mathbf{v} + f(\text{div } \mathbf{v})] \det F\, dV \qquad \text{(after applying (6.18))}$$

$$= \int_{\Omega} [f_t + \text{div } (f\mathbf{v})] \det F\, dV$$

$$= \int_{\mathbf{p}(t,\Omega)} [f_t + \text{div } (f\mathbf{v})]\, dv.$$

which completes the proof.

As an application of Theorem 6.2.2 we consider the mass of an arbitrary subregion Ω of a body of fluid B. With $\rho(t, \mathbf{x})$ representing the density of the fluid at any time t and position $\mathbf{x}$, we note that

$$m(t) = \int_{\mathbf{p}(t,\Omega)} \rho(t, \mathbf{x})\, dv$$

is the total mass of the fluid that occupies $\mathbf{p}(t, \Omega)$ at time t. Since mass is conserved, we have $m'(t) = 0$. Using (6.24), $m'(t)$ is

$$m'(t) = \frac{d}{dt}\left[\int_{\mathbf{p}(t,\Omega)} \rho(t, dv)\, dv\right] = \int_{\mathbf{p}(t,\Omega)} [\rho_t + \text{div } (\rho\mathbf{v})]\, dv, \qquad (6.25)$$

which must vanish. The above integral must vanish for every arbitrary subset Ω of the fluid B, which implies that the integrand must vanish. Hence, conservation of mass reduces to the following PDE for ρ and $\mathbf{v}$:

$$\rho_t + \text{div } (\rho\mathbf{v}) = 0. \qquad (6.26)$$

If the divergence term, div $(\rho\mathbf{v})$, in (6.26) is expanded, we end up with

$$\rho_t + \mathbf{v} \cdot \nabla\rho + \rho\,\text{div } \mathbf{v} = 0. \qquad (6.27)$$

In analogy with the definitions for $\mathbf{v}$ and $\mathbf{V}$, we let $R(\mathbf{X})$ be the density associated with the particle $\mathbf{X}$. When a fluid particle $\mathbf{X}$ is incompressible its density will not change under any deformation $\mathbf{p}$. Hence, we have

$$R(\mathbf{X}) = \rho(t, \mathbf{p}(t, \mathbf{X})).$$

In that case, since $\frac{d}{dt}(R(\mathbf{X})) = 0$, we have $\frac{d}{dt}(\rho(t, \mathbf{p}(t, \mathbf{X})) = 0$. The latter expression reduces to $\rho_t + \nabla\rho \cdot \mathbf{v} = 0$, which when combined with (6.27), leads to

$$\text{div } \mathbf{v} = 0 \qquad (6.28)$$

as the PDE that represents conservation of mass when the flow is incompressible. We summarize the above development as a theorem:

Theorem 6.2.3 (Conservation of Mass)
Let ρ and $\mathbf{v}$ be the density and velocity field associated with a deformation $\mathbf{p}$. Then ρ and $\mathbf{v}$ must satisfy (6.26)

$$\rho_t + \text{div } (\rho\mathbf{v}) = 0.$$

When the flow is incompressible, the latter PDE further reduces to (6.28)

$$\text{div } \mathbf{v} = 0.$$

Before proceeding to a second application of the Transport Theorem, we emphasize a delicate point about the relationship between the density of a fluid particle and the incompressibility of flows: An incompressible flow is characterized mathematically by the PDE div $\mathbf{v} = 0$. As for the fluid density, incompressibility only requires that the density of any given fluid particle $\mathbf{X}$ remains unchanged no matter what deformation this particle undergoes. In particular, incompressibility does not require that all fluid particles have the same density. In fact, as is the case with many oceanic and atmospheric processes, which are often modeled as incompressible flows, the fluid region is stratified, meaning that the density of stable columns of fluid change monotonically with depth or height, decreasing with height in the case of the atmosphere, and increasing with depth in the case of the ocean. Hence the reader will encounter PDE (6.28) as one of the key equations in modeling flows where the assumption of incompressibility is invoked.

A second application of the Transport Theorem is in the context of quantities that are of the form $\rho\phi$, where ϕ is an arbitrary quantity of interest, such as a component of the velocity field, temperature, or salinity. According to this theorem, we have

$$\frac{d}{dt}\left(\int_{\mathbf{p}(t,\Omega)} \rho(t,\mathbf{x})\phi(t,\mathbf{x})\,dv\right) = \int_{\mathbf{p}(t,\Omega)} (\rho\phi)_t + \text{div}\,(\rho\phi\mathbf{v})\,dv. \tag{6.29}$$

After using the conservation of mass relation (6.26), the term $(\rho\phi)_t$ simplifies as

$$(\rho\phi)_t = \rho_t\phi + \rho\phi_t = -\phi\text{div}\,(\rho\mathbf{v}) + \rho\phi_t.$$

Moreover, since $\text{div}\,(\rho\phi\mathbf{v}) = \phi\,\text{div}\,(\rho\mathbf{v}) + \rho\mathbf{v}\cdot\nabla\phi$, the integral on the right side of (6.29) reduces to

$$\int_{\mathbf{p}(t,\Omega)} \rho(\phi_t + \mathbf{v}\cdot\nabla\phi)\,dv. \tag{6.30}$$

Definition 6.2.1 (Material or Total Time Derivative): *Given any differentiable function $\phi(t,\mathbf{x})$, we define the* **Material** *or* **Total Time Derivative** *of ϕ, denoted by $\frac{D\phi}{Dt}$, as*

$$\frac{D\phi}{Dt} = \phi_t + \mathbf{v}\cdot\nabla\phi. \tag{6.31}$$

With this notation, and in light of (6.30), (6.29) is equivalent to

$$\frac{d}{dt}\left(\int_{\mathbf{p}(t,\Omega)} \rho(t,\mathbf{x})\phi(t,\mathbf{x})\,dv\right) = \int_{\mathbf{p}(t,\Omega)} \rho\frac{D\phi}{Dt}\,dv. \tag{6.32}$$

This remarkable identity states that we can simply slip the differentiation $\frac{d}{dt}$, when applied outside of an integral, to the integrand as shown in (6.32), as long as we use the total derivative concept. We have proved the following theorem.

Theorem 6.2.4
Any differentiable quantity $\phi(t, \mathbf{x})$ satisfies the relation (6.32).

When the result of Theorem 6.2.4 is applied to each component of the linear momentum $\rho\mathbf{v}$, we obtain the important identity

$$\frac{d}{dt}\left(\int_{\mathbf{p}(t,\Omega)} \rho(t,\mathbf{x})\mathbf{v}(t,\mathbf{x})\,dv\right) = \int_{\mathbf{p}(t,\Omega)} \rho\frac{D\mathbf{v}}{Dt}\,dv, \tag{6.33}$$

which will play a key role in the derivation of the Navier-Stokes equations.

Problems 6.2

1. Prove the result in Theorem 6.2.1 in three-dimensional setting.
2. Given any two differentiable functions f and g, show that
$$\frac{D(fg)}{DT} = f\frac{Dg}{Dt} + g\frac{Df}{Dt},$$
where $\frac{D}{Dt}$ is the material time derivative defined in this section.
3. Let F be the deformation gradient of a two-dimensional flow. Let λ_1 and λ_2 be the eigenvalues of F. Find relationships between the time-rates of change of λ_i and the components of the associated velocity field $\mathbf{v}$.

6.3 Derivation of Equation of Conservation of Mass—A Heuristic Approach

Theorem 6.2.3 summarizes the mathematical consequences of conservation of mass, whether the fluid is incompressible or not, in terms of the changes in density and the divergence of velocity field. Here we derive the same results (Equations (6.26) and (6.28)) using a heuristic approach based on computing the flux of the fluid that enters and leaves a control volume within the flow.

Let V be a small cube with sides δx, δy and δz, centered at the point $P = (a, b, c)$, a typical but fixed point in the path of the fluid flow. By the flux of the fluid through V we mean the net loss of mass in V per unit time. We measure this quantity in two ways:

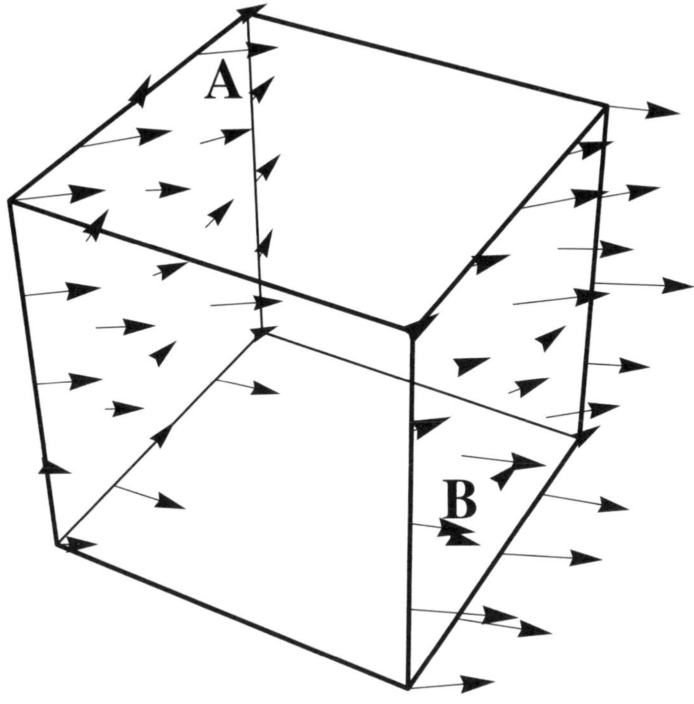

FIGURE 6.2: A schematic to illustrate the conservation of mass equation.

a) first we measure how mass is transported across the boundary of V during the small but fixed time interval $(t, t + \delta t)$,

b) next, we compute the change in the mass in V during the same time interval by observing the dynamics of the density in V.

Since these two approaches must lead to the same value, we will end up obtaining the relation in (6.26).

Without loss of generality, we assume that the faces of V are parallel to the coordinate planes (see Figure 6.2). Let $\mathbf{v} = \langle u, v, w \rangle$ denote the velocity field. Each component of $\mathbf{v}$ is responsible for the transport of the fluid through two of the faces of V—for example, v is responsible for any fluid transport through the faces $y = b - \frac{\delta}{2}$ and $y = b + \frac{\delta}{2}$, denoted by A and B in Figure 6.2. With the density and velocity represented in their Eulerian form, the net transport through these two faces is

$$\int_t^{t+\delta t} \iint_{S_2} \rho(\tau, x, b + \frac{\delta y}{2}, z) v(\tau, x, b + \frac{\delta y}{2}, z)\, dS\, d\tau -$$
$$\int_t^{t+\delta t} \iint_{S_1} \rho(\tau, x, b - \frac{\delta y}{2}, z) v(\tau, x, b - \frac{\delta y}{2}, z)\, dS\, d\tau, \qquad (6.34)$$

where S_1 and S_2 are the two faces of the cube with $y = b - \frac{\delta y}{2}$ and $y = b + \frac{\delta y}{2}$. Since S_1 and S_2 are faces of a cube parallel to the xz-planes, they can be parametrized as

$$\{(x, z) |\, a - \frac{\delta x}{2} < x < a + \frac{\delta x}{2},\, c - \frac{\delta z}{2} < z < c + \frac{\delta z}{2}\}.$$

If the dimensions of the cube are small, we may comfortably apply the Taylor formula to each integrand:

$$\rho(\tau, x, b - \frac{\delta y}{2}, z) v(\tau, x, b - \frac{\delta y}{2}, z) = \rho(\tau, x, b, z) v(\tau, x, b, z) -$$
$$\frac{\partial}{\partial y}(\rho v)\big|_{y=b} \frac{\delta y}{2} + \dots,$$

where dots denote terms in δy^2 and higher. Similarly,

$$\rho(\tau, x, b + \frac{\delta y}{2}, z) v(\tau, x, b + \frac{\delta y}{2}, z) = \rho(\tau, x, b, z) v(\tau, x, b, z) +$$
$$\frac{\partial}{\partial y}(\rho v)\big|_{y=b} \frac{\delta y}{2} + \dots.$$

Hence the net transport across the two faces $y = b - \frac{\delta}{2}$ and $y = b + \frac{\delta}{2}$ reduces to

$$\int_t^{t+\delta t} \int_{a-\delta x/2}^{a+\delta x/2} \int_{c-\delta z/2}^{c+\delta z/2} \left(\frac{\partial}{\partial y}(\rho v)\big|_{y=b} \delta y + \dots \right) dz\, dx\, d\tau.$$

Neglecting the higher order terms from the computation, and assuming that δx, δz, and δt are small enough that the integrand remains essentially constant on each face and during the time interval $(t, t + \delta t)$, we may approximate the integrand by its value at P and replace the above integral by

$$\frac{\partial}{\partial y} (\rho v) \,|_P\, \delta x \delta y \delta z \delta t. \qquad (6.35)$$

Similarly the net loss of mass through the boundaries $x = a - \frac{\delta x}{2}$ and $x = a + \frac{\delta x}{2}$, and $z = c - \frac{\delta z}{2}$ and $z = c + \frac{\delta z}{2}$ are

$$\frac{\partial}{\partial x} (\rho u) \,|_P\, \delta x \delta y \delta z \delta t \qquad \text{and} \qquad \frac{\partial}{\partial z} (\rho w) \,|_P\, \delta x \delta y \delta z \delta t. \qquad (6.36)$$

Hence, the net transport of mass through the boundaries of the cube is the sum of the above three expressions, leading to

$$\text{div}\,(\rho\mathbf{v})|_{P}\,\delta x\delta y\delta z\delta t. \tag{6.37}$$

This concludes the computation of the net transport using the first approach, namely, transport through the boundaries of the cube. Next we compute the transport of mass during $(t, t + \delta t)$ by observing how the changes in the density inside the cube V contribute to the net transport. We begin by noting that

$$\int\!\!\int\!\!\int_V \rho(t, x, y, z)\, dx\, dy\, dz$$

is the mass of the fluid occupying V at time t, and

$$\int\!\!\int\!\!\int_V \rho(t + \delta t, x, y, z)\, dx\, dy\, dz$$

is the total mass at time $t + \delta t$. Naturally, the difference between these expressions

$$\int\!\!\int\!\!\int_V \rho(t, x, y, z)\, dx\, dy\, dz - \int\!\!\int\!\!\int_V \rho(t + \delta t, x, y, z)\, dx\, dy\, dz \tag{6.38}$$

is the net transport of mass. Expression (6.38) can be simplified using the same strategy we applied to (6.34), namely, by applying the Taylor formula to expand

$$\rho(t + \delta t, x, y, z)$$

as

$$\rho(t, x, y, z) + \frac{\partial \rho}{\partial t}\delta t + \ldots,$$

and then approximating the resulting integral in (6.38) by

$$-\frac{\partial \rho}{\partial t}|_P\,\delta x\delta y\delta z\delta t \tag{6.39}$$

by appealing to the smallness of all of the dimensions. Equating (6.39) with (6.37) leads to equation of conservation of mass in (6.26).

Problems 6.3

The heuristic approach presented in this section was developed for a small volume V in the shape of a cube centered at $P = (a, b, c)$. Reconstruct this approach with the point P located instead

1. at a vertex on the lower face of the cube.

2. at a vertex on the upper face of the cube.

6.4 Stream Function and Vector Fields A, B, C, and ABC

As we study the consequences of conservation of mass, and later conservation of linear momentum, it will be instructive to carry with us a few concrete examples of fluid flows whose structures are often representative of the basic features we see in more complicated flows. To that end, we now introduce four vector fields; the first three we refer to by **A**, **B** and **C**, and the fourth by **ABC**, or the Arnold–Beltrami–Childress flow.

Definition 6.4.1
The following vector fields are Vector Fields **A**, **B**, *and* **C**, *respectively:*

$$\mathbf{v}_{\mathrm{A}} = \langle y, -x, 0\rangle, \tag{6.40}$$

$$\mathbf{v}_{\mathrm{B}} = \langle y/\sqrt{x^2+y^2}, -x/\sqrt{x^2+y^2}, 0\rangle, \tag{6.41}$$

and

$$\mathbf{v}_{\mathrm{C}} = \langle y/(x^2+y^2), -x/(x^2+y^2), 0\rangle, \tag{6.42}$$

It is simple to show that all three Vector Fields **A**, **B** and **C** are incompressible. We carry out the computations for vector field **B** and leave the verification for **A** and **C** to the reader. Vector field **B** is defined by (6.41)

$$\mathbf{v}_{\mathrm{B}} = \langle v_1, v_2, v_3\rangle = \langle \frac{y}{\sqrt{x^2+y^2}}, -\frac{x}{\sqrt{x^2+y^2}}, 0\rangle.$$

Its divergence is

$$\begin{aligned}\text{div } \mathbf{v}_{\text{B}} &= \partial v_1/\partial x + \partial v_2/\partial y + \partial v_3/\partial z \\ &= \tfrac{\partial}{\partial x}(y(x^2+y^2)^{-\frac{1}{2}}) - \tfrac{\partial}{\partial y}(x(x^2+y^2)^{-\frac{1}{2}}) \\ &= -xy(x^2+y^2)^{-\frac{3}{2}} + xy(x^2+y^2)^{-\frac{3}{2}} = 0.\end{aligned}$$

As we noted in Theorem 6.2.3, when the flow is incompressible, the equation that describes conservation of mass reduces to the velocity vector field being divergence free. Vector fields **A**, **B** and **C** have the additional property that they are *two-dimensional*, that is, $v_3 = 0$ and that v_1 and v_2 only depend on x and y. As we have seen in Chapter 3, it turns out that we can represent the information in two-dimensional vector fields by a stream function, single function of x and y, denoted by $\psi(x, y, t)$. We recall that a typical incompressible and 2D velocity field $\mathbf{v} = \langle u(x, y, t), v(x, y, t), 0\rangle$ must satisfy the relation

$$\frac{\partial v_1}{\partial x} + \frac{\partial v_2}{\partial y} = 0. \tag{6.43}$$

The expression in (6.43) is satisfied if u and v are related to a scalar function $\psi(x, y, t)$ through the relations

$$u(x, y, t) = \frac{\partial \psi}{\partial y}, \quad v(x, y, t) = -\frac{\partial \psi}{\partial x}, \tag{6.44}$$

where the relations in (6.44) are arranged so that the equation div $\mathbf{v} = 0$ is automatically satisfied, as the reader can verify by direct differentiation of the relations in (6.44). Conversely, if we know *a priori* that the velocity field $\mathbf{v}$ is 2D and incompressible, we will be able to come up with a smooth scalar function $\psi(x, y, t)$ that satisfies (6.44). How one goes about determining ψ requires integrating the relations (6.44), as we will demonstrate shortly with examples. The fact that div $\mathbf{v} = 0$ guarantees that the constants of integration we obtain are compatible and that we end up with a well-defined stream function $\psi(x, y, t)$. We note in passing that this topic is often covered under the title of "exact differential equations" in standard texts on ordinary differential equations.

It is worth remarking at this juncture that the relations in (6.44) could have just as easily been defined as

$$u(x, y, t) = -\frac{\partial \psi}{\partial y}, \quad v(x, y, t) = \frac{\partial \psi}{\partial x}, \tag{6.45}$$

an alternative definition that is adopted in many texts. The definition in (6.44) was especially popular in the early years of the development of physical oceanography, in the 1930s and 1940s, which slowly gave way to the formulation defined in (6.45). The latter definition, on the other hand, is a standard definition in mathematics, where ψ is called a *Hamiltonian* function, and $\mathbf{v}$ is its *symplectic gradient.* We note that the function ψ defined in (6.44) is simply the negative of the stream function defined by (6.45), and hence we reserve the right to use either definition in this text, which will be obvious from context. In this chapter, in particular, we will stick with (6.44), which was H. Stommel's favored formulation.

To see an example of how one determines ψ from $\mathbf{v}$, consider Vector Field **A**, where $\mathbf{v}_{\mathbf{A}} = \langle y, -x, 0\rangle$. Referring to (6.44), the desired stream function $\psi(x, y, t)$ must be a solution of the following system of PDEs:

$$\frac{\partial \psi}{\partial y} = y, \qquad -\frac{\partial \psi}{\partial y} = -x, \tag{6.46}$$

Starting with (6.46)a, we integrate this equation with respect to y:

$$\psi(x, y, t) = \frac{1}{2}y^2 + f(x), \tag{6.47}$$

where f is the constant of integration as far as y is concerned, hence f is a function of x, which is yet to be determined. Next, differentiate (6.47) with respect to x to get

$$\frac{\partial \psi}{\partial x} = f'(x), \tag{6.48}$$

which, when compared with the relation (6.46)a, leads to the equation $f'(x) = x$ for f, resulting in $f(x) = \frac{1}{2}x^2 + C$, where C is a universal constant of integration. Returning to (6.47), we now have the complete expression for ψ:

$$\psi(x, y, t) = \frac{1}{2}(x^2 + y^2) + C. \tag{6.49}$$

The constant C is typically set by a single known data point from the physical setting that ψ and $\mathbf{v}$ model.

The connection that we just established between the stream function ψ of a flow and its velocity field $\mathbf{v}$ has an important geometrical interpretation. At any instant of time t the vector field $\mathbf{v}$ is tangential to instantaneous *contours* or *level curves* of ψ . To see why this is the case consider a point $P = (x^*, y^*)$ in the domain of ψ and an arbitrary but fixed instance t^*. We recall from elementary calculus that the gradient of ψ evaluated at P and at t^*, $\nabla\psi|_{(P,t^*)}$ is perpendicular to the level curve of ψ that passes through P at the instant t^*, which is given by

$$\psi(x, y, t^*) = k.$$

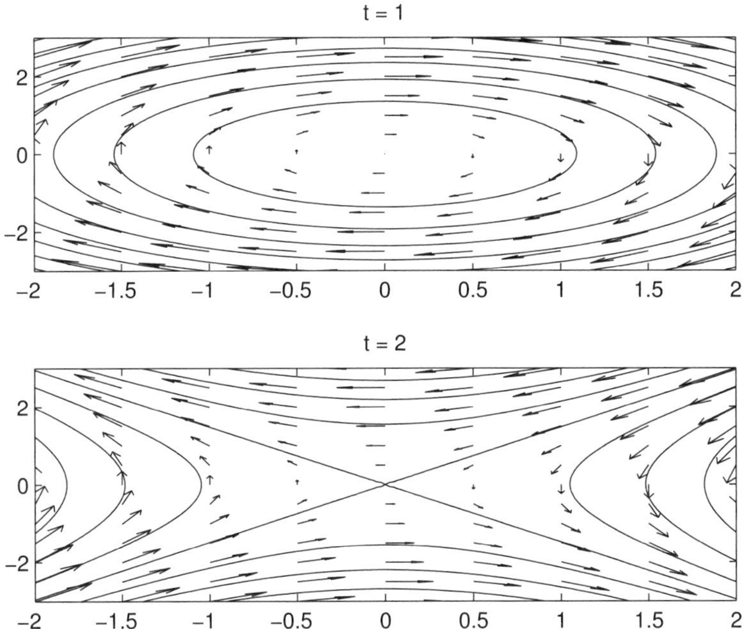

FIGURE 6.3: Contours of the stream function $\psi(x, y, t) = x^2 \sin t + y^2 \cos t$ and its associated vector field at two values of t. Note that the contours of ψ vary with time, as do the velocity vectors at a fixed position (x, y).

The constant k is $\psi(x^*, y^*, t^*)$. On the other hand, recalling the definition of ψ as a stream function, we have

$$\nabla\psi \cdot \mathbf{v} = \langle \frac{\partial \psi}{\partial x}, \frac{\partial \psi}{\partial y} \rangle \cdot \langle \frac{\partial \psi}{\partial y}, -\frac{\partial \psi}{\partial x} \rangle = 0, \tag{6.50}$$

showing that $\mathbf{v}$ is perpendicular to $\nabla\psi$. Since $\nabla\psi$ is perpendicular to both the level curve and $\mathbf{v}$, the vector $\mathbf{v}|_{(x^*, y^*, t^*)}$ must be tangential to the level curve. See Figure 6.3 for an example of a stream function (in this case $\psi(x, y, t) = x^2 \cos t + y^2 \sin t$) that illustrates the relationship between ψ and $\mathbf{v}$ at two instances of t.

The relationship between the contour levels of a stream function ψ and the velocity field associated with ψ is particularly significant in the special case when ψ is independent of t. When ψ is time-independent, the contours of $\psi(x, y)$ remain stationary as time varies, leading us to conclude that the particle paths of the ODE system

$$\left\{ \begin{array}{ll} dx/dt = u(x,y) = \partial\psi(x,y)/\partial y, & x(0) = x_0, \\ dy/dt = v(x,y) = -\partial\psi(x,y)/\partial x, & y(0) = y_0, \end{array} \right. \tag{6.51}$$

in fact coincide with contours of ψ. We summarize this discussion in the following theorem.

Theorem 6.4.1 (Incompressibility and Time-Independent Stream Functions)
Let $\mathbf{v}$ *be a smooth, incompressible, two-dimensional and time-independent vector field. Then there exists a function* $\psi = \psi(x,y)$, *associated with the system of ODEs (6.51), such that the contours of* ψ *coincide with the particle paths of (6.51).*

The fact that the stream function in Theorem 6.4.1 is time-independent is crucial since when ψ is time-dependent, as is the case in the example in Figure 6.3, the contours of ψ are only instantaneously tangential to the associated velocity field and hence do not give a good picture of particle paths as time evolves. On the other hand, when the stream function is time-independent, the contours of ψ can be computed *a priori* and they coincide with the particle paths of the associated system of ODEs. This point is illustrated in Figure 6.4 where we plot the contours of the stream function ψ (which is Vector Field **A**) defined in (6.49) and a single particle path, by solving the associated system of ODEs using `ode45`, through the point $(1.1, 2.3)$. This figure was obtained by running the following lines in MATLAB:

```
clf;
psi=inline('1/2*(x.^2+y.^2)','x','y');
[X,Y]=meshgrid(-3:0.01:3,-3:0.01:3);
contour(X,Y,psi(X,Y));
hold on
[t,z]=ode45(@VectorFieldA, [0 3],[1.2 2.3]);
plot(z(:,1),z(:,2), '*')
```

The M-file `VectorFieldA.m` is

```
function zprime=VectorFieldA(t,z);
%
zprime=[z(2); -z(1)];
```

To recap, if a stream function ψ is explicitly dependent on t, the streamlines do not provide a lot of information about the particle paths of the flow, because the streamlines only show the instantaneous behavior

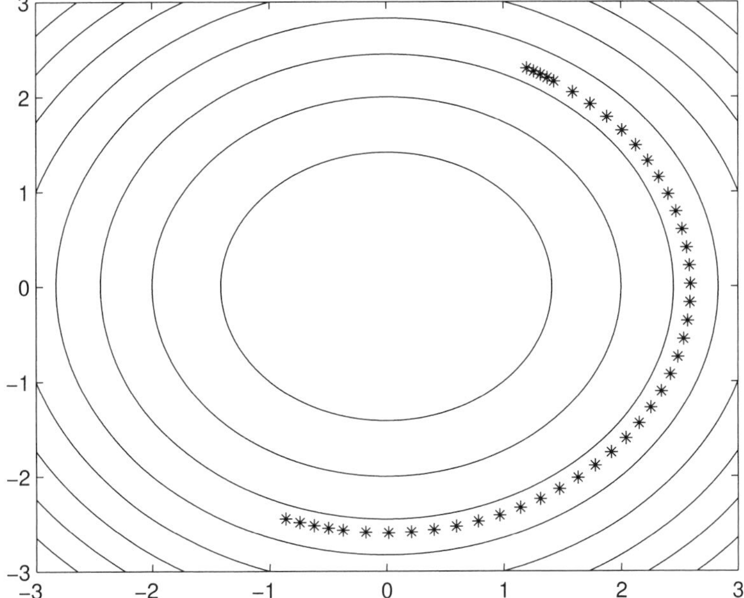

FIGURE 6.4: Streamlines and a trajectory of Vector Field A.

of the vector field and may change substantially from an instant of time to another, as shown in Figure 6.3. By contrast, when ψ is independent of t, the streamlines of ψ coincide with the particle paths of the vector field and hence give us an excellent representation of the fluid flow, as shown in Figure 6.4.

Despite this important qualitative connection between streamlines and particle paths, we lose quantitative information about how fluid particles evolve if we only consider streamlines—a particle path obtained by solving the system of ODEs in (6.51) not only contains spatial information (about the positions occupied by a fluid particle), it also contains temporal information (the amount of time it takes for a fluid particle to travel from one position to another), as shown in Figure (6.4). One of the main features of this approach, i.e., solving the system of ODEs, is that we are able to discover the Lagrangian behavior of solutions, including information as delicate as the chaotic behavior of solutions.

We end this section by recalling the definition of the fourth special flow, namely the ABC flow, that we encountered in an earlier chapter.

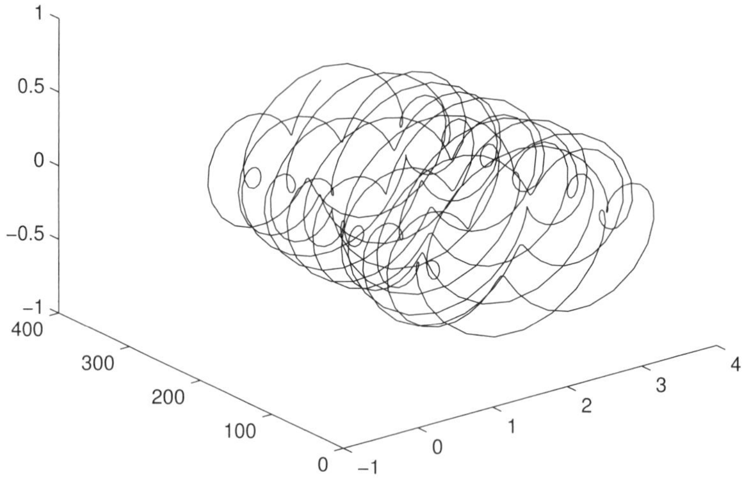

FIGURE 6.5: The particle path of a particle located at (0.1, 0.2, 0.1) at time 0 and otherwise moving under the influence of the ABC flow with $A = 1$, $B = 0.2$ and $C = -0.2$.

The velocity field $\mathbf{v} = \langle u, v, w \rangle$ that gives rise to this flow is as follows:

$$\begin{cases} u &= A \sin z + C \cos y, \\ v &= B \sin x + A \cos z, \\ w &= C \sin y + B \cos x. \end{cases} \tag{6.52}$$

The ensuing fluid motion is called the **ABC** flow, short for the Arnold–Beltrami–Childress, because of the work of the three distinguished researchers, V. I. Arnold, E. Beltrami, and S. Childress, who pointed out several remarkable attributes of the trajectories of this dynamical system. This flow is incompressible, and has the special property that its vorticity, $\nabla \times \mathbf{v}$, and its velocity $\mathbf{v}$ coincide, that is

$$\nabla \times \mathbf{v} = \mathbf{v}. \tag{6.53}$$

Figure 6.5 shows the trajectory of a single particle under the action of the ABC flow. This figure was obtained by executing the following lines in MATLAB:

```
clf;
global A B C
```

```
A=1; B=0.1; C=-0.2;
%
[t,q]=ode45(@ArnoldBeltramiChildress,[0 300],[0.1 0.2 0.1]);
plot3(q(:,1),q(:,2),q(:,3))
```

where the M-file `ArnoldBeltramiChildress.m` is

```
function qprime=ArnoldBeltramiChildress(t,q);
global A B C
%
qprime=[A*sin(q(3))+B*cos(q(2)); B*sin(q(1))+A*cos(q(3));...
    C*sin(q(2))+B*cos(q(1))];
```

Problems 6.4

1. Show that Vector Fields **A** and **C** are incompressible. Determine the stream function of each flow.

2. Compute the vorticity of Vector Fields **A**, **B** and **C**. Consider the particle path of the particle P that is located at $(1, 0)$ at time zero in each of these three flows. In which flow is the rate of rotation of experienced by P largest?

3. Show that the Arnold-Beltrami-Childress flow is incompressible.

4. Consider the vector field

$$\mathbf{v} = \langle u, v, w \rangle = \langle -\frac{\partial \psi_1}{\partial y}, \frac{\partial \psi_1}{\partial x} + \frac{\partial \psi_2}{\partial z}, -\frac{\partial \psi_2}{\partial y} \rangle. \tag{6.54}$$

   (a) Show that $\mathbf{v}$ is incompressible as long as ψ_1 and ψ_2 are sufficiently differentiable.

   (b) Compute the vorticity of $\mathbf{v}$ in terms of ψ_1 and ψ_2.

   (c) Compute the acceleration $\mathbf{a}$ in terms of ψ_1 and ψ_2.

5. Prove the expression given in (6.53).

6. Apply MATLAB's `quiver` to plot the following vector fields. Compute the divergence and the vorticity of each flow and identify which ones are incompressible and/or irrotational, if any.

   (a) $\mathbf{v} = (x + y)\mathbf{i} + (x - y)\mathbf{j}$

   (b) $\mathbf{v} = (x^2y^2 - xy)\mathbf{i} - y\mathbf{j}$

   (c) $\mathbf{v} = \langle \frac{x}{\sqrt{x^2+y^2}}, \frac{1-y}{\sqrt{x^2+y^2}} \rangle$.

   (d) $\mathbf{v} = \langle \sin y, \cos x \rangle$

(e) $\mathbf{v} = \log\sqrt{x^2+y^2}\,\mathbf{i} + (x-y)\mathbf{j}$

7. Let $\mathbf{v} = \nabla p(x,y,z)$ for some scalar function p. Show that if the fluid, whose velocity field is modeled by $\mathbf{v}$, is incompressible, then p satisfies Laplace's equation:

$$\Delta p \equiv \frac{\partial^2 p}{\partial x^2} + \frac{\partial^2 p}{\partial y^2} + \frac{\partial^2 p}{\partial z^2} = 0. \tag{6.55}$$

8. Compute the Laplacian of p where p is described below.

   (a) $p(x,y,z) = x^2+y^2+z^2$
   (b) $p(x,y,z) = \frac{1}{\sqrt{x^2+y^2+z^2}}$
   (c) $p(x,y) = -\frac{y}{x}$
   (d) $p(x,y) = \cos\frac{1}{x^2+y^2}$
   (e) $p(x,y,z) = \ln(x^2+y^2+z^2)$
   (f) $p(x,y) = \frac{1}{\sqrt{x^2+y^2}}$
   (g) $p(x,y) = \arctan\frac{y}{x}$.

9. Find the grad(div $\mathbf{v}$) if $\mathbf{v}$ is given by the following expressions; here the quantity r is the magnitude of the position vector $\langle x,y,z\rangle$:

   (a) $\mathbf{v} = (z-y)\mathbf{i} + (x+z)\mathbf{j} + x\mathbf{k}$
   (b) $\mathbf{v} = 2x^2\mathbf{i} - y^2\mathbf{j} + z^2\mathbf{k}$
   (c) $\mathbf{v} = \frac{1}{r}\langle x,y,z\rangle$
   (d) $\mathbf{v} = \frac{1}{r}\langle y,-x,0\rangle$
   (e) $\mathbf{v} = \frac{1}{r^2}\langle yz+\ln r, -xz-\ln r, z^2\rangle$.

10. Prove the following identities. In each case assume that $\mathbf{v}$, $\mathbf{w}$, p and q are arbitrary smooth functions of x, y and z.

   (a) div $(\rho\mathbf{v}) = \rho$div $\mathbf{v} + \mathbf{v}\cdot\nabla\rho$
   (b) div $(p\nabla q) = p\nabla^2 q + \nabla p\cdot\nabla q$
   (c) div $(p\nabla q) -$ div $(q\nabla p) = p\nabla^2 q - q\nabla^2 p$.
   (d) div $(\nabla\mathbf{v}\times\nabla\mathbf{w}) = 0$.

11. Let $\psi(x,y) = xy\,e^{-(x^2+y^2)}$ be the stream function of a flow. Find the velocity of the particles located at $(1,2)$, $(-1,1)$, $(2,3)$, and $(-1,-2)$. Draw a diagram displaying with the position of the particles and their associated velocity vectors, together with the contours of ψ.

12. Let $\psi(x, y) = \cos(\log(x^2 + y^2))$ be the stream function of a flow. Plot its contours and the velocity vectors for the particles located at $(2, 2)$, $(1, 2)$, $(-2, 2)$, and $(-1, 2)$.

13. Prove that if $\mathbf{v}$ is the velocity field of a 2D and incompressible flow with ψ as its stream function, then

$$\text{div } (\psi\mathbf{v}) = 0. \tag{6.56}$$

14. Find whether each vector field has a stream function. If it does, find ψ and use it to draw the particle paths of the flow.

    (a) $\mathbf{v} = \langle 3y, -x + \sin x\rangle$
    (b) $\mathbf{v} = \langle x^2 + y^2, -2xy\rangle$
    (c) $\mathbf{v} = \langle \sin x \cos y, -\cos x \sin y\rangle$
    (d) $\mathbf{v} = \langle \sinh x \sin y, -\cosh x \cos y\rangle$

15. Let $\mathbf{v}(x, y) = e^{-(x^2+y^2)}\langle y, -x\rangle$. Verify whether this velocity field has a stream function and if it does determine it.

16. Consider the stream function $\psi_c(x, y) = cy + d\sqrt{x^2 + y^2}$, where c and d are nonnegative constants.

    (a) Let $d = 1$ and use the `contour` command of MATLAB and
        i. draw the level curves of ψ_c for $c = 0, 0.1, 0.3, 0.5, 0.7, 0.9$ and 1. Is there any qualitative difference in the level curves as c varies? Explain.
        ii. Use `ode45` and plot several particle paths of each flow for the values of c listed previously. What is the difference between the particle paths when $c = 1$ and the other c's?

    (b) Let $c = 1$ and vary d starting with $d = 0.5$ and increments 0.1. Plot the contours of the resulting stream functions until $d = 2$ and report on any change of geometry of the streamlines as d varies.

17. (The *Double-Gyre Stream Function*) In Reference [2] the authors C. Shadden, F. Lekien and J. Marsden introduce the following stream function:

$$\psi(x, y, t) = A\sin(\pi f(x, t))\sin(\pi y) \tag{6.57}$$

where

$$f(x, t) = a(t)x^2 + b(t)x, \quad a(t) = \epsilon\sin(\omega t),$$
$$b(t) = 1 - 2\epsilon\sin(\omega t). \tag{6.58}$$

Let $A = 0.1$, $\omega = 2\pi$, and $\epsilon = 0.25$. Let $(x, y) \in (0, 2) \times (0, 1)$.

(a) Plot the associated vector fields at $t = 0.01$, $t = 0.3$, $t = 0.65$, and $t = 0.95$ and report on the qualitative behavior of the vector field at these times and in particular the location of two "gyres" (compare your figures to Figure 5 on page 292 of Reference [2] as with Figure 6.6).

(b) Plot the vector field at $t = 0.25$ and report on the location of the two "gyres" generated by (6.58).

(c) Plot the vector field at $t = 0.75$ and report on the location of the two "gyres" now.

(d) Use the animation capabilities of MATLAB and create an animation of the vector field as t varies from 0 to 1 at increments of 0.01. Report on the qualitative behavior of the two gyres' motion as a function of time.

(e) Let $A = 0.1$, $\omega = \frac{\pi}{5}$ and ϵ vary from 0 to 0.3 at increments of 0.1. Use `ode45` and plot the trajectories of the associated vector field for $t \in (0, 25)$ and initial conditions $(0.1 * i, 0.2)$, $i = 1$ to 19. Compare your graphs to Figure 6.7.

18. In [3] the authors M. Branicki and S. Wiggins introduce the following stream function:

$$\psi(x, y, t) = (xy(\sigma(t) - x^2) - \alpha x y^3 + \beta x y^5)e^{-(x^4+y^4)/\delta(t)^2}, \quad (6.59)$$

where σ, α, β and δ are given functions of time.

(a) Determine the vector field associated with this flow.

(b) Let $\alpha = \frac{1}{3}$, $\beta = \frac{0.008}{5}$, $\delta = 5$, and

$$\sigma(t) = 2(\arctan(10t) + \frac{\pi}{2} - 1).$$

(see Page 28 of Reference [3] for more information on the choice of these parameters.

6.5 Acceleration in Rectangular Coordinates

By definition, acceleration is the time rate of change of velocity. In Lagrangian setting, where velocity $\mathbf{V}(t, \mathbf{X})$ is defined in terms of fluid particles, acceleration $\mathbf{A}(t, \mathbf{X})$ is simply

$$\mathbf{A}(t, \mathbf{X}) = \frac{\partial \mathbf{V}}{\partial t}.$$

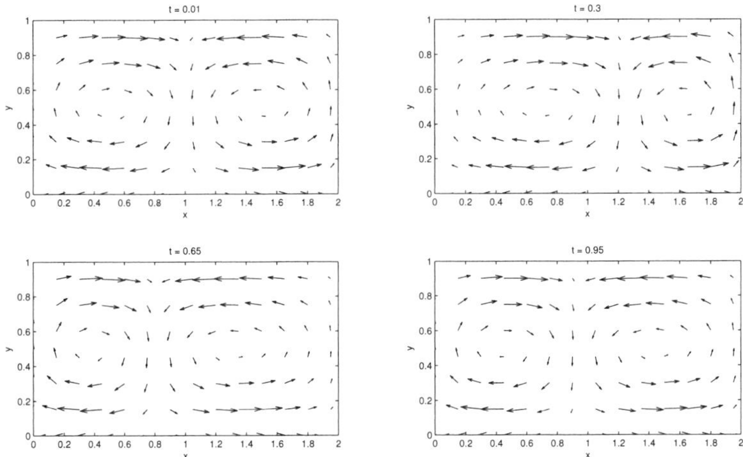

FIGURE 6.6: Four snapshots of ψ when $A = 0.1$, $\omega = 2\pi$ and $\epsilon = 0.25$.

In an Eulerian setting, which is the preferred representation in Fluid Dynamics, $\mathbf{A}$, must be written in terms of position $\mathbf{x}$ rather than particle $\mathbf{X}$. Following the notation we adopted earlier, we denote by $\mathbf{a}$ the representation of acceleration in terms of position, noting the relation

$$\mathbf{a}(t, \mathbf{x}) = \mathbf{A}(t, \mathbf{X}),$$

where $\mathbf{x}$ and $\mathbf{X}$ are related through the deformation $\mathbf{p}$ by $\mathbf{x} = \mathbf{p}(t, \mathbf{X})$. Hence,

$$\mathbf{A}(t, \mathbf{X}) = \mathbf{a}(t, \mathbf{p}(t, \mathbf{x})).$$

The previous expression shows the relation between the Lagrangian and Eulerian representations of acceleration. It does not, however, show how one computes $\mathbf{a}$ from $\mathbf{v}$. To obtain that formula we note that

$$\mathbf{v} = \mathbf{v}(t, \mathbf{x}) = \mathbf{v}(t, \mathbf{p}(t, \mathbf{X})), \tag{6.60}$$

which shows the explicit dependence of $\mathbf{v}$ on t. Differentiating the expression in (6.60) with respect to t leads to

$$\mathbf{a} = \frac{\partial \mathbf{v}}{\partial t} + \mathbf{v} \cdot \nabla \mathbf{v}. \tag{6.61}$$

In terms of the notation $\frac{D}{Dt}$ defined in the previous section, (6.61) states that acceleration $\mathbf{a}$ is simply the material time derivative of $\mathbf{v}$:

$$\mathbf{a} = \frac{D\mathbf{v}}{Dt}.$$

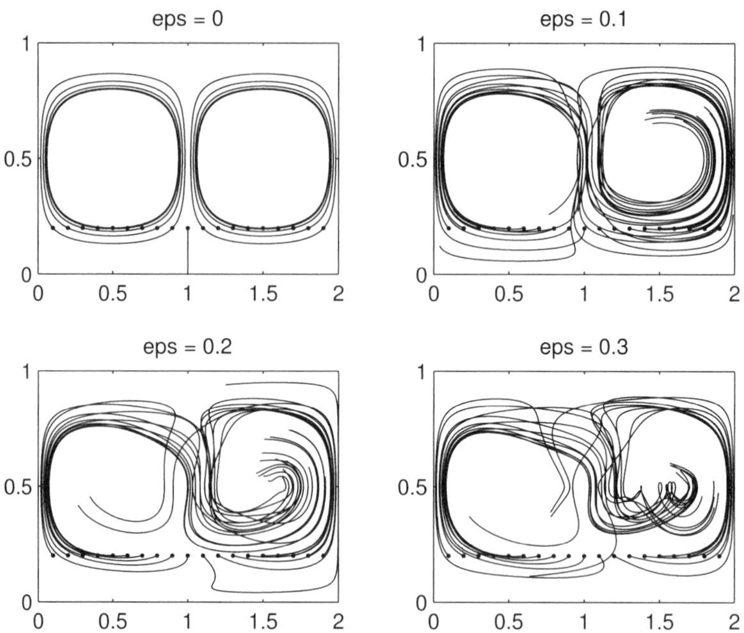

FIGURE 6.7: Particle paths of the Double Gyre stream function for a set of initial conditions located along $y = 0.2$, In each figure $A = 0.1$ and $\omega = \frac{\pi}{5}$ are fixed but ϵ varies. Note that when $\epsilon = 0$ the structure of the double gyre remains intact, while for ϵ positive the line $x = 0$, which is invariant when $\epsilon = 0$ becomes entangled, a sign of the ensuing chaotic behavior.

Moreover, continuing to interpret the notation in (6.61), a good way to think of the vector $\mathbf{v} \cdot \nabla\mathbf{v}$ is to group $\mathbf{v}$ and ∇ together and write the nonlinear part of the acceleration as

$$(\mathbf{v} \cdot \nabla)\mathbf{v}$$

i.e., think of $\mathbf{v}$ and ∇ together as the "scalar" operator

$$(\mathbf{v} \cdot \nabla)$$

so that, with this interpretation, the expression $\mathbf{v} \cdot \nabla\mathbf{v}$ is simply the scalar multiplication of the scalar $(\mathbf{v} \cdot \nabla)$ and the vector $\mathbf{v}$. Hence (6.61) is

$$\mathbf{a} = \frac{\partial \mathbf{v}}{\partial t} + (\mathbf{v} \cdot \nabla\mathbf{v})\mathbf{v}. \tag{6.62}$$

To see how the formula in (6.62) applies, consider the velocity field from Vector Field **B**:

$$\mathbf{v}_{_B} = \langle \frac{y}{\sqrt{x^2+y^2}}, -\frac{x}{\sqrt{x^2+y^2}}, 0\rangle.$$

The first component of the acceleration, a_1, according to formula (6.62), is

$$a_1 = \frac{\partial u}{\partial t} + (\mathbf{v}\cdot\nabla)u = \frac{\partial u}{\partial t} + (u\frac{\partial}{\partial x} + v\frac{\partial}{\partial y} + w\frac{\partial}{\partial z})u.$$

Hence

$$a_1 = u\frac{\partial u}{\partial x} + v\frac{\partial u}{\partial y} = -\frac{x}{x^2+y^2},$$

and similarly, $a_2 = -\frac{y}{x^2+y^2}$, and $a_3 = 0$. Hence

$$\mathbf{a} = -\frac{1}{x^2+y^2}\langle x, y, 0\rangle. \tag{6.63}$$

One of the questions we will be concerned about is if the fluid flows we consider are of the type whose accelerations have potentials. In particular, if ρ is the fluid density and $\mathbf{a}$ is the flow acceleration, is there a function p, which we will refer to as the *pressure* function such that

$$\rho\mathbf{a} = -\nabla p. \tag{6.64}$$

We will refer to (6.64) as the *Euler Equations* of the flow and note that they represent the simplest analog of Newton's law for the balance of linear momentum, that mass times acceleration must be balanced by the sum of all forces acting on the body. In this setting the simplifying assumption we are imposing is that all of the forces that act on the body of fluid can be captured as the gradient of a scalar potential. We will call fluids whose motion can be represented by the Euler Equations *Simple Fluids.*

Returning to (6.64) and recalling the vector identity that

$$\nabla \times \nabla p = \mathbf{0},$$

for any function p, we observe that an acceleration vector $\mathbf{a}$ has a potential if

$$\nabla \times \frac{1}{\rho}\mathbf{a} = \mathbf{0}. \tag{6.65}$$

So if the fluid is homogeneous, so that its density is constant, then the acceleration $\mathbf{a}$ is curl free. When this condition holds, one can integrate the relations in (6.64) to find p. As in the process we outlined in the previous section for obtaining the stream function ψ of an incompressible flow, the curl-free property of $\mathbf{a}$ guarantees that the constants of integration we encounter from integration of (6.64) are compatible, as the following example shows. To see this process at work, consider the acceleration (6.63) of Vector Field **B**. After verifying that this vector is curl free, we set up the equations in (6.64):

$$\partial p/\partial x = x/\sqrt{x^2+y^2},\ \partial p/\partial y = y/\sqrt{x^2+y^2},\ \partial p/\partial z = 0. \tag{6.66}$$

We begin by integrating the first equation in (6.66):

$$p(x,y,z) = \sqrt{x^2+y^2} + f(y,z), \tag{6.67}$$

where f is the constant of integration. Next we differentiate (6.67) with respect to y

$$\partial p/\partial y = \frac{y}{\sqrt{x^2+y^2}} + \frac{\partial f}{\partial y}, \tag{6.68}$$

which, when compared with (6.66)b, shows that $\partial f/\partial y = 0$, or, $f = f(z)$. We infer from this observation that p in (6.67) reduces to

$$p(x,y,z) = \sqrt{x^2+y^2} + f(z). \tag{6.69}$$

Finally, differentiating the above equation with respect to z and comparing the resulting expression with (6.66)c shows that f is in fact independent of z as well. Hence, the pressure experienced by fluid particles in Vector Field **B** is

$$p(x,y,z) = \sqrt{x^2+y^2} + C, \tag{6.70}$$

where C is a universal constant.

The Euler Equations in (6.65) form the precursor to the *Navier-Stokes* equations where fluid particles, in addition to pressure, must endure viscous forces. The latter equations are the subject of the rest of this chapter.

Problems 6.5

1. Complete the calculations that led to (6.61). Start by writing (6.60) in component form $\mathbf{v} = \langle u, v, w\rangle$, then apply the chain rule to differentiate each component to get the components a_1, a_2 and a_3 of **a**. Note that $\mathbf{v}\cdot\nabla = u\frac{\partial}{\partial x} + v\frac{\partial}{\partial y} + w\frac{\partial}{\partial z}$.

2. Complete the calculations that lead to (6.63), the acceleration of the velocity field $\mathbf{v} = \langle \frac{y}{\sqrt{x^2+y^2}}, -\frac{x}{\sqrt{x^2+y^2}}, 0\rangle$.

3. Compute the acceleration of each velocity field below:

   (a) $\mathbf{v} = \langle y, -x, 0\rangle$.

(b) $\mathbf{v} = \langle ax + by, cx + dy, 0 \rangle$.

(c) $\mathbf{v} = \langle x - xz, y - yz, z - xy \rangle$.

(d) The ABC flow, where $\mathbf{v} = \langle A \sin z + C \cos y, B \sin x + A \cos z, C \sin y + B \cos x \rangle$. Here A, B and C are constants.

(e) The Lorenz flow, where $\mathbf{v} = \langle \sigma(y - x), x(\rho - z) - y, xy - \beta z \rangle$. Here σ, ρ and β are constants.

4. Determine if any of the accelerations $\mathbf{a}$ obtained in the previous problem has a potential p, that is, $\mathbf{a} = \nabla p$. Find p in each if appropriate.

5. In each problem below, ψ is the stream function of a velocity field $\mathbf{v}$. Compute $\mathbf{v}$ and its associated acceleration $\mathbf{a}$:

   (a) $\psi(x, y) = \frac{1}{2} \sin(x^2 + y^2)$.

   (b) $\psi(x, y) = \frac{1}{2} y^2 + \sin x$.

6. Compute the accelerations of Vector Fields $\mathbf{A}$, $\mathbf{B}$ and $\mathbf{C}$ and determine if any of these accelerations has a potential p. i.e., demonstrate if any of these vector fields satisfy the Euler equations.

6.6 Strain-Rate Matrix and Vorticity

Fluid motions are not uniform. Fluid particles in different positions have a tendency to move with different velocities. Since discrepancies in velocity often result in changes in acceleration, and acceleration is directly proportional to force, we can expect some form of an internal *frictional* or *viscous* force in the fluid. In our model this force will end up being proportional to the velocity gradient, as we will develop shortly.

Let $\mathbf{v}(t, \mathbf{x}) \in E^3$ be the velocity field of a fluid at time t at position $\mathbf{x}$. We are interested in measuring the relationship between the velocity vectors $\mathbf{v}(t, \mathbf{x})$ when $\mathbf{x}$ is near a fixed but arbitrary position $\mathbf{x}_0$. Let $\mathbf{v}_0$ be the velocity at $\mathbf{x}_0$. We suspect that $\mathbf{v}$ will be relatively close to $\mathbf{v}_0$ if $\mathbf{x}$ is relatively close to $\mathbf{x}_0$. We therefore compute the Taylor series expansion of $\mathbf{v}$ in $\mathbf{x}$ about $\mathbf{x}_0$:

$$\mathbf{v}(t, \mathbf{x}) = \mathbf{v}(t, \mathbf{x}_0) + (\nabla \mathbf{v})\big|_{\mathbf{x}_0} (\mathbf{x} - \mathbf{x}_0) + \cdots \tag{6.71}$$

where the ellipses stand for higher order terms in $\mathbf{x} - \mathbf{x}_0$, and the matrix

$\nabla \mathbf{v}$ in (6.71) is the velocity gradient we saw in the description of acceleration. The standard notation is to let D and A stand for the symmetric and antisymmetric parts of this matrix:

$$\nabla \mathbf{v} = D + A \tag{6.72}$$

where

$$D = \frac{1}{2}(\nabla \mathbf{v} + (\nabla \mathbf{v})^T) \quad \text{and } A = \frac{1}{2}(\nabla \mathbf{v} - (\nabla \mathbf{v})^T). \tag{6.73}$$

The symmetric matrix D is called the *strain-rate* matrix, which has six independent entries related to $\mathbf{v}$ by the relations

$$d_{ij} = \frac{1}{2}(\frac{\partial v_i}{\partial x_j} + \frac{\partial v_j}{\partial x_i}). \tag{6.74}$$

The matrix A, on the other hand, is anti-symmetric and has only three independent entries since $a_{ii} = 0$ for each index i and

$$a_{ij} = \frac{1}{2}(\frac{\partial v_i}{\partial x_j} - \frac{\partial v_j}{\partial x_i}), \tag{6.75}$$

when $i < j$, and $a_{ji} = -a_{ij}$ otherwise. It is interesting to note that A is closely related to the vorticity $\omega = \nabla \times \mathbf{v}$ of the fluid flow. In fact

$$A(\mathbf{x} - \mathbf{x}_0) = \frac{1}{2}\omega \times (\mathbf{x} - \mathbf{x}_0). \tag{6.76}$$

We can therefore summarize the behavior of the velocity field near a fixed but arbitrary position $\mathbf{x}_0$ as follows:

$$\mathbf{v}(t, \mathbf{x}) = \mathbf{v}(t, \mathbf{x}_0) + 2\omega \times (\mathbf{x} - \mathbf{x}_0) + D(\mathbf{x} - \mathbf{x}_0) + \cdots . \tag{6.77}$$

The first term $2\omega \times (\mathbf{x} - \mathbf{x}_0)$ in (6.77) shows one of the impacts of $\mathbf{v}$, which is a rotation about the axis ω with strength $2||\omega||$ radians per second. The second term, $D(\mathbf{x} - \mathbf{x}_0)$, gives us a quantifiable way to measure the rate at which the fluid parcel near $\mathbf{x}_0$ is being sheared, stretched, or compressed.

To gain insight into how the knowledge of D and A, which are local, may provide some understanding about the global behavior of a flow, we now consider an example and refer the reader to the exercises at the end of the section for further examples. Let's consider the Vector Field A, where

$$\mathbf{v} = \langle y, -x, 0 \rangle.$$

In this case

$$\nabla \mathbf{v} = \begin{bmatrix} 0 & 1 & 0 \\ -1 & 0 & 0 \\ 0 & 0 & 0 \end{bmatrix}$$

so that

$$D = \frac{1}{2}(\nabla \mathbf{v}) + (\nabla \mathbf{v})^{\mathrm{T}} = \begin{bmatrix} 0 & 0 & 0 \\ 0 & 0 & 0 \\ 0 & 0 & 0 \end{bmatrix}.$$

Hence, in this example $\nabla \mathbf{v}$ is the same as its anti-symmetric part A. These observations agree with the other calculations we have carried out regarding Vector Field A, since under the influence of this flow, which is a pure rotation, no particle experiences any stretching or shearing. In fact all of the energy of this flow is stored in A.

As a second example, consider the velocity field

$$\mathbf{v} = \langle x, 2.1y, 0 \rangle. \tag{6.78}$$

The vorticity of this flow is identically zero while its strain-rate matrix is

$$D = \begin{bmatrix} 1 & 0 & 0 \\ 0 & 2.1 & 0 \\ 0 & 0 & 0 \end{bmatrix}. \tag{6.79}$$

The system of differential equations that defines the motion is

$$\frac{dx}{dt} = x, \quad \frac{dy}{dt} = 2.1y, \quad \frac{dz}{dt} = 0, \tag{6.80}$$

whose solution is

$$x(t) = x_0 e^t, \quad y(t) = y_0 e^{2.1t}, \quad z(t) = z_0. \tag{6.81}$$

As the diagonal elements of (6.79) show, a parcel of fluid is stretched a little more than twice as far in the y direction after one unit of time has elapsed. Figure 6.8 shows the impact of the deformation on four fluid particles whose initial positions are located near the origin. This figure shows that there is significant stretching in the y-direction. Note that the eigenvalues of D are 1, 2.1 and 0, with eigenvectors $\langle 1, 0, 0 \rangle$, $\langle 0, 1, 0 \rangle$ and $\langle 0, 0, 1 \rangle$, respectively, predicting the amounts of stretching in the three axes directions. Figure 6.8 is the output of the following MATLAB program (note the use of `quiver` to generate a plot of the vector field itself):

```
clf
t=0:0.2:3;
x=-2:).25:2;y=-2::0.25:2;
[X,Y]=meshgrid(x,y);
quiver(X,Y,X,2*Y);
hold on
plot(-0.1*exp(t),-0.01*exp(2.1*t),'*')
```

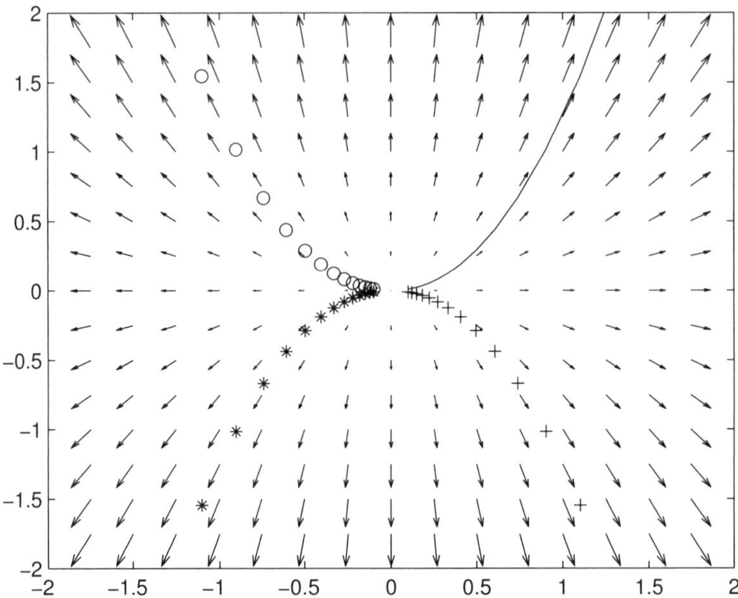

FIGURE 6.8: The motion of four particles, originally located near the origin, under the velocity field $\langle x, 2.1y, 0\rangle$.

```
plot(-0.1*exp(t),0.01*exp(2.1*t),'o')
plot(0.1*exp(t),-0.01*exp(2.1*t),'+')
plot(0.1*exp(t),0.01*exp(2.1*t),'-')
axis([-2 2 -2 2])
```

As a final example, consider the velocity field

$$\mathbf{v} = \langle x - 1.2y, 2.7x + y\rangle. \tag{6.82}$$

The preceding MATLAB program, when appropriately changed for (6.82), leads to Figure 6.9. It is clear from this figure that under this fluid flow parcels of fluid are rotated as well as they are stretched, as can be verified from the values of the vorticity and the strain-rate matrix for (6.82). Figure 6.9 is the output of the following MATLAB program:

```
clear all
clf
t=0:0.1:3;
x=-4:4;y=-4:4;
[X,Y]=meshgrid(x,y);
```

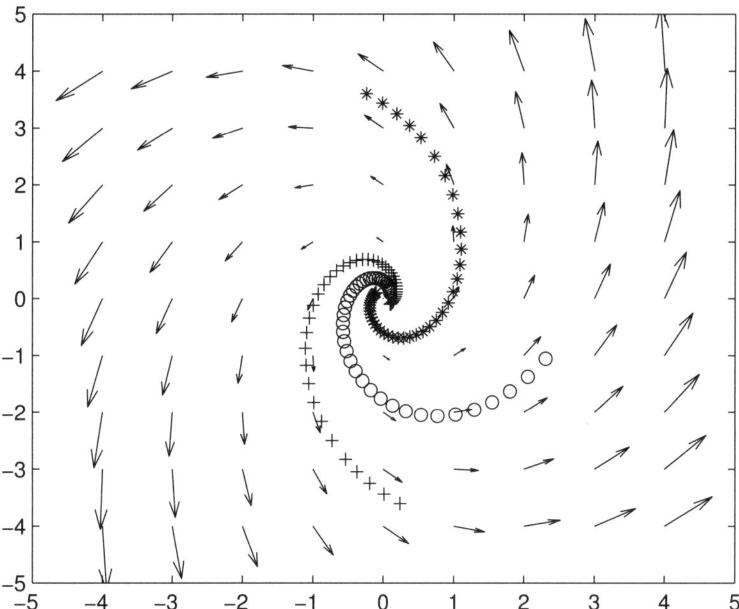

FIGURE 6.9: The motion of three set particles under the velocity field $\langle x - 1.2y, 2.7x + y, 0\rangle$. Note the use of `ode45`.

```
quiver(X,Y,X-1.2*Y,2.7*X+Y);
hold on
[t, y]=ode45(@deform2,[0 3],[0.1 0.1]);
plot(y(:,1),y(:,2),'o')
[t, y]=ode45(@deform2,[0 3],[-0.1 0.1]);
plot(y(:,1),y(:,2),'*')
[t, y]=ode45(@deform2,[0 3],[0.1 -0.1]);
plot(y(:,1),y(:,2),'+')
```

The M-file `deform2.m` is

```
function yprime=deform2(t,y);

yprime=[y(1)-1.2*y(2); 2.7*y(1)+y(2)];
```

The following is the syntax for the MATLAB program that delivers a similar output as the one shown in Figure 6.9. The syntax is written in such a way that it can be extended easily to monitor the evolution of a large set of particles under the action of the system of differential equations that define the flow:

```
clf
clear all
data = [0.1 0.1; -0.1 0.1;0.1 -0.1];
image=[];
for i=1:length(data)
    [t,y]=ode45('deform2',[0 3],data(i,:));
    plot(y(:,1),y(:,2));
    hold on
end
axis([-5 5 -5 5])
```

Problems 6.6

1. Complete the analysis that leads to equations (6.76), that is, let A be the anti-symmetric part of $\nabla \mathbf{v}$ and $\omega = \nabla \times \mathbf{v}$. Show that

$$A\mathbf{x} = \frac{1}{2}\omega \times \mathbf{x},$$

for any vector $\mathbf{x}$.

2. Complete the analysis that leads to the approximation formula (6.77), i.e.,

$$\mathbf{v}(t, \mathbf{x}) = \mathbf{v}(t, \mathbf{x}_0) + 2\omega \times (\mathbf{x} - \mathbf{x}_0) + D(\mathbf{x} - \mathbf{x}_0) + \cdots.$$

3. Find the matrices D and A for the following flows. Describe how you can identify regions of rotation and stretching from the information in these matrices.

   (a) Vector Field B

   (b) Vector Field C

   (c) Vector Field ABC.

4. Draw the graphs of the image of the unit circle centered at the origin as t ranges from 0 to 1 at increments of 0.1 for

   (a) Vector Field A, B and C

   (b) Vector Field ABC with its third component set to zero.

5. In the following problems determine the vorticity vector and strain-rate matrix. Use MATLAB and draw the graphs of the deformation of the unit circle centered at the origin as t ranges from 0 to 1 at increments of 0.1.

(a) $\mathbf{v} = \langle x - 2y + 1, xy, 0 \rangle$

(b) $\mathbf{v} = \langle x^2 y, x - y, 0 \rangle$

(c) $\mathbf{v} = \langle x + y, 2 - 2y, 0 \rangle$

(d) $\mathbf{v} = \langle 2.3x - 3.1y, -3x + 2y, 0 \rangle$

(e) $\mathbf{v} = \langle \frac{x^2}{1+y^2}, -\frac{y^2}{1+x^2}, 0 \rangle$

(f) $\mathbf{v} = \langle e^{-(x^2+y^2)} y, -e^{-(x^2+y^2)}, 0 \rangle$

(g) $\mathbf{v} = \langle \sin(x + y), \cos(x - y), 0 \rangle$.

6. In the following problems determine the vorticity vector and strain-rate matrix. Write a MATLAB program to allow for the three-dimensional graphics aspects of the following vector fields:

   (a) Vector field ABC. Determine the evolution of the unit circle centered at the origin in the xy-plane as t varies from 0 to 1 at increments of 0.1.

   (b) $\mathbf{v}_A(x, z) + \mathbf{v}_B(x, y)$, where $\mathbf{v}_A$ and $\mathbf{v}_B$ are Vector Fields A and B respectively.

6.7 Internal Forces and Cauchy Stress

The strain-rate matrix D provides information about the local deformation of a fluid parcel. The concepts of *pressure* and *viscous forces* are the mathematical entities we introduce to represent how fluid elements respond to these deformations. Our task here is to describe a relationship between the strain-rate D and the internal forces that act on bodies of fluid when these bodies undergo deformations. This discussion will lead to the mathematical definitions of *traction* and *stress* in a fluid.

The concept of the traction $\mathbf{t}$ experienced by a fluid element located at $\mathbf{x}$ is illuminated in terms of the forces exerted by the rest of the body of fluid on $\mathbf{x}$. To be precise, consider a volume V of fluid with $\mathbf{x}$ in its interior, and consider a smooth surface S passing through $\mathbf{x}$—the smoothness of S guarantees that we can define a unique normal vector $\mathbf{n}$ to S at $\mathbf{x}$. Imagine splitting V into two pieces along the surface S. Our task is to place a force $\mathbf{F}$ at $\mathbf{x}$ in such a way that each sliced piece still experiences the same deformation. One of the main assumptions of the discipline of *Continuum Mechanics* is that this force depends continuously on $\mathbf{x}$ and on the normal $\mathbf{n}$. Moreover, it is assumed that the limit of this force, when it is scaled by the area Δ of S, exists as Δ approaches zero and the surface surrounding $\mathbf{x}$ shrinks to $\mathbf{x}$:

$$\lim_{\Delta \to 0} \frac{1}{\Delta} \mathbf{F}(\mathbf{x}, \mathbf{n}) = \mathbf{t}(\mathbf{x}, \mathbf{n}).$$

This limit, which is precisely the force per unit area acting at $\mathbf{x}$, is the traction $\mathbf{t}$ at $\mathbf{x}$.

It follows from the tenets of continuum mechanics that the traction at a point $\mathbf{x}$ is solely determined by the unit normal vector at $\mathbf{x}$. Moreover, a theorem due to Cauchy demonstrates that the dependence of $\mathbf{t}$ on the normal vector is linear. Thus, there is a matrix, denoted by σ, such that

$$\mathbf{t}(\mathbf{x}, \mathbf{n}) = \sigma(\mathbf{x})\mathbf{n}. \tag{6.83}$$

The matrix σ, which is called the *Cauchy Stress* associated with the fluid, depends on the intrinsic properties of the fluid, and not on a specific deformation that the fluid may be undergoing. The reader is encouraged to consult the text by Truesdell and Rajagopal (see Reference [6] at the end of this chapter), which provides an excellent and comprehensive introduction to this subject.

As is the case with any matrix representation, the representation of the Cauchy stress depends on the basis we choose for E^3. Once a basis is selected , the entries in the Cauchy stress are determined and are denoted by σ_{ij}. The physical interpretation of the coefficients σ_{ij} is as follows (again, see Reference [6] for more detailed discussion): Let $\{\mathbf{i}, \mathbf{j}, \mathbf{k}\}$ be the standard orthonormal basis for E^3. Imagine a small cubic block of fluid, centered at $\mathbf{x}$, with sides parallel with the coordinate planes. Each face of the block is perpendicular to one of the coordinate axes, thus the directions of the basis vectors $\pm\mathbf{i}$, $\pm\mathbf{j}$, and $\pm\mathbf{k}$ constitute unit normals to these faces. The vector $\sigma\mathbf{k}$, which is $\langle \sigma_{13}, \sigma_{23}, \sigma_{33} \rangle$, is the traction vector (force per unit area) experienced on this face. The quantity σ_{33} is the component of this vector in the direction normal to the face, while σ_{13} and σ_{23} are the components of the traction vector parallel to the face itself. These quantities measure the extent of shearing the fluid located at $\mathbf{x}$ is undergoing. This discussion, when applied to the other faces of the cube, provides the interpretation for the remaining six σ_{ij}. Because of their inherent physical attributes, the diagonal entries of σ, σ_{ii}, $i = 1, 2, 3$, are called the *normal stresses*, while the off-diagonal entries, σ_{ij}, $i \neq j$ are called the *shear stresses*.

An important property of the Cauchy Stress, which follows directly from the conservation of angular momentum (see Reference [6]), is that the matrix σ is symmetric, that is,

$$\sigma_{ij} = \sigma_{ji}$$

for all i and j.

We define a fluid as an *ideal isotropic incompressible fluid* if its Cauchy stress σ and D, where D is the strain-rate of any arbitrary deformation the fluid may experience, by the relation

$$\sigma = -pI + \mu D. \tag{6.84}$$

The function p is called the *pressure* of the fluid and μ its *kinematic viscosity.* This relation is an example of a *constitutive law* in continuum mechanics, a collection of mathematical rules that help describe how a velocity field $\mathbf{v}$ induces and imparts the resulting internal forces on the fluid particles that are being deformed by the motion.

The constitutive law (6.84) is the last piece of information we need to complete the formulation of the Navier–Stokes equations.

Problems 6.7

1. Suppose that the state of stress in a material is given by the *hydrostatic pressure* p as

$$\sigma = -p(x, y, z)I, \tag{6.85}$$

   where I is the identity matrix. Let $B = (x, y, z)$ be a fixed but arbitrary point in the domain of the deformation. Compute the traction at B along planar surfaces parallel to the coordinate planes (i.e., with unit normals $\mathbf{i}$, $\mathbf{j}$ and $\mathbf{k}$). What are the components of the normal and shear stresses in each case?

2. Consider the state of stress in a material given by the Cauchy Stress

$$\sigma = \begin{bmatrix} x(y+z) - x^2 & \frac{1}{2} - y & z^2 \\ \frac{1}{2} - y^2 x & 1 - yz & 0 \\ 1 + z & -2 & -1 + x + z \end{bmatrix}. \tag{6.86}$$

   Compute the traction and the normal and shear stresses at the following points and directions:

   (a) $\mathbf{n} = \langle 1, 0, 0 \rangle$ and $P = (0, 0, 0)$.

   (b) $\mathbf{n} = \langle 1, 1, 1 \rangle$ and $P = (1, -2, 1)$.

3. Consider the stress matrix of the previous problem and the position $\mathbf{x} = \langle 1, -1, 1 \rangle$. Let $\mathbf{n} = \langle \cos\theta, \sin\theta, 0 \rangle$, with $\theta \in (0, 2\pi]$, be a unit normal vector to a one-parameter family of surfaces passing through $\mathbf{x}$. Compute the normal stress as a function of θ and determine at which θ this quantity achieves its maximum and minimum.

4. Consider the Cauchy stress of a fluid given by

$$\sigma = \begin{bmatrix} \rho g x & 0 & 0 \\ 0 & \rho g y & 0 \\ 0 & 0 & \rho g z \end{bmatrix}, \tag{6.87}$$

where ρ is the density of the fluid and g is the acceleration of gravity. Consider a cube of sides 2 units, centered at the origin, located in this medium. Assume the faces of the cube are parallel to the coordinate planes. Determine the force exerted by the fluid on the side S parametrized by

$$S = \{(x, y, z) | -1 \leq x \leq 1, \ -1 \leq y \leq 1, \ z = -1\}, \tag{6.88}$$

(Hint: Compute $\int\int_S \mathbf{t} \cdot d\mathbf{A}$, where $\mathbf{t}$ is the traction on the surface, whose unit outward normal is $\mathbf{n} = \langle 0, 0, -1 \rangle$)

6.8 Euler and Navier–Stokes Equations

The equations governing the hydrodynamic flow of fluids are derived from the conservations of mass (see (6.26) and (6.28)) and linear momentum. We have already introduced equation (6.26)

$$\rho_t + \text{div } (\rho \mathbf{v}) = 0,$$

and its counterpart

$$\text{div } \mathbf{v} = 0,$$

when the fluid is incompressible. We now derive the equation that results from the conservation of linear momentum. This conservation law states that the rate of change of linear momentum is balanced by the resultant of all forces acting on the fluid, both internal and external. Stating this in mathematical terms (as before, considering a parcel of fluid Ω that occupies the region $\mathbf{p}(t, \Omega)$ at time t)

$$\frac{d}{dt}\left(\int_{\mathbf{p}(t,\Omega)} \rho \mathbf{v} \, dv\right) = \mathbf{G}, \tag{6.89}$$

where $\mathbf{G}$ stands for the resultant forces, and dv denotes volume integration, short for $dx\, dy\, dz$. From Theorem 6.2.4 we have

$$\frac{d}{dt}\left(\int_{\mathbf{p}(t,\Omega)} \rho \mathbf{v} \, dv\right) = \int_{\mathbf{p}(t,\Omega)} \rho \frac{D\mathbf{v}}{Dt} \, dv. \tag{6.90}$$

As for the forces $\mathbf{G}$ acting on a parcel of fluid $\mathbf{p}(t, \Omega)$, they are of two types: the traction forces, which the parcel experiences through its boundary $\partial\mathbf{p}(t, \Omega)$, and body forces. The first force, due to traction, is given by the surface integral

$$\int_{\partial\mathbf{p}(t,\Omega)} \sigma \cdot d\mathbf{A}, \tag{6.91}$$

where σ is the Cauchy stress of the fluid. Body forces, on the other hand, act on the entire parcel of fluid, and are determined from the triple or volume integral

$$\int_{\mathbf{p}(t,\Omega)} \rho\mathbf{F}\, dv. \tag{6.92}$$

The expression in (6.91) can be converted to a volume integral over $\mathbf{p}(t, \Omega)$, by applying the Divergence Theorem (see Theorem 3.6.2), as follows

$$\int_{\partial\mathbf{p}(t,\Omega)} \sigma \cdot d\mathbf{A} = \int_{\mathbf{p}(t,\Omega)} \text{div}\ \sigma\, dv, \tag{6.93}$$

where by div σ, the divergence of a matrix, we mean a vector whose i-th component is the divergence of the i-th row of σ, that is,

$$(\text{div}\ \sigma)_i = \sum_{j=1}^{3} \frac{\partial \sigma_{ij}}{\partial x_j}, \tag{6.94}$$

where x_j is the j-th component of the position $\mathbf{x}$ of a fluid particle. We now combine (6.90), (6.92) and (6.93) to arrive at

$$\int_{\mathbf{p}(t,\Omega)} (\rho\frac{Dv_i}{Dt} - \sum_{j=1}^{3} \frac{\partial \sigma_{ij}}{\partial x_j} - \rho F_i)\, dv = 0, \tag{6.95}$$

for $i = 1, 2$ and 3, and any arbitrary parcel Ω. This discussion results in the following system of PDEs since the parcel Ω is arbitrary and, therefore, the integral in (6.95) vanishes if and only if its integrand vanishes

$$\rho\frac{Dv_i}{Dt} - \sum_{j=1}^{3} \frac{\partial \sigma_{ij}}{\partial x_j} - \rho F_i = 0, \quad \text{for} \quad i = 1, 2, 3, \tag{6.96}$$

where v_i's are the components of $\mathbf{v} = \langle u, v, w\rangle$.

The system of PDEs in (6.96) takes a special form when the material is an ideal isotropic incompressible fluid, i.e., when the velocity field $\mathbf{v}$ and the Cauchy stress σ are related by

$$\sigma = -pI + \mu D.$$

It is not difficult to show that div $(pI) = \nabla p$ and $\text{div}(D) = \Delta \mathbf{v}$, where by $\Delta \mathbf{v}$ we mean $\langle \Delta u, \Delta v, \Delta w \rangle$. With these observations, the PDEs in (6.96) reduce to

$$\rho \left(\frac{\partial \mathbf{v}}{\partial t} + \mathbf{v} \cdot \nabla \mathbf{v} \right) = -\nabla p + \mu \Delta \mathbf{v} + \rho \mathbf{F}. \tag{6.97}$$

Typically the body force $\mathbf{F}$ is the weight of the fluid.

The system of equations (6.97) is called the *Navier–Stokes Equations* equations of the fluid flow. In applications where one studies fluid motions in a revolving body such as the earth, equations (6.97) must also take into account the contribution of the Coriolis and the centripetal forces, which we will develop shortly.

For the special class of fluids for which the kinematic viscosity μ vanishes (or $\Delta \mathbf{v} = \mathbf{0}$), the Navier–Stokes equations reduce to the *Euler* system of PDEs:

$$\rho \left(\frac{\partial \mathbf{v}}{\partial t} + \mathbf{v} \cdot \nabla \mathbf{v} \right) = -\nabla p + \rho \mathbf{F}. \tag{6.98}$$

We summarize the above discussion in the following theorem, which is stated in the context of incompressible flows:

Theorem 6.8.1 (The Euler and Navier–Stokes PDEs for Incompressible Fluids)
The Euler system of PDEs governs motions of incompressible inviscid fluids. These PDEs are

$$\rho \frac{D\mathbf{v}}{Dt} = -\nabla p + \rho F, \quad \text{div}\, \mathbf{v} = 0. \tag{6.99}$$

The Navier–Stokes system of PDEs governs motion of incompressible and viscous fluids. These PDEs are

$$\rho \frac{D\mathbf{v}}{Dt} = -\nabla p + \mu \Delta \mathbf{v} + \rho \mathbf{F}. \tag{6.100}$$

We end this section by pointing out an important connection between the Euler and the Navier-Stokes equations. The proof of this result is left to the reader.

Theorem 6.8.2
Let $\mathbf{v}$ *be the velocity field of an incompressible and viscous material.*

Suppose that $\mathbf{v}$ *is irrotational. Then the equations of motion of* $\mathbf{v}$ *reduce to the Euler Equations*

$$\rho \frac{D\mathbf{v}}{Dt} = -\nabla p + \rho F, \quad \operatorname{div} \mathbf{v} = 0. \tag{6.101}$$

Problems 6.8

1. Show that div $(pI) = \nabla p$ and $\operatorname{div}(D) = \Delta \mathbf{v}$, where $\Delta \mathbf{v} = \langle \Delta u, \Delta v, \Delta w \rangle$.

2. Prove Theorem 6.8.2.

3. Show that the stream function ψ of the flow past the cylinder (with $\mathbf{F} = \mathbf{0}$), which is given by

$$\psi(x, y) = y - \frac{y}{x^2 + y^2},$$

   satisfies $\Delta \mathbf{v} = \mathbf{0}$, and is irrotational. Assuming that the fluid is of constant density, find the pressure p so that the Euler system of PDEs in (6.100) is satisfied.

4. (*Linear Flows*) Consider the flow of an incompressible viscous fluid whose velocity $\mathbf{v}$ satisfies

$$\rho \frac{\partial \mathbf{v}}{\partial t} = -\nabla p + \mu \Delta \mathbf{v}, \qquad \operatorname{div} \mathbf{v} = 0, \tag{6.102}$$

   where ρ is constant. Show that

   (a) p must satisfy $\Delta p = 0$.

   (b) $\omega = \nabla \times \mathbf{v}$ satisfies the PDE

$$\frac{\partial \omega}{\partial t} = \nu \Delta \omega, \tag{6.103}$$

   where $\nu = \frac{\mu}{\rho}$.

6.9 Bernoulli's Equation and Irrotational Flows

In a typical problem the system of PDEs in (6.97) must be augmented by initial conditions, i.e., the state of fluid at time zero, and by boundary conditions, which describe how fluid particles that are located on the boundary interact with the boundary—if the boundary is a solid surface, one often prescribes the *no-slip boundary condition* $\mathbf{v} = \mathbf{0}$, so that fluid particles that are located on the surface remain stationary for all time. Another popular boundary condition, often employed when the model is inviscid (that is, $\mu = 0$ in (6.97)) is to assume that the normal component of the velocity vanishes at the boundary (the so-called slip boundary condition)

$$\mathbf{v} \cdot \mathbf{n} = 0,$$

where $\mathbf{n}$ is a unit outward normal to the boundary. No matter which boundary condition one selects, Navier–Stokes equations are very difficult to solve. Very few exact solutions of this system are available, although the ones we do know of are quite important in that they provide valuable information about the nature of solutions of this system.

The challenges we face when attempting to find solutions to (6.97) stem from two sources: the nonlinear dependence of the acceleration on the velocity, which is due to the presence of $\mathbf{v} \cdot \nabla \mathbf{v}$ in the formula in (6.61), and the complications that arise from the flow domain being geometrically complex. The complication that arises from the nonlinearity of $\mathbf{a}$ prohibits us from building additional solutions of the Navier–Stokes equations once we have two distinct solutions in hand, that is, the *principle of linear superposition*, which is at the heart of obtaining solutions in the case of linear ODEs and PDEs, is not available here. A complex geometry, on the other hand, prohibits us from using well-known analytic functions of mathematics, most notably trigonometric functions and Fourier series, as building blocks for constructing solutions. As a result, our strategy in obtaining exact solutions is to look at cases where the flow and/or the domain are relatively simple. In this section we look at one such special case, the case of irrotational flows.

Consider an irrotational flow, where by definition the velocity field $\mathbf{v}$ satisfies the constraint

$$\nabla \times \mathbf{v} = \mathbf{0}. \tag{6.104}$$

This relation, which is equivalent to stating $\frac{\partial v_i}{\partial x_j} = \frac{\partial v_j}{\partial x_i}$ for all i and j, helps us with rewriting the $\mathbf{v} \cdot \nabla \mathbf{v}$ as

$$\sum_{j=1}^{3} \frac{\partial v_i}{\partial x_j} v_j = \sum_{j=1}^{3} \frac{\partial v_j}{\partial x_i} v_j = \frac{1}{2} \sum_{i=1}^{3} \frac{\partial (v_j^2)}{\partial x_j} = \frac{1}{2} \frac{\partial}{\partial x_i} ||\mathbf{v}||^2. \tag{6.105}$$

Putting it slightly differently, the nonlinear term $(\mathbf{v} \cdot \nabla)\mathbf{v}$ is the gradient of the square of the fluid speed:

$$(\mathbf{v} \cdot \nabla)\mathbf{v} = \frac{1}{2} \nabla ||\mathbf{v}||^2. \tag{6.106}$$

Yet another way of stating this fact is to note that $\frac{1}{2}||\mathbf{v}||^2$ serves as a potential for $(\mathbf{v} \cdot \nabla)\mathbf{v}$. See also Problem 1 at the end of this section.

A second consequence of the irrotationality of $\mathbf{v}$ is that this velocity field must have a potential, that is, there must exist a function ϕ whose gradient is the velocity field:

$$\mathbf{v} = \nabla\phi. \tag{6.107}$$

If we further assume that the fluid is homogeneous and the density ρ remains constant, and that the body force $\mathbf{F}$ itself has a potential f, then the equation in (6.97) takes the form

$$\nabla\left(\frac{\partial\phi}{\partial t} + \frac{1}{2}||\mathbf{v}||^2 + f + \frac{p}{\rho}\right) = 0. \tag{6.108}$$

or equivalently

$$\frac{\partial\phi}{\partial t} + \frac{1}{2}||\mathbf{v}||^2 + f + \frac{p}{\rho} = \text{const.} \tag{6.109}$$

The above expression is called the *Bernoulli* equation for incompressible fluids. When the material is compressible, this expression takes the form

$$\frac{\partial\phi}{\partial t} + \frac{1}{2}||\mathbf{v}||^2 + f + \int\frac{dp}{\rho} = \text{const.} \tag{6.110}$$

Bernoulli's equation has several consequences. For one thing it allows us to compute the pressure function p, say, if we had prior knowledge of the velocity field, and vice versa. Moreover, when the flow has reached a steady-state, so that $\frac{\partial\phi}{\partial t} = 0$, the conserved quantity on the right side of (6.109) reduces to

$$\frac{1}{2}||\mathbf{v}||^2 + f + \int\frac{dp}{\rho}$$

which gives a precise interpretation of how the increase in the speed of a fluid particle must be compensated by an equivalent decrease in the pressure field p. This interpretation is closely related to the concept of "lift" associated with the flow past an airfoil—fluid particles that travel below the wing, as opposed to those that travel above it, experience different pressure fields. The net difference in the pressure values translates into the familiar lift of the airfoil.

Problems 6.9

1. Let $\mathbf{v}$ be an irrotational vector field. Show that $\nabla \times ((\mathbf{v} \cdot \nabla)\mathbf{v}) = \mathbf{0}$, and hence deduce that the vector $(\mathbf{v} \cdot \nabla)\mathbf{v}$ must have a potential.

2. Complete the computations that lead to the compressible version of Bernoulli's equation (6.110).

3. Consider an irrotational flow of an incompressible fluid which has reached its steady-state. Suppose that the body forces are negligible. Show that the velocity and pressure must satisfy

$$\frac{\rho}{2}||\mathbf{v}||^2 + p = \text{const.} \tag{6.111}$$

Use this result and compute the pressure when $\mathbf{v}$ is the velocity field of the flow past the cylinder

---

6.10 Acceleration in Spherical Coordinates

Because of the shape of our planet, the natural setting for studying flows in Geophysical Fluid Dynamics, which will be introduced in the next chapter, is spherical coordinates. Here we begin with the description of these coordinates and develop a basis in terms of unit vectors in the directions of the coordinate curves. We then determine the components of a typical vector, such as $\mathbf{\Omega}$, the vector that defines the axis and magnitude of the Earth's rotation, in this basis. Finally we write down the representation of typical velocity and acceleration fields in spherical coordinates.

6.10.1 Coordinate Curves

Let P be a point having coordinates (x, y, z) in rectangular coordinates and (r, θ, ϕ) in spherical coordinates. Here r is the distance from the origin to P, θ measures the longitude and ranges between 0 and 2π, and ϕ is the latitude, ranging between $-\frac{\pi}{2}$ and $\frac{\pi}{2}$. Note that this definition is different from the traditional definition of spherical coordinates in most mathematical texts where θ stands for the co-latitude angle. Our definition is consistent with how these coordinates are introduced in most oceanography texts because of the natural significance of latitude in this discipline.

The rectangular and spherical descriptions of P are related through the following relations:

$$x = r\cos\theta\cos\phi, \quad y = r\sin\theta\cos\phi, \quad z = r\sin\phi. \tag{6.112}$$

These relations are readily reversed to write r, θ and ϕ in terms of their rectangular counterparts:

$$r = \sqrt{x^2 + y^2 + z^2}, \quad \theta = \text{Arctan}\frac{y}{x}, \quad \phi = \text{Arcsin}\frac{z}{\sqrt{x^2 + y^2 + z^2}}. \tag{6.113}$$

In any coordinate system the expression *coordinate curve* is a curve along which only one of the three coordinate parameters varies while the other two are kept constant. For example, the x-axis is a coordinate curve in rectangular coordinates along which the coordinate x varies while y and z remain constant. Because of the special importance of the three axes in rectangular coordinates, we are interested in identifying the corresponding coordinates curves in a spherical coordinates.

To that end, let P have spherical coordinates (r_0, θ_0, ϕ_0). By keeping r and θ fixed at r_0 and θ_0, respectively, while allowing ϕ to take on all values between $-\frac{\pi}{2}$ and $\frac{\pi}{2}$, we obtain a coordinate curve, a semi-circle in this case, which is part of a great circle (a *meridian* circle) that passes through P and the two poles. We will refer to this curve as the ϕ-curve through P. Similarly, a θ-curve is obtained by fixing $r = r_0$ and $\phi = \phi_0$, while allowing θ to take on all values between 0 and 2π, which defines the familiar *parallel* circle through P. Finally, fixing $\theta = \theta_0$ and $\phi = \phi_0$, while allowing r take on all nonnegative values, we construct an r-curve, a straight line which passes through the origin and P and defines the *radial* direction at P. These three coordinate curves play a role similar to the role that the x-, y- and z-axes play in rectangular coordinates. Figure 6.10 shows the three coordinate curves associated with the point P with rectangular coordinates $(-2, 1, 3)$. This figure is obtained in MATLAB as follows:

```
clf;
clear all;
lon=-0:(2*pi)/100:2*pi;
lat=-pi/2:pi/100:pi/2;
x=-2; y=1; z=3;
r0=sqrt(x^2+y^2+z^2);
theta0=atan(y/x);
phi0=asin(z/r0);
% Meridian
plot3(r0*cos(theta0)*cos(lat),r0*sin(theta0)*cos(lat),...
    r0*sin(lat),'b');
hold on
% Parallel
plot3(r0*cos(phi0)*cos(lon),r0*cos(phi0)*sin(lon),...
    r0*sin(phi0)*ones(length(lon),1)', 'r');
hold on
```

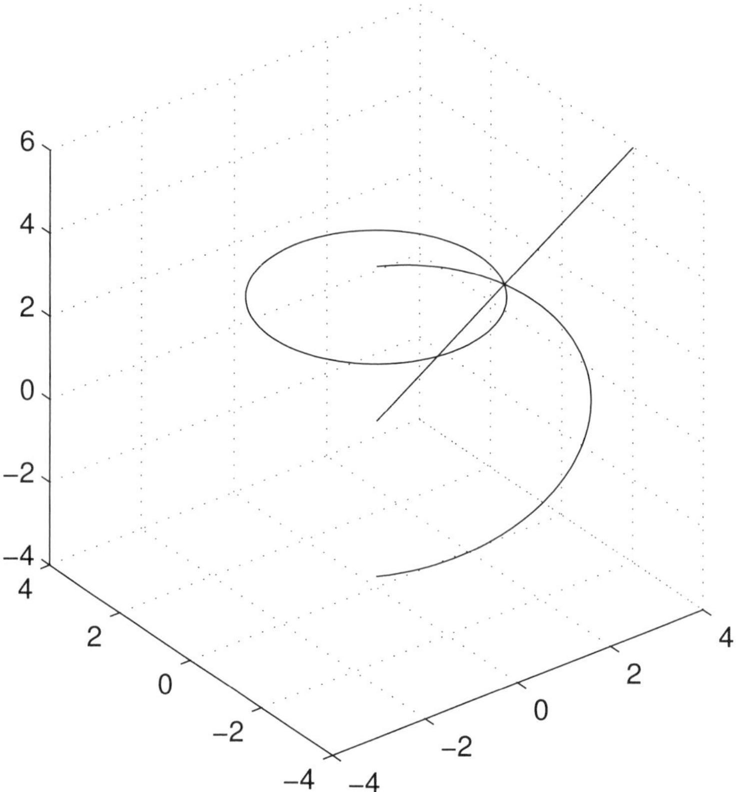

FIGURE 6.10: The spherical coordinate curves passing through $P = (-2, 1, 3)$.

```
% Radial
r=0:0.01:2*r0;
plot3(cos(phi0)*cos(theta0)*r,cos(phi0)*sin(theta0)*r,...
    sin(phi0)*r,'g')
grid on
axis square
```

6.10.2 Spherical Basis

Given a specific point P on a sphere, we now determine three vectors, denoted by $\mathbf{e}_\theta(P)$, $\mathbf{e}_\phi(P)$ and $\mathbf{e}_r(P)$, which play a similar role to $\mathbf{i}$, $\mathbf{j}$ and $\mathbf{k}$ of rectangular coordinates in that they will be mutually orthogonal and have magnitude one. By definition, $\mathbf{e}_\theta$ is a unit tangent vector to the θ-

curve through P, while $\mathbf{e}_\phi$ is a unit tangent vector to the corresponding ϕ-curve, and $\mathbf{e}_r$ is a unit tangent vector to the r-curve.

We begin by determining $\mathbf{e}_\theta$ by first parametrizing the θ-curve through the point P. Let P have coordinates (r_0, θ_0, ϕ_0) in spherical coordinates. Then the θ-curve through P has the parametrization

$$\mathbf{r}(\theta) = \langle r_0 \cos\theta \cos\phi_0, r_0 \sin\theta \cos\phi_0, r_0 \sin\phi_0 \rangle.$$

Since $\mathbf{e}_\theta$ is a *unit tangent* vector to the θ-curve, we find $\mathbf{e}_\theta$ by differentiating the above expression with respect to θ, followed by dividing the result by its magnitude, to get

$$\mathbf{e}_\theta(P) = -\sin\theta_0 \mathbf{i} + \cos\theta_0 \mathbf{j}. \tag{6.114}$$

As expected $\mathbf{e}_\theta$ does not have a component in the north-south direction. A similar consideration leads to the following formulas for $\mathbf{e}_\phi$ and $\mathbf{e}_r$:

$$\mathbf{e}_\phi(P) = -\cos\theta_0 \sin\phi_0 \mathbf{i} - \sin\theta_0 \sin\phi_0 \mathbf{j} + \cos\phi_0 \mathbf{k}. \tag{6.115}$$

and

$$\mathbf{e}_r(P) = \cos\theta_0 \cos\phi_0 \mathbf{i} + \sin\theta_0 \cos\phi_0 \mathbf{j} + \sin\phi_0 \mathbf{k}. \tag{6.116}$$

Note that $\mathbf{e}_\phi$, unlike $\mathbf{e}_\theta$, depends on the longitude and the latitude of P. Also, as expected, $\mathbf{e}_r$ is in the radial direction and is therefore perpendicular to the sphere of radius r_0. Moreover, these vectors are mutually orthogonal, that is

$$\mathbf{e}_\theta \cdot \mathbf{e}_\phi = \mathbf{e}_\theta \cdot \mathbf{e}_r = \mathbf{e}_\phi \cdot \mathbf{e}_r = 0. \tag{6.117}$$

The expressions in (6.114), (6.115) and (6.116) show the relationship between $\{\mathbf{e}_\theta, \mathbf{e}_\phi, \mathbf{e}_r\}$ and $\{\mathbf{i}, \mathbf{j}, \mathbf{k}\}$. These relations are easily inverted to give us

$$\begin{cases} \mathbf{i} &= -\sin\theta_0\, \mathbf{e}_\theta - \cos\theta_0 \sin\phi_0\, \mathbf{e}_\phi + \cos\theta_0 \cos\phi_0\, \mathbf{e}_r, \\ \mathbf{j} &= \cos\theta_0\, \mathbf{e}_\theta - \sin\theta_0\, \sin\phi_0\, \mathbf{e}_\phi + \sin\theta_0\, \cos\phi_0\, \mathbf{e}_r, \\ \mathbf{k} &= \cos\phi_0\, \mathbf{e}_\phi + \sin\phi_0\, \mathbf{e}_r. \end{cases} \tag{6.118}$$

What we have accomplished so far is to introduce the concept of *spherical basis vectors* $\mathbf{e}_\theta$, $\mathbf{e}_\phi$ and $\mathbf{e}_r$. The significance of this set of mutually orthogonal unit vectors is that any vector $\mathbf{v}$ can be represented in terms of these three vectors as

$$\mathbf{v} = u\mathbf{e}_\theta + v\mathbf{e}_\phi + w\mathbf{e}_r. \tag{6.119}$$

The coefficients u, v and w are the coordinates of $\mathbf{v}$ in spherical coordinates. The same vector $\mathbf{v}$ has a similar representation in terms of the rectangular basis vectors $\{\mathbf{i}, \mathbf{j}, \mathbf{k}\}$: There are scalars v_1, v_2 and v_3 such that

$$\mathbf{v} = v_1\mathbf{i} + v_2\mathbf{j} + v_3\mathbf{k}. \tag{6.120}$$

However, in oceanography and meteorology it is the spherical representation (6.119) that is most natural when one studies ocean currents or pressure fronts, especially when the study involves large-scale structures.

Because the spherical and rectangular bases involve mutually orthogonal vectors, it is an easy task to write the coordinates of a vector $\mathbf{v}$ expressed in one basis in terms of its coordinates in another. Since the spherical basis vectors are orthonormal, we have

$$u = \mathbf{v} \cdot \mathbf{e}_\theta, \quad v = \mathbf{v} \cdot \mathbf{e}_\phi, \quad w = \mathbf{v} \cdot \mathbf{e}_r. \tag{6.121}$$

One can now deduce the relations among u, v, w in (6.119) and v_1, v_2 and v_3 in (6.120) by using (6.114), (6.115) and (6.116).

We summarize the above discussion in the following theorem.

Theorem 6.10.1 (Basis in Spherical Coordinates)
Given an arbitrary point P with coordinates (x, y, z) in rectangular coordinates and (r, θ, ϕ) in spherical coordinates, the vectors $\mathbf{e}_\theta$, $\mathbf{e}_\phi$ and $\mathbf{e}_r$ defined by

$$\begin{aligned} \mathbf{e}_\theta(P) &= -\sin\theta\mathbf{i} + \cos\theta\mathbf{j}, \\ \mathbf{e}_\phi(P) &= -\cos\theta\sin\phi\mathbf{i} - \sin\theta\sin\phi\mathbf{j} + \cos\phi\mathbf{k}, \\ \mathbf{e}_r(P) &= \cos\theta\cos\phi\mathbf{i} + \sin\theta\cos\phi\mathbf{j} + \sin\phi\mathbf{k}, \end{aligned} \tag{6.122}$$

form an orthonormal basis constructed at P. Given any vector $\mathbf{v}$, it can be represented as

$$\mathbf{v} = u\mathbf{e}_\theta + v\mathbf{e}_\phi + w\mathbf{e}_r. \tag{6.123}$$

The component u is the contribution of $\mathbf{v}$ in the east-west direction, v its contribution in the north-south direction, and w is its radial component. Conversely, the standard rectangular basis $\{\mathbf{i}, \mathbf{j}, \mathbf{k}\}$ is related to the spherical basis by the relations in (6.118).

6.10.3 The Eulerian Formulation of Velocity and Acceleration Revisited

Consider a particle of fluid P and its trajectory C, consisting of a curve in the three-dimensional space R^3. Let us assume that the position of P at any time t can be specified by a set of differentiable functions $x(t)$, $y(t)$ and $z(t)$ so that

$$\mathbf{r}(t) = x(t)\mathbf{i} + y(t)\mathbf{j} + z(t)\mathbf{k} \tag{6.124}$$

defines the position vector $\mathbf{r}$, or equivalently, the parametrization of the curve C. The velocity $\mathbf{v}$ of P is then determined by direct differentiation of $\mathbf{r}$:

$$\mathbf{v}(t) = x'(t)\mathbf{i} + y'(t)\mathbf{j} + z'(t)\mathbf{k},$$

where prime denotes differentiation with respect to time t. The above expression defines the Lagrangian representation of the velocity, which was introduced earlier. The above components of velocity (i.e., x', y' and z') are converted to functions of position and time, so typically

$$\mathbf{v} = v_1\mathbf{i} + v_2\mathbf{j} + v_3\mathbf{k}, \tag{6.125}$$

where each component v_i is a function of position and time

$$v_i = v_i(x, y, z, t),$$

so that the expression is (6.125) is the familiar Eulerian formulation of the velocity field.

As we saw in Section 6.5, the Eulerian representation of velocity implies that the acceleration $\mathbf{a}$ is determined from (6.125) by the formula

$$\mathbf{a} = \frac{\partial \mathbf{v}}{\partial t} + (\mathbf{v} \cdot \nabla)\mathbf{v}. \tag{6.126}$$

In rectangular coordinates the components of $\mathbf{a}$ are

$$a_i = \frac{\partial v_i}{\partial t} + \sum_{j=1}^{3} v_j \frac{\partial v_i}{\partial x_j}, \quad i = 1, \ldots, 3. \tag{6.127}$$

Here we are adopting the convention that $x_1 = x$, $x_2 = y$ and $x_3 = z$. Recall that the operator

$$\frac{D}{Dt} = \frac{\partial}{\partial t} + \sum_{j=1}^{3} v_j \frac{\partial}{\partial x_j} \tag{6.128}$$

is the *total* or the *material* derivative and that (6.128) can be recast as

$$a_i = \frac{Dv_i}{Dt}. \tag{6.129}$$

6.10.4 Velocity in Spherical Basis

In order to write down an expression for acceleration in spherical coordinates, we first need to write the expression (6.128) in spherical coordinates. Since the positions $\mathbf{p}(t)$, occupied by P, have coordinates

$(x(t), y(t), z(t))$ in rectangular coordinates, and $(\theta(t), \phi(t), r(t))$ in spherical coordinates, and since we know the relations among the rectangular and spherical coordinates, we can rewrite the position vector $\mathbf{r}(t)$ in (6.124) as

$$\mathbf{r} = r(t)\cos\theta(t)\cos\phi(t)\,\mathbf{i} + r(t)\sin\theta(t)\cos\phi(t)\,\mathbf{j} + r(t)\sin\phi(t)\,\mathbf{k} \tag{6.130}$$

in spherical coordinates. Differentiating (6.130) with respect to t yields

$$\begin{aligned}\mathbf{v} = {} & (r'\cos\theta\cos\phi - r\theta'\sin\theta\cos\phi - r\phi'\cos\theta\sin\phi)\,\mathbf{i} + \\ & (r'\sin\theta\cos\phi + r\theta'\cos\theta\cos\phi - r\phi'\sin\theta\sin\phi)\,\mathbf{j} + \\ & (r'\sin\phi + r\phi'\cos\phi)\,\mathbf{k}.\end{aligned} \tag{6.131}$$

Using the formulas (6.114), (6.115) and (6.116), it is easy to see that (6.131) is equivalent to

$$\mathbf{v} = r\theta'\cos\phi\,\mathbf{e}_\theta + r\phi'\mathbf{e}_\phi + r'\mathbf{e}_r. \tag{6.132}$$

The coefficients of $\mathbf{e}_\theta$, $\mathbf{e}_\phi$ and $\mathbf{e}_r$ in the above expressions are the components of velocity in spherical coordinates. We denote them by v_θ, v_ϕ and v_r respectively, i.e.,

$$\mathbf{v} = v_\theta\mathbf{e}_\theta + v_\phi\mathbf{e}_\phi + v_r\mathbf{e}_r \tag{6.133}$$

where

$$v_\theta = r\theta'\cos\phi, \quad v_\phi = r\phi', \quad v_r = r'. \tag{6.134}$$

We note that v_θ is the component of the velocity in the east-west direction, v_ϕ is the component in the north-south direction, and v_r is the component in the radial direction. As is common in most oceanography texts, v_θ is denoted by u, v_ϕ by v and v_r by w.

We summarize the above discussion in the following theorem.

Theorem 6.10.2 (Velocity Fields in Spherical Coordinates) *When the velocity vector* $\mathbf{v}$ *is represented in spherical coordinates as in (6.133), its components* v_θ, v_ϕ *and* v_r *(or equivalently* u, v, *and* w*) are related to* $\theta(t)$, $\phi(t)$ *and* $r(t)$ *through the relations (6.134). In particular, particle trajectories can be obtained from the system of differential equations*

$$\frac{d\theta}{dt} = \frac{u}{r\cos\phi}, \quad \frac{d\phi}{dt} = \frac{v}{r}, \quad \frac{dr}{dt} = w. \tag{6.135}$$

6.10.5 Dynamics of Basis Vectors

To compute the acceleration **a** in spherical coordinates we need to differentiate $\mathbf{v} = u\mathbf{e}_\theta + v\mathbf{e}_\phi + w\mathbf{e}_r$ with respect to t. Unlike the rectangular basis $\{\mathbf{i}, \mathbf{j}, \mathbf{k}\}$, where each vector is independent of t, the spherical basis $\{\mathbf{e}_\theta, \mathbf{e}_\phi, \mathbf{e}_r\}$ varies with t because this basis depends on position, and the particle P, whose acceleration we seek, occupies different positions at different values of t. This time dependence will additionally contribute to the computation of acceleration.

Recall from (6.114) that $\mathbf{e}_\theta$ is related to the standard rectangular basis through the relation $\mathbf{e}_\theta = -\sin\theta\,\mathbf{i} + \cos\theta\,\mathbf{j}$. Differentiating this relation with respect to t yields

$$\frac{d\mathbf{e}_\theta}{dt} = -\theta'\cos\theta\,\mathbf{i} - \theta'\sin\theta\,\mathbf{j}.$$

But from (6.135) we have $\theta' = \frac{u}{r\cos\phi}$ so the above expression takes the form

$$\frac{d\mathbf{e}_\theta}{dt} = \frac{u}{r\cos\phi}(-\cos\theta\,\mathbf{i} - \sin\theta\,\mathbf{j}). \tag{6.136}$$

Recall that we derived the relationship between the standard and spherical bases in (6.118). In particular, the latter expressions relate the vectors **i** and **j** to their spherical counterparts, which we use to replace **i** and **j** in (6.136):

$$\frac{d\mathbf{e}_\theta}{dt} = \frac{u}{r\cos\phi}(\sin\phi\,\mathbf{e}_\phi - \cos\phi\,\mathbf{e}_r). \tag{6.137}$$

Similarly, we derive the following expressions for $\mathbf{e}_\phi$ and $\mathbf{e}_r$:

$$\frac{d\mathbf{e}_\phi}{dt} = -\frac{u\tan\phi}{r}\mathbf{e}_\theta - \frac{v}{r}\mathbf{e}_r, \qquad \frac{d\mathbf{e}_r}{dt} = \frac{u}{r}\mathbf{e}_\theta + \frac{v}{r}\mathbf{e}_\phi. \tag{6.138}$$

6.10.6 Formula for Acceleration in Spherical Coordinates

Returning to (6.133), we differentiate this relation with respect to t to get

$$\mathbf{a} = \frac{du}{dt}\mathbf{e}_\theta + v\frac{d\mathbf{e}_\theta}{dt} + \frac{dv}{dt}\mathbf{e}_\phi + v\frac{d\mathbf{e}_\phi}{dt} + \frac{dw}{dt}\mathbf{e}_r + w\frac{d\mathbf{e}_r}{dt}. \tag{6.139}$$

Next we substitute (6.137), (6.138) into (6.139) to get

$$\mathbf{a} = (\frac{du}{dt} - \frac{uv}{r}\tan\phi + \frac{uw}{r})\mathbf{e}_\theta + (\frac{dv}{dt} + \frac{u^2}{r}\tan\phi + \frac{vw}{r})\mathbf{e}_\phi + (\frac{dw}{dt} - \frac{u^2+v^2}{r})\mathbf{e}_r. \tag{6.140}$$

Equation (6.140) determines the acceleration when the velocity is given in spherical coordinates. We summarize this finding in the following theorem.

Theorem 6.10.3 (Acceleration in Spherical Coordinates) *The expression in (6.140) defines the acceleration* **a** *when expressed in spherical coordinates.*

Problems 6.10

1. Consider the point P whose coordinates are $(1, 2, 3)$ in Cartesian coordinates.

   (a) Find the spherical coordinates of P.

   (b) Plot the three spherical coordinate curves that pass through P.

2. Verify (6.117), that the spherical basis vectors are mutually orthogonal.

3. Verify the following relations:

$$\mathbf{e}_\theta \times \mathbf{e}_\phi = \mathbf{e}_r, \quad \mathbf{e}_\phi \times \mathbf{e}_r = \mathbf{e}_\theta, \quad \mathbf{e}_r \times \mathbf{e}_\theta = \mathbf{e}_\phi. \tag{6.141}$$

4. Derive (6.118). Hint: Start with (6.115) and (6.116) and eliminate $\mathbf{k}$ between them. Then consider the resulting equation with (6.114) and solve for $\mathbf{i}$ and $\mathbf{j}$.

5. Use the orthogonality properties of the spherical basis vectors to show that v_1, v_2 and v_3 in (6.119) are given by

$$v_1 = \mathbf{v} \cdot \mathbf{e}_\theta, \quad v_2 = \mathbf{v} \cdot \mathbf{e}_\phi, \quad v_3 = \mathbf{v} \cdot \mathbf{e}_r. \tag{6.142}$$

6. Show that $\mathbf{a}$ and $\frac{d\mathbf{a}}{dt}$ are orthogonal, where $\mathbf{a}$ is any of the three vectors in the spherical basis (6.122).

7. Use the identities in (6.141) to arrive an alternative derivation of the equations in (6.137), (6.138).

6.11 Project A: Inviscid Linear Fluid Motions and Surface Gravity Waves

The main goal of this project is to develop a linear two-dimensional model based on the Navier–Stokes equations which is capable of supporting surface gravity waves. For more discussion regarding surface gravity waves and the questions posed below see pages 343–348 of [1]. In addition, surface gravity waves are treated extensively in many oceanography texts, an excellent source being Adrian Gill's book, Reference[7]. See, in particular, pages 95 through 105 of this reference.

1. Consider an inviscid fluid (i.e., $\mu = 0$ in (6.97)), occupying an infinite region bounded by the planes $z = 0$ and $z = -h$, so that the domain is
$$D = \{(x, y, z) | -h \leq z \leq 0\}. \tag{6.143}$$
First, consider the *stationary flow* $\mathbf{v} = \mathbf{0}$. Let $\mathbf{F} = \begin{bmatrix} 0 & 0 & -g \end{bmatrix}^{\mathrm{T}}$ be the force acting on the fluid, i.e., the only external force acting on the body of the fluid is its weight. By looking at the first two equations in (6.97), show that the pressure p is independent of x and y.

2. Show that the third equation in (6.97) reduces to $\frac{\partial p}{\partial z} = -\rho g$, from which deduce the expression
$$p(z) = -\rho g z. \tag{6.144}$$
To get the above result choose the constant of integration so that the surface $z = 0$ is pressure free and corresponds to the free surface of the domain. Equation (6.144) defines the *hydrostatic pressure* in the equilibrium flow.

3. So far we have obtained the simplest solution of the Navier–Stokes equations, namely, one where every fluid particle is standing still and the only pressure a fluid particle feels is induced by the column of fluid resting above it (see (6.144)). Now consider a two-dimensional perturbation of this flow by perturbing the free surface $z = 0$ in the form
$$z = \epsilon\eta(x, y, t). \tag{6.145}$$
Let $\mathbf{v} = (\epsilon U, \epsilon V, \epsilon W)$ denote the perturbed velocity—note that when $\epsilon = 0$ we end up with the basic stationary solution $\mathbf{v} = \mathbf{0}$. The pressure field in the perturbed motion also deviates from the hydrostatic pressure expressed in (6.144). Let ϵP denote this deviation:
$$p = -\rho g z + \epsilon P. \tag{6.146}$$
Show that $\{U, V, W, P\}$ satisfy
$$\rho\frac{\partial U}{\partial t} = -\frac{\partial P}{\partial x} + \text{h.o.t.}, \quad \rho\frac{\partial V}{\partial t} = -\frac{\partial P}{\partial y} + \text{h.o.t.},$$

$$\rho \frac{\partial W}{\partial t} = -\frac{\partial P}{\partial z} - \rho g + \text{h.o.t.}, \tag{6.147}$$

where h.o.t., the "higher order terms," denotes terms that depend on ϵ or its higher powers.

4. Show that the conservation of mass equation, div $\mathbf{v} = 0$, takes the form

$$\frac{\partial U}{\partial x} + \frac{\partial V}{\partial y} + \frac{\partial W}{\partial z} = 0. \tag{6.148}$$

5. Show that on the free surface $z = \epsilon\eta(x, y, t)$ the following relation holds

$$W = \frac{\partial \eta}{\partial t} + \epsilon U \frac{\partial \eta}{\partial x} + \epsilon V \frac{\partial \eta}{\partial y}. \tag{6.149}$$

(Hint: Differentiate the expression $z(t) = \epsilon\eta(x(t), y(t), t)$ with respect to t.)

6. The main assumption we impose now is that we can neglect terms that depend on ϵ throughout equations (6.147)–(6.149). With this assumption invoked, the nonlinear terms in (6.147)–(6.149) drop out. Show that the resulting linearized equations are

$$\rho \frac{\partial U}{\partial t} = -\frac{\partial P}{\partial x}, \quad \rho \frac{\partial V}{\partial t} = -\frac{\partial P}{\partial y}, \quad \rho \frac{\partial W}{\partial t} = -\frac{\partial P}{\partial z} - \rho g, \tag{6.150}$$

together with the boundary conditions

$$W = \frac{\partial \eta}{\partial t}, \quad \text{when } z = 0. \tag{6.151}$$

7. Assume that the bottom boundary, $z = -h$, is impenetrable. Hence the vertical component of the velocity must vanish, that is,

$$W = 0, \quad \text{when } z = -h. \tag{6.152}$$

8. By manipulating the equations in (6.150) show that P satisfies the Laplace equation

$$\frac{\partial^2 P}{\partial x^2} + \frac{\partial^2 P}{\partial y^2} + \frac{\partial^2 P}{\partial z^2} = 0. \tag{6.153}$$

9. Apply the separation of variables

$$P(x, y, z, t) = X(x)Y(y)Z(z)T(t), \tag{6.154}$$

to (6.153) and arrive at

$$X'' + \mu_1^2 X = 0, \quad Y'' + \mu_2^2 Y = 0, \quad Z'' - \lambda^2 Z = 0, \tag{6.155}$$

where

$$\mu_1^2 + \mu_2^2 = \lambda^2. \tag{6.156}$$

10. Show that (6.155) have the following solutions

$$\begin{aligned} X(x) &= c_1 \sin(\mu_1 x) + c_2 \cos(\mu_1 x), \\ Y(y) &= c_3 \sin(\mu_2 y) + c_4 \cos(\mu_2 y), \\ Z(z) &= c_5 \sinh(\lambda z) + c_6 \cosh(\lambda z). \end{aligned} \tag{6.157}$$

11. Show that Z must satisfy the boundary condition

$$Z'(-h) = 0. \tag{6.158}$$

Use this information to conclude that P must be

$$P(x, y, z, t) = (c_1 \sin(\mu_1 x) + c_2 \cos(\mu_1 x))(c_3 \sin(\mu_2 y) +$$

$$c_4 \cos(\mu_2 y)) \cosh \lambda(z + h) T(t), \tag{6.159}$$

12. Returning to the free surface equation, show that

$$\eta(x, y, t) = \frac{\cosh(\lambda h)}{\rho g}(c_1 \sin(\mu_1 x) + c_2 \cos(\mu_1 x))(c_3 \sin(\mu_2 y) +$$

$$\cos(\mu_2 y))T(t). \tag{6.160}$$

13. Show that T must satisfy the relation

$$T'' + \lambda g \tanh(\lambda h) T = 0. \tag{6.161}$$

14. Conclude that the final solution to the perturbation problem is

$$P(x, y, z, t) = P_0 \cos(\mu_1 x + \mu_2 y - \omega t) \cosh \lambda(z + h), \tag{6.162}$$

$$\eta(x, y, z, t) = \frac{P_0}{\rho g} \cos(\mu_1 x + \mu_2 y - \omega t), \tag{6.163}$$

$$W(x, y, z, t) = \frac{P_0 \lambda}{\omega} \sin(\mu_1 x + \mu_2 y - \omega t) \sinh \lambda(z + h), \tag{6.164}$$

$$\lambda^2 = \mu_1^2 + \mu_2^2, \quad \text{and } \omega^2 = \lambda g \tanh(\lambda h). \tag{6.165}$$

The above function η is called a *Surface Gravity Wave.* The relation

$$\omega^2 = \lambda g \tanh(\lambda h)$$

is called the *dispersion relation* of these waves, providing a constraint between the wave numbers (μ_1, μ_2), the direction of propagation of waves, and ω, the frequency of the waves.

6.12 Project B: Internal Gravity Waves

We considered surface gravity waves in the previous project. These are waves that typically arise at the interface between two fluids with vastly different densities, such as at the air-sea interface. Here we would like to develop the basic mathematical equations that govern waves that are generated in regions where two fluids with relatively similar densities are present. This scenario often happens in the case of the ocean and the atmosphere, where the fluid is stratified and the density varies, albeit slowly, with depth or height. The waves generated in a stratified domain are usually referred to as *internal gravity waves.*

A simple mathematical setting where we can explore the effect of change in density is when we have multi-layer fluids, where the fluid density experiences (small) jumps across interfaces. In this project we consider the case of two such incompressible fluids whose densities differ slightly and study the behavior of the interface. See [1] and [7] for more background on thi topic, as well as, many other texts, including [9], which treat internal waves, not only in the linear setting described here, but also for considerably more complicated and realistic domains.

Consider the domain D containing two fluids with densities ρ_1 and ρ_2, with $\rho_2 > \rho_1$. Suppose D is the union of two regions D_1 and D_2 where

$$D_1 = \{\mathbf{x} \in R^3 | -h_1 \leq z < 0\}$$

and

$$D_2 = \{\mathbf{x} \in R^3 | -h_2 - h_1 \leq z < -h_1\}.$$

Here h_1 and h_2 are constant, so that the initial, equilibrium interfaces are flat. The surface defined by $H = -h_1 - h_2$ is a solid, flat boundary. The regions D_1 and D_2 contain fluids with densities ρ_1 and ρ_2, respectively. We assume that the two interfaces $z = 0$ and $z = -h_1$ are perturbed and are now represented by

$$z = \eta_1(x, y, t), \qquad \text{and} \qquad z = -h_1 + \eta_2(x, y, t).$$

Follow the strategy described in the previous project in what follows. In particular, assume that pressure remains hydrostatic through the water column, a reasonable assumption if η_1 and η_2 remain small.

1. Show that in the upper layer, where $-h_1 + \eta_2 \leq z < \eta_1$, that the perturbed velocity $\mathbf{V}_1 = \langle U_1, V_1, W_1 \rangle$ satisfies

$$\frac{\partial U_1}{\partial t} = -g\frac{\partial \eta_1}{\partial x}, \quad \frac{\partial V_1}{\partial t} = -g\frac{\partial \eta_1}{\partial y}, \quad \frac{\partial W_1}{\partial t} = -g\frac{\partial \eta_1}{\partial z} + \rho_1 g. \quad (6.166)$$

2. Show that $\mathbf{V}_1$ satisfies div $\mathbf{V}_1 = 0$. Integrate this equation in the interval $z \in (-h_1 + \eta_2, \eta_1)$ and arrive at

$$\frac{\partial^2 \eta_1}{\partial t^2} = gh_1 \Delta \eta_1 + \frac{\partial^2 \eta_2}{\partial t^2}. \tag{6.167}$$

   Note how the dynamics from the lower layer influences the dynamics of the upper layer through the "forcing" term $\frac{\partial^2 \eta_2}{\partial t^2}$.

3. Show that the pressure P_1 in the upper layer is

$$P_1 = \rho_1 g(\eta_1 + h_1 - \eta_2).$$

4. We now turn our focus to the lower layer. We assume that the pressure is continuous across the interface. Use this boundary condition to show that P_2 is

$$P_2 = \rho_1 g(\eta_1 + h_1 - \eta_2) + \rho_2 g(z - h_1 + \eta_2).$$

5. Let $\mathbf{V}_2 = \langle U_2, V_2, W_2 \rangle$ be the velocity vector field in the lower layer. As before, apply the conservation of mass equation to conclude that η_2 is related to U_2 and V_2 by

$$\frac{\partial \eta_2}{\partial t} + h_2 (\frac{\partial U_2}{\partial x} + \frac{\partial V_2}{\partial y}) = 0.$$

6. Next, begin with the equations for U_2 and V_2, eliminate P_2 from these equations, followed by differentiating the above equation for η_2 with respect to t to arrive at the following second-order PDE for η_2:

$$\frac{\partial^2 \eta_2}{\partial t^2} = \frac{\rho_1}{\rho_2} gh_2 \Delta \eta_1 + \frac{\rho_2 - \rho_1}{\rho_2} gh_2 \Delta \eta_2. \tag{6.168}$$

   The behavior of the solutions (η_1, η_2) of the system of PDEs in (6.167) and (6.168) determine what an internal wave is. Of particular interest is the term

$$\frac{\rho_2 - \rho_1}{\rho_2} g$$

   which is usually referred to as *reduced gravity* and denoted by g'. This value is usually small when ρ_1 and ρ_2 are close to each other, which could result in generation of relatively large amplitude waves relative to the surface gravity waves we observe at the air-sea interface.

7. Write a MATLAB program to solve the system of PDEs in (6.167) and (6.168). Begin with values for ρ_1, ρ_2, h_1, and h_2, as well as for initial conditions for η_1 and η_2, of your own choosing. Assume periodic boundary conditions in the x and y directions. Experiment with values of ρ_1 and ρ_2 and report on the relative sizes of η_1 and η_2. In particular, consider the special case when $\eta_1 = \mu\eta_2$ at time zero and explore the conditions under which the two perturbations remain proportional for all time.

6.13 Project C: Equation for Bubble Dynamics

The goal of this project is to apply Bernoulli's equation to derive a differential equation for the motion of a spherical bubble immersed in an incompressible fluid. See Reference [1], pages 335–336, for more discussion and some hints to the questions posed below.

1. Consider a compressible fluid, say air, occupying a region $D = \{\mathbf{x} | |\mathbf{x}| \leq R(t)\}$, embedded in an incompressible fluid, say water, outside of D having density ρ. Assume that the motions of the bubble and the fluid are spherically symmetric so that the velocity field of the fluid is in the form

$$\mathbf{v}(\mathbf{x}, t) = w(r, t)\mathbf{e}_r, \tag{6.169}$$

where $w(r, t)$ is unknown and $\mathbf{e}_r$ is the unit radial vector $\mathbf{e}_r = \mathbf{x}/||\mathbf{x}||$.

2. Recall the formula for the divergence operator in spherical coordinates (see the expression in (3.39), and also Problem 9 in Chapter 3). Start with the equation of conservation of mass, the formula for divergence in spherical coordinates, and show that $w(r, t)$ must be of the form

$$w(r, t) = \frac{T(t)}{r^2}, \tag{6.170}$$

where f is an arbitrary function of t.

3. Apply the formula in (6.140) to (6.169) to arrive the formula for the acceleration $\mathbf{a}$ as a function of w and T.

4. Assuming that R', the velocity of the gas-fluid interface, is equal

to $\mathbf{v}|_{r=R}$, the velocity of the fluid adjacent to it, show that the function f in (6.170) satisfies

$$T(t) = R^2 R'. \tag{6.171}$$

5. Show that the velocity field (6.169) is irrotational and that

$$\phi = -\frac{1}{r}T = -\frac{1}{r}R^2 R' \tag{6.172}$$

is a potential for this field.

6. Assume that the body forces are negligible so that the function f in the Bernoulli equation (6.110) is zero. Show that under this assumption equation (6.110) takes the form

$$\frac{1}{r}(R^2 R')' - \frac{1}{2r^4}R^4 R'^2 - p = \text{const.} \tag{6.173}$$

where the above constant could be a function time. The differential equation (6.173) can be solved for $R(t)$ once appropriate boundary conditions are specified.

7. Consider the case of a bubble immersed in a fluid of density ρ with constant pressure p_0 far away from the bubble. Use this boundary condition to determine the constant in (6.173):

$$\text{const} = \frac{p_0}{r}. \tag{6.174}$$

8. Substitute $r = R$ in (6.173) to get the ODE

$$-\frac{1}{R}(R^2 R')' + \frac{1}{2}R'^2 + \frac{p}{r} = \frac{p_0}{r}, \tag{6.175}$$

for R, the radius of the bubble at time t.

9. Use `ode45` of MATLAB to study (6.175) after setting p, the internal pressure inside the bubble, equal to zero, acknowledging that this pressure is negligible in comparison to the fluid pressure. Solve the initial value problem

$$RR'' + R'^2 = -\frac{2}{3}\frac{p_0}{r}, \quad R(0) = R_0,\ R'(0) = 0, \tag{6.176}$$

with parameter values $r = 1$, $R_0 = 1$, $p_0 = 10^6$ to see if the bubble ever collapses, that is, if there is a time $T > 0$ such that $\lim_{t\to T} R(t) = 0$.

6.14 Project D: Chaotic Transport

This project is motivated by a paper of H. Yang and Z. Liu (see [8]) where the authors present an analysis of a three-dimensional oceanic model and introduce the concept of the "Great Ocean Barrier."

1. Read the abstract and the introduction to [8] and write up a summary of the issues that are addressed in this paper. Specifically,

   (a) what do the mathematical symbols L, H, $\operatorname{curl} \tau(y)$, δ_s and δ_B represent?

   (b) What are the physical (and simplifying) assumptions under which this work is undertaken?

   (c) What are the mathematical tools being employed (for example, what is the "Lyapunov Analysis" described in Section 5 of the paper)?

2. Consider two stream functions $\psi_W(x, y)$ and $\psi_B(y, z)$, as yet not explicitly specified, in terms of which we define the velocity field $\mathbf{v}$ as

$$\mathbf{v} = \langle u, v, w \rangle = \langle -\frac{\partial \psi_W}{\partial y}, \frac{\partial \psi_W}{\partial x} + \frac{\partial \psi_B}{\partial z}, -\frac{\partial \psi_B}{\partial y} \rangle. \tag{6.177}$$

   Show that $\mathbf{v}$ is incompressible. Compute the vorticity of this flow in terms of the stream functions ψ_W and ψ_B.

3. Begin with the following definitions for ψ_W and ψ_B:

$$\psi_W(x, y) = (x - 2)\sin(\pi y)(1 - \exp(-\frac{x}{\delta_s})),$$

$$\psi_B = \epsilon(y + 1)\sin(\frac{\pi z}{b})(1 - \exp(\frac{y - 1}{\delta_B})). \tag{6.178}$$

   Find formulas for the velocity field $\mathbf{v}$ in (6.11).

4. Using the results of the velocity field from the previous part, consider the following system of ODEs for the particle trajectories:

$$\frac{dx}{dt} = -\frac{\partial \psi_W}{\partial y}, \qquad \frac{dy}{dt} = \frac{\partial \psi_W}{\partial x} + \frac{\partial \psi_B}{\partial z}, \qquad \frac{dz}{dt} = -\frac{\partial \psi_B}{\partial y}, \tag{6.179}$$

   subject to the initial conditions $x(0) = x_0$, $y(0) = y_0$ and $z(0) =$

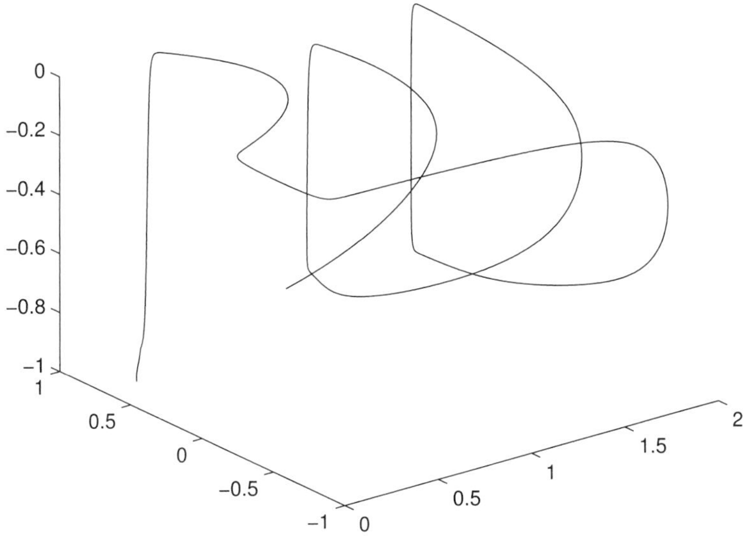

FIGURE 6.11: The trajectory of (6.179) with initial position $(0.3, -0.2, -0.5)$.

z_0. Use `ode45` to solve (6.179) for $t \in (0, 4)$ and initial position $(0.3, -0.2, -0.5)$ for the following set of parameter values:

$$\delta_s = 0.025,\ \delta_B = 0.01,\ b = 1,\ \epsilon = 1. \tag{6.180}$$

Plot the graph of the trajectory you obtain and compare your graph to Figure 6.11 and to Figure 2(a), page 1262 of [8]. To get a graph that resembles these two figures, first apply `ode45` with its default values for relative and absolute errors—the values you will obtain will unfortunately be only accurate for a short period of time and you will obtain `NaN` value for the coordinates of $(x(t), y(t), z(t))$ as soon as the trajectory reaches its northern most destination and must dip down in the z direction. To improve on accuracy we must appeal to the `odeset` function in MATLAB to set the options in `ode45` to something like

```
options = odeset('RelTol',1e-10,'AbsTol', ...
       [1e-10 1e-10 1e-10]);
[t,y]=ode45(@YangLiu,[0 4],[0.3 -0.2 -0.5], options);
```

where `YangLiu.m` is the M-file in which the ODEs for this problem are defined.

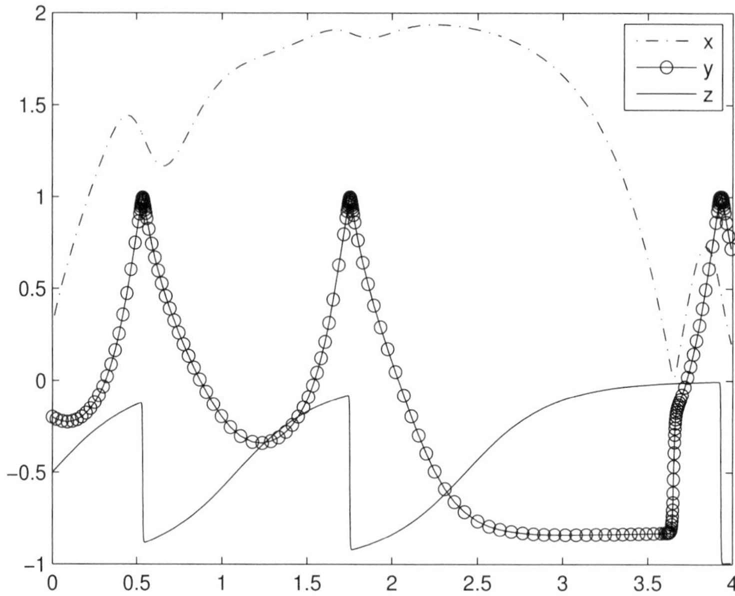

FIGURE 6.12: The graphs of x, y, and z versus t for the trajectory with initial position $(0.3, -0.2, -0.5)$.

5. Apply `ode45` to (6.179) to obtain the other three figures in Figure 2 of [8]. The initial positions are $(1, -0.5, -0.2)$, $(0.01, 0.8, -0.05)$, and $(0.01, -0.5, -0.01)$.

6. To gain insight into the chaotic character of (6.179) apply the above analysis to a small neighborhood of each of the four initial positions. To be precise, consider a spherical neighborhood of 0.01 radius about each one of the four initial positions described above. Select 100 random positions on each one of these spheres and solve (6.179) for each of these points and plot their positions after 4 units of time have expired. Report on the level of dispersion you observe on the final positions.

6.15 References

1. Malek-Madani, R., *Advanced Engineering Mathematics with Mathematica® and MATLAB®*, 1998, Addison-Wesley.

2. Shadden, C., Lekien, F., Marsden, J., "Definition and properties of Lagrangian coherent structures from finite-time Lyapunov exponents in two-dimensional aperiodic flows", *Physica D*, Vol 212, 2005, pp. 271–304.

3. Branicki, M., Wiggins, S., "Finite-time Lagrangian transport analysis: stable and unstable manifolds of hyperbolic trajectories and finite-time Lyapunov exponents", *Nonlinear Processes in Geophysics*, 2010, pp. 1–36.

4. Majda, A., Bertozzi, A., *Vorticity and Incompressible Flow*, Cambridge, 2002.

5. Majda, A., Wang, X., *Nonlinear Dynamics and Statistical Theories for Basic Geophysical Flows*, Cambridge, 2006.

6. Truesdell, C., Rajagopal, K. *An Introduction to the Mechanics of Fluids*, Birkhauser, 1999.

7. Gill, A., *Atmosphere-Ocean Dynamics*, Academic Press, 1982.

8. Yang, H., Liu, Z, "The three-dimensional chaotic transport and the great ocean barrier", *J. of Phys. Ocean.*, Vol 27, 1997, pp. 1258–1273.

9. Pedlsoky, J., *Geophysical Fluid Dynamics*, Springer-Verlag, New York, 1986.

Chapter 7

Equations of Geophysical Fluid Dynamics

7.1 Introduction

In the previous chapters we concentrated on the derivation of equations of motion of fluids in stationary, non-rotating frames and obtained these equations in rectangular as well as spherical coordinate systems. In this chapter our goal is to extend the derivation of equations of motion to rotating frames. Moreover, we begin the discussion of the mathematical relationship between measuring quantities when confined to the rotating frame itself, which is typical of the measurements we make when we collect data about velocity, pressure or salinity on our planet, and the same quantities when viewed from an inertial frame, say a coordinate frame that is stationary relative to our planet. The key new idea now is the mathematical description of the entity called the *Coriolis* force that appears in the equations of balance linear momentum. The new equations of motion are referred to as the equations of *Geophysical Fluid Dynamics*, or *GFD* for short.

As we will see shortly, the appearance of the Coriolis force will have an enormous impact on the properties of solutions of the PDEs that govern geophysical flows, especially in modeling large-scale motions, as in the Gulf Stream and the Kuroshio. It was the discovery of how Coriolis counterbalances other geophysical forces, such as the prevailing winds and friction or viscous forces, that resulted in the genesis of geophysical fluid dynamics as a distinct discipline from the fluid flows whose temporal and spatial scales confine them to laboratory settings.

7.2 Coriolis

For geophysical fluid flows on our planet the formula (6.140) of the previous chapter for acceleration $\mathbf{a}$ needs to be augmented to include the influence of the Earth's rotation. Assuming the axis of rotation passes through the poles, this rotation induces an angular velocity represented by

$$\boldsymbol{\Omega} = \Omega\,\mathbf{k},$$

where, assuming that it takes 24 hours for our planet to complete one rotation about its axis, we have

$$\Omega = \frac{2\pi\,\text{rad}}{24\,\text{hrs}} = 0.000072722 \quad \text{rad/s}.$$

Consider now a fluid particle P that remains stationary relative to the rotating planet. To an observer outside of the planet this particle undergoes a motion, tracing a path in the shape of a parallel circle, where the latitude remains constant while the longitude changes according to Ωt:

$$\mathbf{r}(t) = \langle a\cos\phi\cos\Omega t, a\cos\phi\sin\Omega t, a\sin\phi\rangle. \tag{7.1}$$

The velocity vector associated with this motion is computed by differentiating (7.1) with respect to t:

$$\mathbf{v} = \mathbf{r}' = \Omega\langle -a\cos\phi\sin\Omega t, a\cos\phi\cos\Omega t, 0\rangle, \tag{7.2}$$

or equivalently

$$\mathbf{v} = \boldsymbol{\Omega} \times \mathbf{r}. \tag{7.3}$$

The rotational motion represented by (7.3) induces the acceleration

$$\mathbf{a} = \frac{d\mathbf{v}}{dt} = \frac{d\boldsymbol{\Omega}}{dt} \times \mathbf{r} + \boldsymbol{\Omega} \times \frac{d\mathbf{r}}{dt} = \boldsymbol{\Omega} \times \mathbf{v}, \tag{7.4}$$

assuming the rate of rotation $\boldsymbol{\Omega}$ of our planet is time independent. Formula (7.4) describes the apparent acceleration of a stationary particle P in a rotating frame when this particle is viewed and measured in a non-rotating coordinate system.

Particles, however, typically move relative to the rotating planet itself. Denoting this relative velocity by $\mathbf{v}_r$, we note that a particle's absolute velocity, which we now denote by $\mathbf{v}_a$, when measured in a non-rotating frame, will be the sum of its relative velocity $\mathbf{v}_r$ and the velocity (7.3) induced by the planet:

$$\mathbf{v}_a = \mathbf{v}_r + \boldsymbol{\Omega} \times \mathbf{r}. \tag{7.5}$$

It is worth emphasizing that $\mathbf{v}_r$ is what our instruments measure when we measure the velocity of a particle on our planet.

A good way to view the expression in (7.5) is to use its content to define the time-rate of change of any quantity (denoted by an "$\circ$" in the following formula) in the non-rotating coordinate system:

$$\frac{D\circ}{Dt} = \frac{d\circ}{dt} + \mathbf{\Omega} \times \circ. \tag{7.6}$$

Hence, we are using $\frac{D}{Dt}$ to denote the absolute time differentiation, in a non-rotating cartesian coordinate system, while $\frac{d}{dt}$ denotes the time differentiation relative to the rotating frame. With this notation the absolute acceleration of a particle is simply the second derivative $\frac{D^2}{Dt^2}$, which, using the symbol operator in (7.6), reduces to

$$\frac{D^2\circ}{Dt^2} = (\frac{d\circ}{dt} + \mathbf{\Omega} \times \circ)(\frac{d\circ}{dt} + \mathbf{\Omega} \times \circ) =$$

$$\frac{d^2\circ}{dt^2} + 2\mathbf{\Omega} \times \frac{d\circ}{dt} + \mathbf{\Omega} \times (\mathbf{\Omega} \times \circ). \tag{7.7}$$

For example, when we apply the formula in (7.7) to the position vector $\mathbf{r}$ of a particle, we obtain the important formula that describes the absolute acceleration of a particle:

$$\frac{D^2\mathbf{r}}{Dt^2} = \frac{d^2\mathbf{r}}{dt^2} + 2\mathbf{\Omega} \times \frac{d\mathbf{r}}{dt} + \mathbf{\Omega} \times (\mathbf{\Omega} \times \mathbf{r}). \tag{7.8}$$

The three terms on the right side (7.8) all have significant physical interpretations. The first term, $\frac{d^2\mathbf{r}}{dt^2}$, is the relative acceleration $\mathbf{a}$, which we should now denote by $\mathbf{a}_r$, that is,

$$\mathbf{a}_r = \frac{d^2\mathbf{r}}{dt^2} = \frac{\partial \mathbf{v}_r}{\partial t} + \mathbf{v}_r \cdot \nabla \mathbf{v}_r. \tag{7.9}$$

We recall that in Section 6.10 we presented the spherical representation of $\mathbf{a}_r$.

The second term in (7.8), $2\mathbf{\Omega} \times \mathbf{v}_r$, is the *Coriolis force*. We will have more to say about this term shortly, but would like to comment now that this "force" is in reality part of the acceleration of the particle and not a force acting on it.

The third term, $\mathbf{\Omega} \times (\mathbf{\Omega} \times \mathbf{r})$, is the *centripetal acceleration* of the particle. While this expression is quite significant when the rate of rotation, Ω, is large, in most geophysical flows its magnitude is small relative to the magnitude of $-g\mathbf{k}$, the acceleration due to gravity. To see this, note that the largest value

$$||\mathbf{\Omega} \times (\mathbf{\Omega} \times \mathbf{r})||$$

can assume occurs when the particle is located at the equator, and that value is $||\Omega||^2 a$, where a is the Earth's radius. With $\Omega = 7.2 \times 10^{-5}$ and $a = 6,400$ km, the magnitude of the centripetal acceleration is approximately $0.03\,\mathrm{m}/s^2$, more than two orders of magnitude smaller than g's value, which is $9.8\,\mathrm{m}/s^2$. For this reason the contribution of the centripetal acceleration is often ignored.

It is also worth noting that $\mathbf{\Omega} \times (\mathbf{\Omega} \times \mathbf{r})$ can be rewritten as

$$-\frac{\Omega^2}{2}\nabla(||\mathbf{R}||^2), \tag{7.10}$$

where $\mathbf{R} = \langle x, y, 0\rangle$ is the projection of the position vector $\mathbf{r}$ onto the xy-plane. So if the contribution of this term needs to be taken into account, one can alter the potential of the conservative forces in the problem by adding $-\frac{\Omega^2}{2}||\mathbf{R}||^2$ to this potential.

In summary, in what follows in this book we will use the expression

$$\frac{D^2\mathbf{r}}{Dt^2} = \frac{d^2\mathbf{r}}{dt^2} + 2\mathbf{\Omega} \times \mathbf{v}_r \tag{7.11}$$

for absolute acceleration. In the next section we will rewrite the Coriolis force $2\mathbf{\Omega} \times \mathbf{v}_r$ in spherical coordinates.

Problems 7.2

1. Verify the relationship between the expressions in (7.2) and (7.3).
2. Verify the statement in (7.10).
3. The Earth's rotation vector $\mathbf{\Omega}$ has the form $\Omega\mathbf{k}$ in Cartesian coordinates. Find the components of $\mathbf{\Omega}$ in spherical coordinates, i.e., find a, b and c such that

$$\mathbf{\Omega} = a\,\mathbf{e}_\theta + b\,\mathbf{e}_\phi + c\,\mathbf{e}_r.$$

Answer: $a = 0$, $b = \Omega\cos\phi$, $c = \Omega\sin\phi$. Is it intuitively clear why Ω does not have a component in the $\mathbf{e}_\theta$ direction?

7.3 Coriolis Acceleration: $2\mathbf{\Omega} \times \mathbf{v}_r$

The term $2\mathbf{\Omega} \times \mathbf{v}_r$ plays a crucial role in the equations of motion of geophysical fluid flows, especially when the goal is to understand large and medium scale behavior of large bodies of fluids. In order to compare its impact to the relative acceleration, $\mathbf{a}_r$, we first write this vector in spherical coordinates. Recall that $\mathbf{\Omega} = \Omega\,\mathbf{k}$ and $\mathbf{v}_r = u\mathbf{e}_\theta + v\mathbf{e}_\phi + v_r\mathbf{e}_r$. The relation

$$\mathbf{k} = \cos\phi\,\mathbf{e}_\phi + \sin\phi\,\mathbf{e}_r$$

(see (6.118)) enables us to recast the Coriolis contribution as

$$2\mathbf{\Omega} \times \mathbf{v}_r = 2(\cos\phi\,\mathbf{e}_\phi + \sin\phi\,\mathbf{e}_r) \times (u\,\mathbf{e}_\theta + v\,\mathbf{e}_\phi + w\,\mathbf{e}_r),$$

which, after taking advantage of the orthogonality properties of the spherical basis (see (6.141)), reduces to

$$2\Omega \times \mathbf{v}_r = (-2\Omega\,v\sin\phi + 2\Omega\,w\cos\phi)\,\mathbf{e}_\theta +$$

$$2\Omega\,u\sin\phi\,\mathbf{e}_\phi - 2\Omega\,u\cos\phi\,\mathbf{e}_r. \tag{7.12}$$

Combining this formula with (6.140) and (7.11) we obtain the following working formula for absolute acceleration written in spherical coordinates:

$$\frac{D^2\mathbf{r}}{Dt^2} = (-2\Omega\,v\sin\phi + 2\Omega\,w\cos\phi + \frac{du}{dt} - \frac{uv}{r}\tan\phi + \frac{v_\theta v_r}{r})\,\mathbf{e}_\theta +$$

$$(2\Omega\,v_\theta\sin\phi + \frac{dv_\phi}{dt} + \frac{v_\theta^2}{r}\tan\phi + \frac{v_\phi v_r}{r})\,\mathbf{e}_\phi +$$

$$(-2\Omega\,v_\theta\cos\phi + \frac{dv_r}{dt} - \frac{v_\theta^2 + v_\phi^2}{r})\,\mathbf{e}_r. \tag{7.13}$$

The above formula is the fundamental result we will end up relying on when we study the various reduced models of geophysical fluid flows, notably the f-plane and the β-plane approximations, when the scales of the flow allow us to ignore some of the nonlinearities in (7.13).

In the next section we develop the gradient operator in spherical coordinates, which, together with (7.13), leads to the equivalent of the Navier-Stokes equations in a rotating frame.

7.4 Gradient Operator in Spherical Coordinates

Consider a function p, represented as $p(x, y, z)$ in rectangular coordinates, and its equivalent representation $P(r, \theta, \phi)$ in spherical coordinates. These expressions satisfy the relation

$$p(x, y, z) = P(r, \theta, \phi). \tag{7.14}$$

The gradient operator ∇, when applied to p , can be written as

$$\nabla \mathrm{p} = \frac{\partial p}{\partial x}\mathbf{i} + \frac{\partial p}{\partial y}\mathbf{j} + \frac{\partial p}{\partial z}\mathbf{k} \tag{7.15}$$

and equivalently as

$$\nabla \mathrm{p} = a\,\mathbf{e}_\theta + b\,\mathbf{e}_\phi + c\mathbf{e}_r. \tag{7.16}$$

Our task is to determine the terms a, b and c in terms of the various derivatives of P. Note that the basis vectors $\{\mathbf{i},\mathbf{j},\mathbf{k}\}$ and $\{\mathbf{e}_\theta,\mathbf{e}_\phi,\mathbf{e}_r\}$ are related by the expressions listed in (6.118). In order to complete (7.16) we will need the following fact (recall that $r^2 = x^2+y^2+z^2$, $\theta = \tan^{-1}\frac{y}{x}$ and $\phi = \frac{z}{\sqrt{x^2+y^2+z^2}}$)

$$\begin{bmatrix} r_x & r_y & r_z \\ \theta_x & \theta_y & \theta_z \\ \phi_x & \phi_y & \phi_z \end{bmatrix} = \begin{bmatrix} \cos\theta\cos\phi & \sin\theta\cos\phi & \sin\phi \\ -\frac{1}{r}\sin\theta\sec\phi & \frac{1}{r}\cos\theta\sec\phi & 0 \\ -\frac{1}{r}\cos\theta\sin\phi & -\frac{1}{r}\sin\theta\sin\phi & \frac{1}{r}\sec\phi \end{bmatrix}, \tag{7.17}$$

which shows the relationships between the rates of change of the spherical coordinate variables with respect to the corresponding rectangular ones. Since $\frac{\partial p}{\partial x} = \frac{\partial P}{\partial r}r_x + \frac{\partial P}{\partial \theta}\theta_x + \frac{\partial P}{\partial \phi}\phi_x$, it follows from (7.17) that

$$\frac{\partial p}{\partial x} = \cos\theta\cos\phi\frac{\partial P}{\partial r} - \frac{1}{r}\sin\theta\sec\phi\frac{\partial P}{\partial \theta} - \frac{1}{r}\cos\theta\sin\phi\frac{\partial P}{\partial \phi}. \tag{7.18}$$

Similar expressions follow for $\frac{\partial p}{\partial y}$ and $\frac{\partial p}{\partial z}$. Once the latter expressions are substituted into (7.15) and use is made of the relations in (6.118), we have

$$\nabla p = \frac{1}{r\cos\phi}\frac{\partial P}{\partial \theta}\mathbf{e}_\theta + \frac{1}{r}\frac{\partial P}{\partial \phi}\mathbf{e}_\phi + \frac{\partial P}{\partial r}\mathbf{e}_r. \tag{7.19}$$

Problems 7.4

1. Verify the assertions in (7.17).

2. Complete the calculation that leads to (7.19).

3. Consider each function p defined below. First compute its gradient in rectangular coordinates, and next in spherical coordinates, by transforming p to $P(r,\theta,\phi)$:

   (a) $p(x,y,z) = xyz$.

   (b) $p(x,y,z) = \frac{1}{\sqrt{x^2+y^2+z^2}}$.

   (c) $p(x,y,z) = \sin(x^2+y^2)$.

7.5 Navier–Stokes Equation in a Rotating Frame

We have now obtained formulas for the absolute acceleration (see (7.13)) and pressure gradient in a rotating frame (in (7.19)). Hence the equations that express the balance of linear momentum in the θ, ϕ and r directions are

$$\frac{du}{dt} - \frac{uv}{r}\tan\phi + \frac{uw}{r} + (-2\Omega v\sin\phi + 2\Omega w\cos\phi) = -\frac{1}{r\rho\cos\phi}\frac{\partial P}{\partial\theta} + F_\theta,$$

$$\frac{dv}{dt} + \frac{u^2}{r}\tan\phi + \frac{vw}{r} + 2\Omega u\sin\phi = -\frac{1}{r\rho}\frac{\partial P}{\partial\phi} + F_\phi,$$

$$\frac{dw}{dt} - \frac{u^2+v^2}{r} - 2\Omega u\cos\phi = -\frac{1}{\rho}\frac{\partial P}{\partial r} - g - F_r, \qquad (7.20)$$

where the F terms are the components of external and viscous forces. These equations are complemented by the conservation of mass equation (i.e., div $\mathbf{v} = 0$), which has the following form

$$\frac{1}{r\cos\phi}\frac{\partial u}{\partial\theta} + \frac{1}{r\cos\phi}\frac{\partial(\cos\phi\, v)}{\partial\phi} + \frac{\partial w}{\partial r} + \frac{2w}{r} = 0 \qquad (7.21)$$

in spherical coordinates. Note that the expression in (7.21) differs from the expression we gave for the divergence operator in spherical coordinates in (3.39), the reason being that here the angle ϕ represents the latitude, whereas in (3.39) it represented the co-latitude.

The above four equations constitute the fundamental set of PDEs of dynamics in geophysical flows in a rotating frame. The models we will study in the subsequent chapters are all based on various simplifications of this system of equations. In the next section, we will take up the derivation of the β–plane approximation, which is considered a good model for studying large–scale features in the ocean and the atmosphere.

7.6 β-Plane Approximation

The equations in (7.20)–(7.21) are exact and suited well for modeling flows that may visit large segments of the planet. In cases where the extent of domain is small enough where the planetary surface may reasonably be approximated by a tangent plane, it seems prudent to look into simplifying these equations to remove the terms that do not contribute substantially to the behavior of the solutions we are seeking. This is the motivation behind the so-called "β plane" approximation, where the domain is replaced by a tangent plane while keeping the impact of Coriolis, which varies with latitude, intact. An excellent treatment of this approach is given by G. Veronis in [1], which we now outline.

The strategy behind the β plane approximation is to concentrate attention at the fixed point P with longitude-latitude (θ_0, ϕ_0) and replace the domain with the tangent plane at P. We can think of x now as being the east-west distance relative to P along the direction where longitude varies, but projected or confined to the tangent plane, and define it as (let a denote the radius of the planet)

$$x = (a \cos \phi_0)(\theta - \theta_0).$$

Similarly, we think of y as the north-south distance, measured from P, in the latitude direction, and define it as

$$y = a(\phi - \phi_0).$$

And finally z, defined by $r = a + z$, is the deviation from the planet's radius. In doing so, we note that the various derivatives in the (7.20)–(7.21) are now replaced by derivatives in x, y and z because

$$\frac{\partial}{\partial x} = \frac{1}{a \cos \phi_0} \frac{\partial}{\partial \theta}, \quad \frac{\partial}{\partial y} = \frac{1}{a} \frac{\partial}{\partial \phi}, \quad \frac{\partial}{\partial z} = \frac{\partial}{\partial r}.$$

In particular, we note that this approximation introduces the length scale a, the radius of the planet, into the equations. Therefore, if the length scale L of the domain is small compared with a, one may be persuaded to ignore terms that have the term a in their denominators in deference to ones that do not. We do not carry out the details of this calculation and refer the reader to [1], pages 144–145 for the details, but state the resulting reduced equations, often called the *Quasi-Geostrophic Equations in the β-plane*:

$$\rho(\frac{Du}{Dt} - fv) = -\frac{\partial p}{\partial x} + F_x, \tag{7.22}$$

$$\rho(\frac{Dv}{Dt} + fu) = -\frac{\partial p}{\partial y} + F_y, \tag{7.23}$$

$$\rho\frac{Dw}{Dt} = -\frac{\partial p}{\partial z} - \rho g z + F_z, \tag{7.24}$$

and

$$\frac{\partial u}{\partial x} + \frac{\partial v}{\partial y} + \frac{\partial w}{\partial z} = 0. \tag{7.25}$$

Some variation of the equations (7.22)– (7.25) will be the subject of our study in the next chapters.

Problems 7.6

1. Derive the Quasi-Geostrophic equation (7.22)–(7.25) by completing the calculations on pages 144-145 of [1]. This material is available from MIT's free online course materials.

7.7 References

1. Veronis, G., "Dynamics of large-scale circulation," Chapter 5 of *Evolution of Physical Oceanography*, edited by B. Warren and C. Wunsch, 1981, pp. 140–183. Also available at MIT's free online course materials.

Chapter 8

Shallow Water Equations (SWE)

8.1 Introduction

In Chapters 6 and 7 we derived the Navier–Stokes equations in non-rotating and rotating frames. In this chapter we concentrate on the PDEs in the non-rotating frame and derive the Shallow Water Equations (SWE) as a perturbation of the Navier-Stokes equations. Shallow Water Equations constitute one of the fundamental systems of equations in fluid dynamics, typically applied to settings where horizontal scales are considerably larger than the vertical one, a common occurrence in oceans and the atmosphere. The presentation here is motivated by those in the books *An Introduction to Fluid Dynamics* by G. K. Batchelor, *Water Waves*, by J. J. Stoker, and in the paper "Derivation of viscous Saint-Venant system for laminar shallow water; numerical validation," by J.- F. Gerbeau and Benoit Perthame.

We present the derivation here for the simpler case of a two-dimensional flow. Later in the chapter we derive the scalar wave equation as a special of SWE and apply the Fourier series method to derive the solution to a typical initial-boundary value problem for this equation. We derive the D'Alembert solution to this IBVP from its Fourier series solution, and then proceed to introduce its finite-difference solution.

8.2 Derivation of Equations

The derivation we present here will be confined to two-dimensional basins. The extension of the methodology to three dimensional basins is straightforward.

Consider a flow of a homogeneous fluid, so that ρ is constant throughout all deformations, in a domain Ω defined by

$$B = \{(x, z) |\, b(x) < z < b(x) + h(x, t),\, x \in R\} \tag{8.1}$$

where $z = b(x)$ defines the bathymetry (bottom surface) of the basin and $z = b(x) + h(x, t)$ is the free surface of the fluid—$h(x, t)$ is the fluid height (column) at any time t and point x. See Figure 8.1. We consider the two-dimensional velocity field $\mathbf{v} = \langle u, w \rangle$ (setting $v \equiv 0$) which satisfies the Navier–Stokes equations (6.97)

$$\frac{\partial u}{\partial x} + \frac{\partial w}{\partial z} = 0 \tag{8.2}$$

and

$$u_t + uu_x + wu_z = -\frac{1}{\rho} p_x + \nu \Delta u, \tag{8.3}$$

$$w_t + uw_x + ww_z = -\frac{1}{\rho} p_z - g + \nu \Delta w. \tag{8.4}$$

The symbol Δ stands for the Laplacian, $\frac{\partial^2}{\partial x^2} + \frac{\partial^2}{\partial z^2}$. The constant $\nu = \frac{\mu}{\rho}$ is viscosity of the fluid.

Equations (8.2)–(8.4) are supplemented by the boundary conditions we need to impose on the two boundaries of the region B: We assume that both surfaces $z = b(x)$ and $z = b(x) + h(x, t)$ are Lagrangian-invariant, that is, if a fluid particle is located on either surface at one time, it continues to remain on that surface for all time. For the bathymetry $z = b(x)$, this assumption may be interpreted by stating that this stationary surface is impenetrable. We impose an additional boundary condition on the free surface $z = b(x) + h(x, t)$, since it the interface between the fluid under study and the outside environment, that the pressure function $p(t, x, z)$ remains continuous on this surface, whether measured from the atmospheric side or the fluid side.

The fact that the free surface $z = b(x) + h(x, t)$ is Lagrangian-invariant imposes certain conditions on the functions b and h. To see this let $\mathbf{r}(t) = \langle \hat{x}(t), \hat{z}(t) \rangle$ be the trajectory of a fluid particle that remains on this surface for all time. Then $\hat{x}(t)$ and $\hat{z}(t)$ must satisfy the equation

$$\hat{z}(t) = b(\hat{x}(t)) + h(\hat{x}(t), t).$$

Since this expression holds for all t, the identity must also hold for its derivative. Differentiating this equation yields (recall $u = \hat{x}'(t)$ and $w = \hat{z}'(t)$)

$$w(\hat{x}, b + h, t) = h_t + (b' + h_x) u(\hat{x}, b + h, t). \tag{8.5}$$

This relation between u and w on the free-surface of the flow will play a key role in the reduced model we are about to develop.

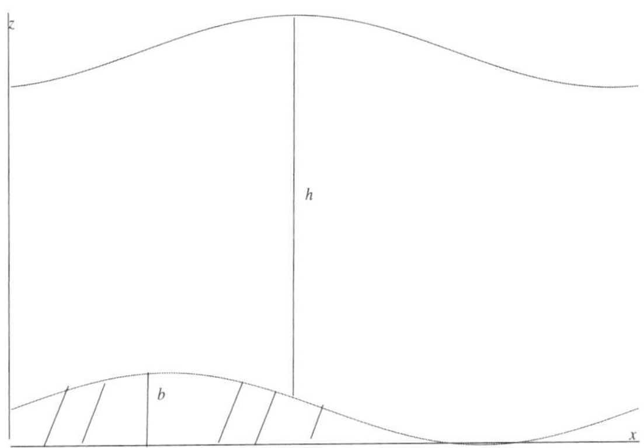

FIGURE 8.1: Schematic of a 2D flow with bathymetry.

The constraints on the bathymetry are similar. Since the bottom surface $z = b(x)$ is Lagrangian-invariant, we arrive at

$$w(x, b, t) = b' u(x, b, t). \tag{8.6}$$

The last two constraints are summarized in the following theorem.

Theorem 8.2.1 (Free and Bottom Surface Conditions)
Equations (8.5) and (8.6) express the boundary conditions that the free surface/water column $z = b(x) + h(x, t)$ and the bottom surface $z = b(x)$ must satisfy.

To derive the our first model of fluid flows in shallow waters, we begin by integrating the conservation of mass equation, (8.2), in z over the entire water column $b < z < b + h$. We get

$$\int_b^{b+h} u_x(x, \eta, t)\, d\eta = -w(x, b + h, t)) + w(x, b, t). \tag{8.7}$$

The term on the left side of (8.7) can be rewritten as

$$\int_b^{b+h} u_x(x, \eta, t)\, d\eta = \frac{\partial}{\partial x}(\int_b^{b+h} u(x, \eta, t)\, d\eta) -$$

$$(b' + h_x)u(x, b + h, t) + b' u(x, b, t). \tag{8.8}$$

After applying (8.5) and (8.6), the non-integral terms in (8.8) become equivalent to

$$h_t - w(x, b + h, t) + w(x, b, t). \tag{8.9}$$

Thus (8.8) reduces to

$$\int_b^{b+h} u_x(x,\eta,t)\,d\eta = \frac{\partial}{\partial x}(\int_b^{b+h} u(x,\eta,t)\,d\eta)+$$

$$h_t - w(x,b+h,t) + w(x,b,t). \tag{8.10}$$

Compare (8.10) with (8.7). It is clear now that

$$\frac{\partial}{\partial x}\left(\int_b^{b+h} u(x,\eta,t)\,d\eta\right) + h_t = 0 \tag{8.11}$$

Define the new quantity $U(x,t)$, the horizontal velocity $u(x,z,t)$ averaged over the water column $(b, b+h)$, by

$$U(x,t) = \frac{1}{h}\int_b^{b+h} u(x,\eta,t)\,d\eta, \tag{8.12}$$

in terms of which the expression (8.11) takes the form

$$h_t + (hU)_x = 0. \tag{8.13}$$

This equation is the first equation in the SWE system. We note in passing that if u is independent of z, the quantity U reduces to u and Equation (8.13) takes the form

$$h_t + (hu)_x = 0. \tag{8.14}$$

So far all calculations have been exact and no approximations have been imposed on the governing equation. Our first assumption, which has a significant simplifying impact, is to replace the equation in (8.4) by

$$0 = -\frac{1}{\rho}\frac{\partial p}{\partial z} - g. \tag{8.15}$$

The rationale in weighing the terms in (8.15) more relative to the remaining terms in (8.4) is the underlying shallowness assumption of the basin, that horizontal processes in general have more impact on the dynamics in a shallow basin. Hence, when viewing the balance of linear momentum in the vertical direction, the acceleration term in the z-direction (the term $w_t + uw_x + ww_z$), and the viscous dissipation in that direction, $\nu\Delta w$, end up being negligible relative to the pressure gradient term, $\frac{1}{\rho}\frac{\partial p}{\partial z}$, and the fluid's weight (represented by the acceleration due to gravity g). This assumption, which as we have seen earlier is referred to as the hydrostatic approximation, can be borne out by making a back-of-the-envelope calculation of the relative size of each term in (8.4), as done in many texts listed at the end of this chapter.

Returning to (8.15), and recalling that ρ is constant, we integrate this equation with respect to z in the interval $(z, b+h)$:

$$p(x,z,t) = p(x, b(x) + h(x,t), t) + \rho g(b + h - z). \tag{8.16}$$

The first term on the right side of (8.16) is the same as the atmospheric pressure, which we assume to be constant, p_0, for convenience. In that case (8.16) reduces to

$$p(x,z,t) = p_0 + \rho g(b(x) + h(x,t) - z). \tag{8.17}$$

We eliminate pressure p from (8.3) using (8.17) to get

$$u_t + uu_x + wu_z = -g(b' + h_x) + \nu \Delta u. \tag{8.18}$$

Following the strategy we applied to the conservation of mass equation, we integrate (8.18) with respect to z in the water column $(b, b+h)$:

$$\int_b^{b+h} u_t \, d\eta + \int_b^{b+h} uu_x \, d\eta + \int_b^{b+h} wu_z \, d\eta =$$

$$-g(b' + h_x)h + \nu \int_b^{b+h} \Delta u \, d\eta. \tag{8.19}$$

We begin simplifying the expressions in (8.19) by first addressing the third integral. Integrating this expression by parts yields

$$\int_b^{b+h} wu_z \, d\eta = (wu)\big|_{b+h} - (wu)\big|_b - \int_b^{b+h} w_z u \, d\eta.$$

But $u_x + w_z = 0$, hence the above integral reduces to

$$\int_b^{b+h} wu_z \, d\eta = (wu)\big|_{b+h} - (wu)\big|_b + \int_b^{b+h} uu_x \, d\eta. \tag{8.20}$$

Substituting (8.20) back into (8.19) yields

$$\int_b^{b+h} u_t \, d\eta + \int_b^{b+h} 2uu_x \, d\eta + (wu)\big|_{b+h} - (wu)\big|_b =$$

$$-g(b' + h_x)h + \nu \int_b^{b+h} \Delta u \, d\eta.$$

Note that $\int_b^{b+h} 2uu_x \, d\eta = \int_b^{b+h} (u^2)_x \, d\eta$ so the above expression now takes the form

$$\int_b^{b+h} u_t \, d\eta + \int_b^{b+h} (u^2)_x \, d\eta + (wu)\big|_{b+h} - (wu)\big|_b =$$

$$-g(b' + h_x)h + \nu \int_b^{b+h} \Delta u \, d\eta. \tag{8.21}$$

The first two integrals in (8.21) have the following alternative forms once we apply the chain rule to them:

$$\int_b^{b+h} u_t \, d\eta = \frac{\partial}{\partial t}(\int_b^{b+h} u \, d\eta) - u|_{b+h} h_t, \tag{8.22}$$

and

$$\int_b^{b+h} (u^2)_x \, d\eta = \frac{\partial}{\partial x}(\int_b^{b+h} u^2 \, d\eta) - u^2|_{b+h} (b' + h_x) + u^2|_b b. \tag{8.23}$$

Expressions (8.22) and (8.23), when substituted in (8.21), reduce the latter expression to

$$\frac{\partial}{\partial t}(\int_b^{b+h} u \, d\eta) + \frac{\partial}{\partial x}(\int_b^{b+h} u^2 \, d\eta) - u|_{b+h} h_t - u^2|_{b+h} (b' + h_x) +$$

$$u^2|_b b + (wu)|_{b+h} - (wu)|_b = -g(b' + h_x)h + \nu \int_b^{b+h} \Delta u \, d\eta. \tag{8.24}$$

Finally, using the values of w on the free surface and on the bottom from (8.5) and (8.6), we note that the non-integral expressions on the left side of (8.24) vanish. We are left with

$$\frac{\partial}{\partial t}(\int_b^{b+h} u \, d\eta) + \frac{\partial}{\partial x}(\int_b^{b+h} u^2 \, d\eta) =$$

$$-g(b' + h_x)h + \nu \int_b^{b+h} \Delta u \, d\eta. \tag{8.25}$$

Equations (8.13) and (8.25) constitute the Shallow Water Equations, which we now state as a theorem. Recall the relation between u and U as defined in (8.12).

Theorem 8.2.2 (SWE)
The system of Shallow Water Equations, which governs the dynamics of h, the water column, and u, the horizontal velocity, is

$$h_t + (hU)_x = 0,$$

$$(hU)_t + \frac{\partial}{\partial x}(\int_b^{b+h} u^2 \, d\eta) = -g(b' + h_x)h + \nu \int_b^{b+h} \Delta u \, d\eta, \tag{8.26}$$

$$U(x,t) = \frac{1}{h} \int_b^{b+h} u(x, \eta, t) \, d\eta.$$

The system of equations in (8.26) simplifies considerably when u is independent of z. As noted before, when the horizontal velocity is depth-independent, the variables u and U are identical and all integrals in (8.26) become trivial to evaluate. In fact, this system reduces to

$$h_t + (hu)_x = 0, \quad (hu)_t + (hu^2 + \frac{1}{2}gh^2)_x = -gb'h + \nu h \Delta u. \tag{8.27}$$

which we refer to as the *reduced* system of Shallow Water Equations.

The reduced SWE have received quite a bit of analytical and computational treatment. We refer to the book by R. LeVeque, [4], for further illuminating discussion concerning the mathematical challenges one faces in obtaining solutions to the initial-boundary value problem for this system. Additionally, the article in [3] is concerned with the derivation of SWE (referred to as the Saint-Venant system in this paper) and the numerical validation of the model. An interesting application of SWE is presented in [5] where a single-layer system of Shallow Water Equations is studied in the context of tsunami propagation and inundation. In this work finite volume numerical methods are developed for SWE, together with numerical schemes based on Riemann solvers when SWE is viewed from the perspective PDEs as Conservation Laws. Of special interest in [5] is the introduction of the software package *CLAWPACK*, which consists of a comprehensive collection of numerical schemes and tools for solving conservation laws, and, in particular, systems of the type (8.49). This package can be downloaded from the web site

`www.amath.washington.edu/~claw`

Problems 8.2

1. The derivation of the SWE system, (8.26), was based on the two-dimensional Navier–Stokes equations. Apply this derivation to the three-dimensional system and arrive at the following system:

$$h_t + \nabla \cdot (h\mathbf{u}) = 0, \quad \mathbf{u}_t + \mathbf{u} \cdot \nabla \mathbf{u} + g\nabla h = \mathbf{0}, \tag{8.28}$$

where $\mathbf{u} = \langle u, v\rangle$ is the horizontal velocity. Here, the effect of viscosity and bathymetry have been ignored.

2. Find the equivalent of (8.28) when the bottom topography is not flat. Let $z = H_B(x, y)$ stand for the bathymetry. Compare your result with (8.26).

3. Find the equivalent of (8.28) when Coriolis parameter f is not ignored. Show that $\mathbf{u}$ must satisfy

$$h_t + \nabla \cdot (h\mathbf{u}) = 0, \quad \mathbf{u}_t + \mathbf{u} \cdot \nabla \mathbf{u} + g\nabla h - f\mathbf{u}^{\perp} = \mathbf{0}, \tag{8.29}$$

where $\mathbf{u}^\perp = \langle v, -u \rangle$. With regard to this system see page 69 of the paper in [6].

4. One of the main assumptions in our derivation of the SWE in (8.26) was the homogeneity of the fluid, the fact that the density was assumed constant. Both the oceans and the atmosphere are quite stratified, with the density varying with depth in both settings, inversely with depth in the case of the atmosphere, and directly with depth in the case of oceans. Returning to the technique used in arriving at (8.26), explore how this method needs to be modified in order to accommodate $\rho(\mathbf{x})$, and the special case of $\rho(z)$.

8.3 Rotating Shallow Water Equations (RSWE)

The above derivation is just one approach that takes advantage of the shallowness of the basin in order to reduce the complexity of the original full set of equations to a more manageable lower order system. There are several ways we can generalize the approach. One is to consider the full three-dimensional system to begin with, that is, allow $\mathbf{v} = \langle u, v, w \rangle$ and continue with treating v as equal to u in all of the calculations. This is relatively routine and was assigned to an exercise (see Problem 1 in Section 8.2).

Another important feature we ignored in (8.2)–(8.4) is the Coriolis effect. When the balance of linear momentum equations are augmented by

$$-f\langle v, -u \rangle$$

and the same technique of averaging in the z direction is applied, we end up with the *Rotating Shallow Water Equations*, or RSWE, which are

$$\begin{aligned} \frac{Du}{Dt} - fv &= -gh_x, \\ \frac{Dv}{Dt} + fu &= -gh_y, \\ \frac{Dh}{Dt} + h\left(\frac{\partial u}{\partial x} + \frac{\partial v}{\partial y}\right) &= 0, \end{aligned} \tag{8.30}$$

where $\frac{D}{Dt}$ is the usual total time derivative

$$\frac{D}{Dt} = \frac{\partial}{\partial t} + u\frac{\partial}{\partial x} + v\frac{\partial}{\partial y}.$$

The above equations correspond to a single homogeneous fluid viewed as a shallow-water flow. The atmosphere and the oceans, however, are not generally homogeneous and are often quite stratified. Allowing the density ρ to vary spatially, at a minimum, creates quite a few new mathematical difficulties, but also introduces a necessary richness to the models, new structures that are capable of capturing features observed in nature.

Concentrating here only on the effect of stratification, which is arguably the most significant attribute missing from the above RSWE, we adopt a mathematical strategy that has paid off well in the past few decades, leading to the introduction of *multi-layer* models. In this approach reduced models are obtained by approximating the density, which is assumed to vary primarily in the depth direction z, by step functions (see also Project B in Chapter 6 where a linear multi-layer model was introduced in order to study internal gravity waves). This strategy leads to reduced models that retain enough of the physics from the original set of equations and, at the same time, allow a researcher to explore some of the physical parameters of interest to arrive at meaningful conclusions as to which parameters are responsible for the presence of a particular feature or phenomenon under study.

As an example of a multi-layer model consider the following two-dimensional two-layer generalization of the domain (8.1): We let B consist of two segments B_1 and B_2

$$B = B_1 \cup B_2$$

where B_1 is the upper layer and defined by

$$B_1 = \{(x, z)|\, H_2(x, t) < z < H_1(x, t),\ x \in R\}, \tag{8.31}$$

where

$$H_1(x, t) = H_2(x, t) + h_1(x, t)$$

with H_2, defined below, is the sum of the bathymetry and the water column height in the lower layer, h_1 is the height of the water column in the first layer, $u_1 = u_1(x, t)$ and the density ρ_1 is constant in this layer. The second layer B_2 is defined as follows

$$B_2 = \{(x, z)|\, b(x) < z < H_2(x, t),\ x \in R\} \tag{8.32}$$

where

$$H_2(x, t) = b(x) + h_2(x, t)$$

with $z = b(x)$ is the bathymetry, h_2 is the height of the water column in the lower layer, and where $u_2 = u_2(x, t)$ and the density $\rho_2 \neq \rho_1$ is also constant. With this in mind, the governing equations for the pair (u_1, u_2) is

$$(h_1)_t + (h_1 u_1)_x = 0, \tag{8.33}$$

$$(h_1 u_1)_t + (h_1 u_1^2 + \frac{1}{2} g h_1^2)_x = -g h_1 (h_2)_x - g h_1 b_x, \tag{8.34}$$

$$(h_2)_t + (h_2 u_2)_x = 0, \tag{8.35}$$

$$(h_2 u_2)_t + (h_2 u_2^2 + \frac{1}{2} g h_2^2)_x = -g \frac{\rho_1}{\rho_2} h_2 (h_1)_x - g h_2 b_x. \tag{8.36}$$

Note that the coupling between the layers manifests itself only in the balance of linear momentum equations. Also, typically $\rho_1 < \rho_2$ so that the fluid in the upper layer is lighter, leading to a stable stratification of the fluid column.

The equations in (8.33)–(8.36), and their natural generalization to multi-layer fluid columns, are the governing equations studied by many investigators who are seeking to understand phenomena such as storm surges and tsunami in coastal regions. Among these investigations are the ones in [7] and [8]. In [7] the authors present finite-volume numerical schemes for the above equations with application to submarine landslides and the onset of tsunami propagation. In [8] the authors present a Lagrangian-based numerical scheme for the single-layer Shallow Water Equations in the presence of bathymetry. One of the features of the latter work is their use of exact solutions of SWE, a topic we take up in the next section, to validate their numerical approach.

Problems 8.3

1. Consider the original Navier–Stokes equations of motion in a rotating frame:

$$u_x + v_y + w_z = 0 \tag{8.37}$$

$$\frac{Du}{Dt} - fv = -\frac{1}{\rho}\frac{\partial p}{\partial x} + \nu \Delta u, \tag{8.38}$$

$$\frac{Dv}{Dt} + fu = -\frac{1}{\rho}\frac{\partial p}{\partial y} + \nu \Delta v, \tag{8.39}$$

$$\frac{Dw}{Dt} = -\frac{1}{\rho}\frac{\partial p}{\partial z} - g + \nu \Delta w. \tag{8.40}$$

Let T, L, H, U and W be typical values for the scales in time, length, depth, the horizontal velocity (in x and y directions) and the vertical velocity directions. Non-dimensionalize (8.30) as follows: Let the new independent variables $\bar{t}$, $\bar{x}$, $\bar{y}$ and $\bar{z}$, and the new dependent variables $\bar{u}$, $\bar{v}$, $\bar{w}$ and $\bar{p}$ be related to their dimensional counterparts by

$$\bar{t} = \frac{t}{T},\ \bar{x} = \frac{x}{L},\ \bar{y} = \frac{y}{L},\ \bar{z} = \frac{z}{H},$$

$$\bar{u} = \frac{u}{U},\ \bar{v} = \frac{v}{U},\ \bar{w} = \frac{w}{W},\ \bar{p} = \frac{p}{P}.$$

Show that the resulting non-dimensional equations are as follows:

(a) Equation (8.37) becomes

$$\frac{\partial \bar{u}}{\partial \bar{x}} + \frac{\partial \bar{v}}{\partial \bar{y}} + \epsilon_1 \frac{\partial \bar{w}}{\partial \bar{z}} = 0, \tag{8.41}$$

where ϵ_1 is

$$\epsilon_1 = \frac{LW}{HU}.$$

According to (8.41), assuming that the spatial rates of change of $\bar{u}$, $\bar{v}$, $\bar{w}$ are similar to each other, the term ϵ_1 must then remain in the neighborhood of unity, the coefficients of $\frac{\partial \bar{u}}{\partial \bar{x}}$ and $\frac{\partial \bar{v}}{\partial \bar{y}}$. Note that ϵ_1 is the product of two ratios, one the ratio of the vertical velocity scale to the horizontal one, $\frac{W}{U}$, and the other the ratio of the horizontal scale to the vertical, $\frac{L}{H}$. For this expression to remain near one, the four terms L, H, U and W must balance each other delicately. Since in the case of "shallow" waters we expect that the ratio $\frac{L}{H}$ is quite large, the fact that ϵ_1 must stay near one implies that the vertical velocity scale, W, must be considerably smaller than U, the horizontal velocity scale.

(b) Equation (8.38) becomes

$$\frac{U}{T}\frac{\partial \bar{u}}{\partial \bar{t}} + \frac{U^2}{L}(\bar{u}\frac{\partial \bar{u}}{\partial \bar{x}} + \bar{v}\frac{\partial \bar{u}}{\partial \bar{y}}) + \frac{UW}{H}\bar{w}\frac{\partial \bar{u}}{\partial \bar{z}} - fU\bar{u} =$$

$$-\frac{P}{\rho L}\frac{\partial \bar{p}}{\partial \bar{x}} + \frac{\nu U}{L^2}(\frac{\partial^2 \bar{u}}{\partial \bar{x}^2} + \frac{\partial^2 \bar{u}}{\partial \bar{y}^2}) + \frac{\nu U}{H^2}\frac{\partial^2 \bar{u}}{\partial \bar{z}^2}. \tag{8.42}$$

(c) Find the equivalent equations for (8.39) and (8.40).

2. Returning to the equation (8.42), consider the following special scaling:

$$W = \frac{UH}{L}, \quad P = \rho f U L. \tag{8.43}$$

Show that now (8.42) reduces to

$$\epsilon_2(\frac{\partial \bar{u}}{\partial t} + \bar{u}\frac{\partial \bar{u}}{\partial x} + \bar{v}\frac{\partial \bar{u}}{\partial y} + \bar{w}\frac{\partial \bar{u}}{\partial z}) - v =$$

$$-\frac{\partial \bar{p}}{\partial \bar{x}} + \frac{\nu}{fL^2}(\frac{\partial^2 \bar{u}}{\partial \bar{x}^2} + \frac{\partial^2 \bar{u}}{\partial \bar{y}^2}) + \frac{\nu}{fH^2}\frac{\partial^2 \bar{u}}{\partial \bar{z}^2}, \quad (8.44)$$

where

$$\epsilon_2 = \frac{U}{fL} \quad (8.45)$$

is called the *Rossby* number (see page 200 of [9] for more details on the choice of scaling and the interpretation of the Rossby number).

It is common in oceanography to distinguish between the horizontal viscosity and the vertical one, thus replacing ν by A_H and A_V in (8.44). With this new notation, the two coefficients

$$\frac{A_H}{fL^2} \quad \text{and} \quad \frac{A_V}{fH^2}$$

are called the *horizontal* and *vertical Ekman* numbers.

8.4 Some Exact Solutions of the RSWE

A novel and interesting mathematical aspect of the Coriolis effect in (8.30) is that this system of nonlinear PDEs supports nontrivial exact analytical solutions. In a series of papers between 1960s and 1990s several investigators were able to identify some of these exact solutions, solutions that seems to have significant physical counterparts in nature. At the same time, these exact solutions are invaluable when we attempt to build approximate solutions for initial-boundary value problems for the PDEs in (8.30).

We concentrate here on the special solution sought in [10], but the reader is urged to study any of the papers cited at the end of this chapter, not only for the historical value of how the ideas presented evolved, but for the sake of observing the small details in presentation and emphasis that each author brings to his development. What was observed ultimately was that these solutions, which all have an elliptic vortex structure to them, have strong resemblances to some of the large-scale structures we see near the Gulf Stream and elsewhere in the world's oceans.

These analytical solutions all have very simple spatial structures; the velocity fields are linear in x and y while the height h is quadratic in these variables. The time dependence, in both velocity and height, are quite complicated and are obtained only after solving a nonlinear system of ordinary differential equations. Since today we can solve nonlinear ODEs extremely accurately, we consider these solutions as "exact."

The special solution sought in [10] is of the following form:

$$\begin{aligned} u(x, y, t) &= U_0(t) + U_1(t)x + U_2(t)y, \\ v(x, y, t) &= V_0(t) + V_1(t)x + V_2(t)y. \\ h(x, y, t) &= A(t)x^2 + 2B(t)xy + C(t)y^2 + \\ &\quad 2D(t)x + 2E(t)y + F(t), \end{aligned} \tag{8.46}$$

The triple $\langle u, v, h \rangle$ in (8.46) is designed to be a solution of the PDEs in (8.30). Substitution of the template in (8.46), which has twelve time-dependent unknowns A, B, ..., through U_2, into (8.30) will result in a system of twelve coupled nonlinear ordinary differential equations in the unknowns. If the reader has access to a symbolic manipulator such as *Mathematica*, it would be ideal for the type of manipulation that leads to this system of ODEs. We state the first equation in this system and refer the reader to Project D as well as page 236 of [10] for the complete system:

$$\frac{dA}{dt} = -(3U_1 + V_2)A - 2V_1 B.$$

Note how the rate of change of A, the leading coefficient in the definition of h, depends on A and B to begin with (i.e., depends on how h itself varies in the x and y directions), it depends on U_1, a coefficient in the template for u, and on V_1 and V_2, the coefficients in v. The other eleven equations have a similar complexity. Project D in this chapter provides details of the derivation of the twelfth-order system in (8.46), as well as the construction of a MATLAB program to solve the initial-value problem for this system.

8.5 Linearization of SWE

In the next few sections we take a more careful look at the details of the system of PDEs in (8.26)

$$h_t + (hU)_x = 0,$$

$$(hU)_t + \frac{\partial}{\partial x}(\int_b^{b+h} u^2\, d\eta) = -g(b' + h_x)h + \nu \int_b^{b+h} \Delta u\, d\eta, \qquad (8.47)$$

$$U(x,t) = \frac{1}{h}\int_b^{b+h} u(x,\eta,t)\, d\eta,$$

by looking at some of its special solutions. As we have noted earlier, these PDEs are nonlinear and, other than some of the exact solutions we alluded to in the previous section, the main approach to understanding these PDEs is to compute their solutions by numerical methods. A different approach, however, is to linearize these equations about time-independent (stationary) equilibrium solutions, much like the approach we introduced earlier in the context of ODEs, when we studied the stability of equilibrium solutions of systems of ODEs by applying the Taylor series method.

We begin by considering the simplest basin geometry, namely a basin with a flat bottom ($b' = 0$), containing an inviscid (i.e., $\nu = 0$) fluid that is standing still at height H. We observe that the pair

$$(u, h) = (0, H), \qquad (8.48)$$

describing the state of the stationary fluid, is a solution of the equations (8.47). Our goal in this section is to study solutions (u_ϵ, h_ϵ) of the full system (8.47) which remain close to the stationary solution (8.48). To that end we consider the small perturbations of the form

$$(u_\epsilon, h_\epsilon) = (0, H) + \epsilon(\hat{u}(x,t), \hat{\eta}(x,t)), \qquad (8.49)$$

i.e., the sum of the stationary solution and a small addition that may vary in time and space. Here ϵ is a small positive number. Also, note the important assumption that $\hat{u}$ is independent of z.

Beginning with (8.47)c, and noting that $\hat{u}$, defined in (8.49), is independent of z, we have

$$U(x,t) = \frac{1}{h}\int_b^{b+h} u(x,\tau,t)\, d\tau = \epsilon\hat{u}(x,t).$$

The second variable in (8.47), h, is $H + \epsilon\hat{\eta}(x,t)$, with H constant. Hence, (8.47)a reduces to

$$\epsilon\hat{\eta}_t + \epsilon(H\hat{u})_x + \epsilon^2(\hat{\eta}\,\hat{u})_x = 0.$$

Divide the above expression by ϵ to get $\hat{\eta}_t + H\hat{u}_x + \epsilon(\hat{\eta}\hat{u})_x = 0$. Assuming that ϵ is a small number to the extent that the first two terms dominate the term multiplying ϵ, we neglect the $\epsilon(\hat{\eta}\hat{u})_x$ term and arrive at the reduced and linear equation

$$\hat{\eta}_t + H\hat{u}_x = 0, \qquad (8.50)$$

in place of (8.47)a.

We treat (8.47)b similarly. Substituting $u = U = \epsilon\hat{u}$ and $h = H + \epsilon\hat{\eta}$ into (8.47)b yields (recall that by assumption $b' = \nu = 0$)

$$\epsilon H \hat{u}_t + \epsilon^2(\hat{\eta}\,\hat{u})_x + \epsilon^2(\hat{u}^2)_x = -\epsilon g H \hat{\eta}_x - g\epsilon^2\hat{\eta}\hat{\eta}_x.$$

Divide this expression by ϵ and neglect terms with powers of ϵ to get

$$\hat{u}_t = -g\hat{\eta}_x. \tag{8.51}$$

The system of equations in (8.50) and (8.51) constitute the linearization of the SWE (8.47) about its trivial solution $(u, h) = (0, H)$. These equations can be combined, by differentiating (8.51) with respect to t, (8.50) with respect to x, to get the linear *wave equation*

$$\hat{u}_{tt} - gH\hat{u}_{xx} = 0, \tag{8.52}$$

with a similar equation for $\hat{\eta}$. In the next few sections we study the properties of this equation and discover some of its general characteristics in relation to wave propagation in shallow basins and channels.

Problems 8.5

1. Complete the calculations that leads to the wave equation (8.52).
2. Relax the assumption that the fluid is inviscid, i.e., allow $\nu \neq 0$, and derive the generalization of the equation in (8.52).
3. Return to the inviscid case, but now consider the small perturbation where the bathymetry is a small perturbation of a flat bottom, i.e.,
$$b(x) = b_0 + \epsilon B(x).$$
What complications, if any, does this additional observation contribute the derivation of the linearized equations?

8.6 Linear Wave Equation

Equation (8.52), with u replacing $\hat{u}$ and $c^2 = gH$, is rewritten in the form

$$u_{tt} - c^2 u_{xx} = 0. \tag{8.53}$$

This equation involves two derivatives in time and two derivatives in space, so naturally we expect that four additional conditions, two in time and two in space, will be required to determine a unique solution to (8.53). A typical initial-boundary value problem for (8.53) will have additional side constraints in the form of boundary conditions

$$u(0,t) = u(L,t) = 0, \tag{8.54}$$

and initial conditions

$$u(x,0) = f(x),\ u_t(x,0) = g(x). \tag{8.55}$$

Here L, a constant, represents the length of the basin, and f and g, known functions of x, represent the initial states of u.

In what follows we develop several techniques for determining the solution to (8.53)–(8.55). Some of the methods, such as the Fourier and the Characteristics methods will lead to the exact solution of the problem, while others, such as the finite difference method and the Galerkin Method give us approximate solutions.

Problems 8.6

1. Let u and v be two solutions of (8.53). Show that $w = c_1 u + c_2 v$ is also a solution of (8.53), where c_1 and c_2 are constants.

2. Let u be a solution of (8.53)–(8.55).

   (a) (*Conservation of Energy*) Show that the quantity

   $$E = \frac{1}{2}\int_0^L u_t^2 dx + \frac{c^2}{2}\int_0^L u_x^2 dx, \tag{8.56}$$

   which is the total energy of the wave u, remains constant. Hint: Differentiate the above expression with respect to t, apply an integration by parts, and use (8.53).

   (b) Show that

   $$E = \frac{1}{2}\int_0^L g^2(x)\, dx + \frac{c^2}{2}\int_0^L f'^2(x)\, dx.$$

3. Let c in (8.53) be a function of x. Let E be defined as in (8.56). Is it still true that E is constant?

8.7 Separation of Variables and the Fourier Method

The Fourier method is predicated on our ability to construct solutions of (8.53) in terms of *basis functions* or *normal modes*, which are the natural building blocks for solutions of linear PDEs such as the wave equation. We will not give a complete introduction to the Fourier series and instead refer the reader to texts that provide such an introduction, including several listed at the end of this chapter (see, for example, Chapter 14 of [15] for information on Fourier series and how they apply to general PDEs). We will, however, review here the concept of *separation of variables*, which is the enabling factor behind the use of Fourier series, and later provide the connection between this method and the *Galerkin* method and the *Method of Lines*, which are two general purpose methods for obtaining approximate solutions of evolution equations like the wave equation, but powerful enough that they also apply to a large class of nonlinear PDEs.

The method of separation of variables seeks solutions of (8.53) that treat the dependence of u on x and t separately. We look for u in the form

$$u(x,t) = F(x)G(t).$$

Substituting this template into (8.53) yields

$$G''F - c^2GF'' = 0,$$

or, after division by c^2FG,

$$\frac{G''}{c^2G} - \frac{F''}{F} = 0. \tag{8.57}$$

Since $\frac{G''}{c^2G}$ is only a function of t (recall that $c^2 = gH$ is a constant) and $\frac{F''}{F}$ only a function of x, each must be a constant, a fact that can readily be verified by differentiating (8.57) with respect to t, say. Assuming for the time being that this constant is negative, we denote it by $-\lambda^2$ and rewrite (8.57) equivalently as

$$\frac{G''}{c^2G} = -\lambda^2, \qquad \frac{F''}{F} = -\lambda^2.$$

These equations reduce to the two familiar ordinary differential equations

$$F'' + \lambda^2F = 0, \quad G'' + c^2G = 0, \tag{8.58}$$

which have the general solutions

$$F(x) = c_1 \sin\lambda x + c_2 \cos\lambda x, \quad G(t) = c_3 \sin c\lambda t + c_4 \cos c\lambda t. \tag{8.59}$$

Recall that $u(x,t) = G(t)F(x)$. We have therefore succeeded in finding a general solution of the linear wave equation, $u_{tt} - c^2u_{xx} = 0$, in the form

$$u(x,t) = (c_3 \sin c\lambda t + c_4 \cos c\lambda t)(c_1 \sin\lambda x + c_2 \cos\lambda x). \tag{8.60}$$

To complete the solution of the initial-boundary value problem (8.54)–(8.55), we need to determine the constants c_1 through c_4 and λ in such a way that the four constraints $u(0,t) = u(L,t) = 0$, the two boundary conditions, and $u(x,0) = f(x)$ and $u_t(x,0) = g(x)$, the two initial conditions, hold.

We begin by applying the first boundary condition in (8.54), that $u(0,t) = 0$, to the expression for u in (8.60). This results in

$$0 = u(0,t) = c_2(c_3 \sin c\lambda t + c_4 \cos c\lambda t),$$

which must hold for all values of t, so naturally we select $c_2 = 0$. This choice of c_2 reduces the expression for u in (8.60) to

$$u(x,t) = (A \sin c\lambda t + B \cos c\lambda t) \sin \lambda x, \tag{8.61}$$

where now A and B stand for c_1c_3 and c_1c_4, respectively. Next we apply the second boundary condition in (8.54), namely $u(L,t) = 0$, which, when applied to (8.61), results in

$$0 = u(L,t) = (A \sin c\lambda t + B \cos c\lambda t) \sin \lambda L. \tag{8.62}$$

This expression must hold for all t, hence we choose λ so that $\sin \lambda L = 0$. Since the sine function is 2π-periodic, there are quite a few angles θ, namely $\theta = n\pi$, at which $\sin\theta$ vanishes. Hence we select λ such that $\lambda L = n\pi$ or

$$\lambda_n = \frac{n\pi}{L}. \tag{8.63}$$

The quantities λ_n are the *eigenvalues* of the linear wave equation. They represent the natural frequencies at which the *normal modes* $\sin \lambda_n x$ end up being part of a solution to the wave equation when this equation is supplemented by the boundary conditions in (8.54).

The choice of the eigenvalues $\lambda_n = \frac{n\pi}{L}$, when substituted into (8.61), leads to infinitely many natural modes or solutions to (8.53) and (8.54). They are

$$u_n(x,t) = (A_n \sin \frac{n\pi ct}{L} + B_n \cos \frac{n\pi ct}{L}) \sin \frac{n\pi x}{L}. \tag{8.64}$$

Since the wave equation is linear, the superposition of any two solutions in (8.64) results in another solution to the wave equation. In fact,

$$u(x,t) = \sum_{n=1}^{\infty} (A_n \sin \frac{n\pi ct}{L} + B_n \cos \frac{n\pi ct}{L}) \sin \frac{n\pi x}{L} \tag{8.65}$$

constitutes a general solution to (8.53) where the boundary conditions (8.54) are automatically satisfied.

To complete obtaining the solution to the initial-boundary value problem (8.53)–(8.55) we need to determine the coefficients A_n's and B_n's in (8.65) so that the initial conditions (8.55) hold. The first of these conditions, that $u(x,0) = f(x)$, requires that u in (8.65) satisfy the relation

$$f(x) = u(x,0) = \sum_{n=1}^{\infty} B_n \sin \frac{n\pi x}{L}. \tag{8.66}$$

In other words, the coefficients B_n in (8.65) must also be the Fourier sine coefficients of f when f is expanded in terms of $\sin \frac{n\pi x}{L}$.

We recall the formula for computing the Fourier sine coefficients of a function f:

$$B_n = (f, \sin \frac{n\pi x}{L})/(\sin \frac{n\pi x}{L}, \sin \frac{n\pi x}{L}), \tag{8.67}$$

where the notation (f, g), called the *inner product* of the two functions f and g in the interval $(0, L)$, is defined as

$$(f, g) = \int_0^L f(x)g(x)\, dx. \tag{8.68}$$

With this definition, the expression in (8.67) reduces to the familiar formula

$$B_n = \frac{2}{L} \int_0^L f(x) \sin \frac{n\pi x}{L}. \tag{8.69}$$

A similar argument applies to A_n's. Since $u_t(x,0) = g(x)$, we have

$$g(x) = u_t(x,0) = \sum_{n=1}^{\infty} \frac{n\pi c}{L} A_n \sin \frac{n\pi x}{L},$$

which states that $\frac{n\pi c}{L} A_n$ is the Fourier sine coefficient of g, that is

$$\frac{n\pi c}{L} A_n = (g, \sin \frac{n\pi x}{L})/(\sin \frac{n\pi x}{L}, \sin \frac{n\pi x}{L}).$$

The above expression simplifies to

$$A_n = \frac{2}{n\pi c} \int_0^L g(x) \sin \frac{n\pi x}{L}. \tag{8.70}$$

Several exercises at the end of this section involve applying the Fourier method to various initial-boundary value problems. In the next section we introduce a MATLAB program to carry out all of the underlying computations.

Before leaving this topic we make one important observation about the formula in (8.65): Because $\sin \frac{n\pi ct}{L}$ and $\cos \frac{n\pi ct}{L}$ have the common fundamental period of $T = \frac{2L}{c}$ for all n, the function u inherits this period as well. Hence, any disturbance (wave) supported by the linear wave equation and the initial-boundary conditions (8.54)–(8.55) will travel in time periodically with period $T = \frac{2L}{c}$. We summarize the above discussion in the following theorem:

Theorem 8.6.1 (Fourier Series and the Wave Equation) *The solution to the initial-boundary value problem (8.53)–(8.55) is given by*

$$u(x,t) = \sum_{n=1}^{\infty} (A_n \sin \frac{n\pi ct}{L} + B_n \cos \frac{n\pi ct}{L}) \sin \frac{n\pi x}{L}$$

where A_n and B_n are given by

$$A_n = \frac{2}{n\pi c} \int_0^L g(x) \sin \frac{n\pi x}{L}, \quad B_n = \frac{2}{L} \int_0^l f(x) \sin \frac{n\pi x}{L}.$$

This solution is unique and is periodic with period $\frac{2L}{c}$.

See the exercises for a proof of the uniqueness of the solutions.

Problems 8.6

1. Let $\phi_n(x) = \sin \frac{n\pi x}{L}$ be the normal modes of the wave equation. Show that ϕ_n and ϕ_m are *orthogonal*, that is

$$(\phi_n, \phi_m) = 0,$$

the inner product (,) is defined in (8.68).

2. Consider the expression $f(x) = \sum_{n=1}^{N} a_n \phi_n(x)$ where a_n's are scalars (constants in the set of real numbers R). Suppose that $f(x) \equiv 0$ for all $x \in (0, L)$. Use the orthogonality property of ϕ_n to show that the coefficients a_n's must all vanish.

3. Consider the expression $f(x) = \sum_{n=1}^{N} a_n \phi_n(x)$ where $\phi_n(x) = \sin \frac{n\pi x}{L}$. Show that

$$\frac{2}{L} \int_0^L ||f(x)||^2 \, dx = \sum_{n=1}^{N} a_n^2.$$

4. In the analysis we presented we assumed that the constant of separation of variables was negative, i.e., $\frac{F''}{F} = -\lambda^2$. Consider now the two alternative cases:

(a) Suppose that $\frac{F''}{F} = \lambda^2$. Show that this equation's general solution is $F(x) = c_1 e^{\lambda x} + c_2 e^{-\lambda x}$. Apply the boundary conditions $u(0,t) = u(L,t) = 0$ to this expression to show that the only solution that satisfies both boundary conditions requires that $c_1 = c_2 = 0$. Thus the only solution we obtain by assuming that the constant of separation of variables is positive is the trivial solution $F(x) \equiv 0$.

(b) Suppose that $\frac{F''}{F} = 0$. Show that this equation's general solution is $F(x) = c_1 x + c_2$. Apply the boundary conditions $u(0,t) = u(L,t) = 0$ to this expression to show that, again, the only solution that satisfies both boundary conditions requires that $c_1 = c_2 = 0$, that is, the trivial solution $F(x) \equiv 0$.

5. Find the solution the following initial-boundary value problems for the wave equation $u_{tt} = c^2 u_{xx}$, subject to boundary conditions $u(0,t) = u(L,t) = 0$ and initial data $u(x,0) = f(x)$ and $u_t(x,0) = g(x)$.

   (a) $c^2 = 4$, $L = 5$, $f(x) = x(5-x)$, $g(x) \equiv 0$.

   (b) $c^2 = 16$, $L = 1$, $f(x) \equiv 0$, $g(x) \equiv 1$.

   (c) $c^2 = 25$, $L = 2$, $f(x) = \begin{cases} x, & \text{if } 0 < x \le 1, \\ 2-x & \text{otherwise} \end{cases}$, $g(x) \equiv 0$.

   (d) $c^2 = 1, L = 5$, $f(x) = \sin\frac{\pi x}{5}$, $g(x) = \begin{cases} x, & \text{if } 0 < x \le \frac{5}{2}, \\ 5-x & \text{otherwise} \end{cases}$.

6. (*Uniqueness of Solutions*) Consider the initial-boundary value problem (8.53)–(8.55). Let $u_1(x,t)$ and $u_2(x,t)$ be two solutions of this system. Let $v(x,t) = u_1(x,t) - u_2(x,t)$. Show that

   (a) v satisfies the wave equation, that is, $v_{tt} - c^2 v_{xx} = 0$, and the boundary conditions $v(0,t) = v(L,t) = 0$.

   (b) v satisfies the initial data $v(x,0) \equiv 0$ and $v_t(x,0) \equiv 0$.

   (c) v must then be identically zero, by determining its Fourier series solution.

   Thus we conclude that the solution to the initial-boundary value problem (8.53)–(8.55) is unique.

7. Consider the wave equation (8.53) with boundary conditions $u_x(0,t) = u_x(L,t) = 0$. Show that applying the method of separation of variables to this boundary value problem leads to solutions of the form

$$u_n(x,t) = (A_n \cos\frac{n\pi ct}{L} + B_n \sin\frac{n\pi ct}{L}) \cos\frac{n\pi x}{L}, \tag{8.71}$$

where $n = 0, 1, \cdots$.

8. Starting with the result of the previous problem, find the formula for the approximate solution of the initial-boundary value problem

$$u_{tt} = 4u_{xx}, \quad u_x(0,t) = u_x(3,t) = 0,$$

$$u(x,0) = f(x), u_t(x,0) = g(x).$$

9. Repeat Problems 1) to 3) but now with the normal modes $\phi_n(x) = \cos \frac{n\pi x}{L}$, with $n = 0, 1, 2, \ldots$.

8.8 Fourier Method in MATLAB

We now develop a MATLAB program to solve the initial-boundary value problem (8.53)–(8.55). The specific set of data we consider is

$$c^2 = 4, \quad L = 3,$$

$$f(x) = \begin{cases} x, & \text{if } 0 < x \le 1, \\ -\frac{1}{2}x + \frac{3}{2}, & \text{if } 1 < x \le 3, \end{cases} \tag{8.72}$$

$$g(x) \equiv 0.$$

Since g is identically zero, all of the coefficients A_n in (8.70) vanish. To determine the B_n's in (8.69) we need to split the domain of integration according to the definition of f and carry out the integrations separately as follows:

$$B_n = \frac{2}{3}(\int_0^1 x \sin \frac{n\pi x}{3}\, dx + \int_1^3 (-\frac{1}{2}x + \frac{3}{2}) \sin \frac{n\pi x}{3}\, dx) =$$

$$\frac{12}{n^2\pi^2} \sin^3 \frac{n\pi}{3}. \tag{8.73}$$

Thus the exact solution of the initial-boundary value problem (see (8.65)) is

$$u(x,t) = \sum_{n=1}^{\infty} \frac{12}{n^2\pi^2} \sin^3 \frac{n\pi}{3} \cos \frac{2n\pi t}{3} \sin \frac{n\pi x}{3}. \tag{8.74}$$

The main program below consists of a segment where the Fourier coefficients are computed using MATLAB's `quad`, `quadl`, `quadgk` or `quadv`, and a segment that computes u from (8.74) and then plots the graph of the various snapshots. Here is the syntax that computes the first ten coefficients B_n's using `quadv`, which integrates an array of functions (such as $f(x) \sin \frac{n\pi x}{L}$ with n ranging between 1 and 10, say):

```
b=2/3*(quadv('x.*sin((1:10)*pi*x/3)',0,1)+ ...
     quadv('(-1/2*x+3/2).*sin((1:10)*pi*x/3)',1,3));
b
```

MATLAB returns

```
b =

  Columns 1 through 9

    0.7897    0.1974    0.0000   -0.0494   -0.0316    0.0000
    0.0161    0.0123   -0.0000

  Column 10

   -0.0079
```

To compare these results with the exact value of B_n's (from 8.73) we compute the exact B_n's and use the `max` and `abs` commands within MATLAB to measure the error incurred in using `quadv`:

```
exactb=12./((1:10).^2*pi^2).*sin((1:10)*pi/3).^3;
exactb
```

which results in

```
exactb =

  Columns 1 through 9

    0.7897    0.1974    0.0000   -0.0494   -0.0316   -0.0000
    0.0161    0.0123    0.0000

  Column 10

   -0.0079
```

These values agree very well with the values from `quadv`, at least with the accuracy embedded in the first four significant digits used to display the results. The absolute error in this computation is measured as follows:

```
max(abs(b-exactb))
```

or

```
6.5904e-009
```

The relative error, computed by

```
max(abs(b-exactb))/max(abs(b))
```

is

```
8.3453e-009.
```

The following program combines `inline` and `quadv` with the plotting capabilities of MATLAB to produce the graph of the snapshots seen in Figure 8.2:

```
clf;
c=2;
L=3;
n=50;   % Number of terms in the Fourier Sine series
f1=inline('x.*sin((1:n)*pi*x/3)','n','x');
f2=inline('(-1/2*x+3/2).*sin((1:n)*pi*x/3)','n','x');
b=2/3*(quadv(@(x)f1(n,x),0,1)+quadv(@(x)f2(n,x),1,3));
x=0:0.01:L;
sine=sin(pi*(1:n)'*x/3);
for i=1:8
    t=0.3*(i-1);
    coeff=b.*cos(2*pi*(1:n)*t);
    u=coeff*sine;
    subplot(4,2,i)
    plot(x,u)
    title(['u at t = ', num2str(t)]);
    axis([0 3 -1 1]);
    hold on
end
```

This figure demonstrates one of the key features of the wave equation, how discontinuities (shock waves) propagate. In this example the initial condition f is a continuous function in its domain, the interval $(0, 3)$, but is not differentiable at $x = 1$ where the derivative f' experiences a jump. This discontinuity in the derivative of f manifests itself in u_x, propagates in both directions, as seen in the snapshots in Figure 8.2: at $t = 0.3$ the discontinuity that was originally located at $x = 1$ can now be found around $x = 0.3$ and 1.6, approximately. It turns out, as we will demonstrate in the next section, that the two shock waves propagate with wave speed c, in this example $c = 2$, until they reach the endpoints of the domain, at which time they reverse their course and propagate toward each other and reunite at time $t = \frac{L}{c}$, in this example $\frac{3}{2}$, which is half of the period of oscillation of u.

So far we have seen a graphical evidence of how the wave operator propagates discontinuities. In the next section, where the method

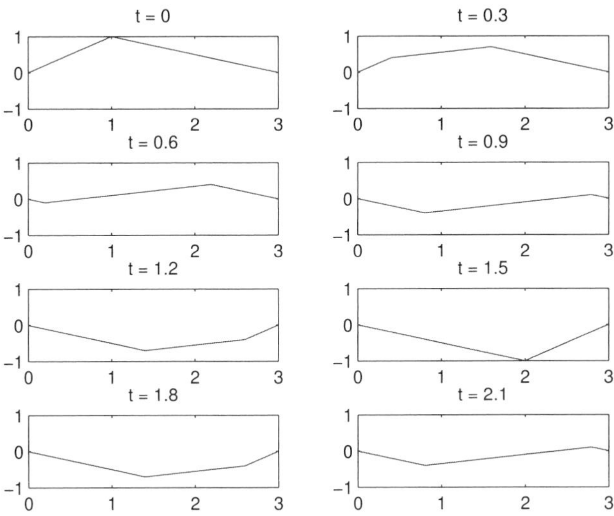

FIGURE 8.2: Snapshots of the function u where u is the solution to the initial-value problem $u_{tt} - 4u_{xx} = 0$, $u(0,t) = u(3,t) = 0$, $u(x,0) = f(x)$, with f given in (8.72), and $g \equiv 0$.

of characteristics is introduced, we will revisit this feature of the wave equation once we are able to rewrite the analytic solution in a different and more illuminating form.

We end this section by pointing out that the above MATLAB code may be altered slightly, by introducing the `drawnow` command, to animate the snapshots:

```
clf;
c=2;     % Speed of propagation
L=3;     % Domain size
period=2*L/c;
count = 100; % Number of Snapshots
n=50;   % Discretization along the x axis
f1=inline('x.*sin((1:n)*pi*x/3)','n','x');
f2=inline('(-1/2*x+3/2).*sin((1:n)*pi*x/3)','n','x');
b=2/3*(quadv(@(x)f1(n,x),0,1)+quadv(@(x)f2(n,x),1,3));
x=0:0.01:3;
sine=sin(pi*(1:n)'*x/3);
for i=1:count
    t=period/count*(i-1);
```

```
    coeff=b.*cos(2*pi*(1:n)*t/3);
    u=coeff*sine;
    drawnow
    plot(x,u)
    title(['t = ', num2str(t)]);
    axis([0 3 -1 1]);
end
```

Problems 8.8

1. Execute the various MATLAB programs in this section and obtain the results and figures cited.

2. Write a MATLAB program to plot the snapshots of the solution to each initial-boundary value problem stated in Problem 5, Section 8.7.

8.9 Method of Characteristics

One of the features of the wave equation we discussed in the previous sections was how a discontinuity at a point x_0 in the initial data propagates by the wave operator with speed c. We observed this attribute of the wave equation after we were able to determine the solution to the initial-boundary value problem and visualize the snapshots of the solution at various time values. The method we describe here, the *method of characteristics*, is another way where this property of the wave equation becomes apparent. This method has a second important property in that it can be applied to nonlinear problems, almost as easily as we apply it to linear problems such as the wave operator in (8.53).

We now develop and apply the method of characteristics to (8.53)–(8.55). For background information on this method see Chapter 13 of [15]. Also, in a project at the end of this chapter we develop this method in its traditional way. Here, instead, we appeal to the formula in (8.65) to show that there is an entirely different way of representing the solution to (8.53)–(8.55), which in light of the uniqueness of solutions to this system (see Problem 6, Section 8.7) is in fact the same solution we obtained using the Fourier method.

For convenience we repeat formula (8.65), written slightly differently and with ω_n replacing $\frac{n\pi}{L}$:

$$u(x,t) = \sum_{n=1}^{\infty} B_n \cos(c\,\omega_n t)\sin(\omega_n x) + \sum_{n=1}^{\infty} A_n \sin(c\,\omega_n t)\sin(\omega_n x). \quad (8.75)$$

Recall the following elementary formulas:

$$\begin{cases} 2\cos t \sin x &= \sin(x-t) + \sin(x+t) \\ 2\sin t \sin x &= \cos(x-t) - \cos(x+t) \end{cases} \quad (8.76)$$

The first identity in (8.76) allows us to cast the first summation in (8.75), $\sum_{n=1}^{\infty} B_n \cos(c\,\omega_n t)\sin(\omega_n x)$, in the following form:

$$\frac{1}{2}\sum_{n=1}^{\infty} B_n(\sin\omega_n(x-ct) + \sin\omega_n(x+ct)). \quad (8.77)$$

Recall that B_n's are the Fourier sine coefficients of f, the initial data in (8.55), that is,

$$f(x) = \sum_{n=1}^{\infty} B_n \sin(\omega_n x), \qquad \text{for all } x \in (0, L). \quad (8.78)$$

In view of the identity in (8.78), we can restate the expression in (8.77) as

$$\sum_{n=1}^{\infty} B_n \cos(c\,\omega_n t)\sin(\omega_n x) = \frac{1}{2}(f(x-ct) + f(x+ct)). \quad (8.79)$$

The expression in (8.79) is valid as long as $x - ct$ and $x + ct$ remain in the domain of f, which is the interval $(0, L)$. What is remarkable is that the expression on the left side of (8.79) provides us with the correct representation of f outside of its original domain, i.e, this expression defines a function *for all* x *and* t, if we are willing or need to extend the definition of function f outside of the interval $(0, L)$, to the entire set of real numbers. In fact, this is the strategy we adopt, that is, we extend f to the entire real line by assigning values from the left side of (8.79) when x and t cause $x - ct$ or $x + ct$ to land outside of the interval $(0, L)$. We will denote this extension of f by $\tilde{f}$ and write

$$\tilde{f}(x) = \sum_{n=1}^{\infty} B_n \sin\frac{n\pi x}{L}, \qquad \text{for all } x \in R. \quad (8.80)$$

The function $\tilde{f}$ has two important properties that will help us in constructing its image geometrically. First, because each sine function is an odd function, it follows that $\tilde{f}$, as a sum of odd functions, is itself an odd function:

$$\tilde{f}(-x) = -\tilde{f}(x). \tag{8.81}$$

Thus to construct an image of $\tilde{f}$ on the interval $(-L, 0)$, we simply reflect the image of f on the interval $(0, L)$ about the origin. We now have $\tilde{f}$ defined on an interval of length $2L$. Next we note that each sine function in (8.80) is $2L$ periodic. Therefore, the function $\tilde{f}$ must be $2L$ periodic. Since we already have an image of $\tilde{f}$ on an interval of length $2L$, we can construct its image everywhere along the real line. This completes the construction of $\tilde{f}$. With this definition of $\tilde{f}$ we have

$$\sum_{n=1}^{\infty} B_n \cos(c\,\omega_n t)\sin(\omega_n x) = \frac{1}{2}(\tilde{f}(x-ct) + \tilde{f}(x+ct)). \tag{8.82}$$

The second summation in (8.75) is treated in exactly the same way using the second trigonometric identity in (8.77). This summation ends up being related to $\tilde{g}$, the odd and $2L$-periodic extension of the initial function g from (8.55), as follows:

$$\sum_{n=1}^{\infty} A_n \sin(c\,\omega_n t)\sin(\omega_n x) = \frac{1}{2c}\int_{x-ct}^{x+ct} \tilde{g}(\tau)\,d\tau. \tag{8.83}$$

We leave the details of this derivation to an exercise.

Combining now the formulas in (8.82) and (8.83), we have a closed form solution, known as the *D'Alembert* solution, to the initial-boundary value problem (8.53)–(8.55):

$$u(x,t) = \frac{1}{2}(\tilde{f}(x-ct) + \tilde{f}(x+ct)) + \frac{1}{2c}\int_{x-ct}^{x+ct} \tilde{g}(\tau)\,d\tau. \tag{8.84}$$

We state this result as a theorem.

Theorem 8.9.1 (D'Alembert's Solution)
The formula in (8.84) gives a solution to the initial-boundary value problem (8.53)–(8.55).

One of the consequences of (8.84) is that it demonstrates why c is the speed of propagation of any disturbance (wave) in (8.53)–(8.55). To see this, we note that the graphs of $y = \tilde{f}(x)$ and $y = \tilde{f}(x-ct)$ are identical except for a shift, that the graph of $\tilde{f}(x-ct)$ is the same as the graph of $\tilde{f}(x)$ except that it is shifted to the right by the amount ct. Similarly, the graph of $\tilde{f}(x+ct)$ is identical with the graph of $\tilde{f}(x)$ except for a shift to the left by the amount ct. We reach the same conclusion for the integral in (8.84) because we can break it up into a sum of two integrals, for example by writing it as

$$\int_{x-ct}^{x+ct} \tilde{g}(\tau)\, d\tau = \int_{x-ct}^{0} \tilde{g}(\tau)\, d\tau + \int_{0}^{x+ct} \tilde{g}(\tau)\, d\tau,$$

and applying the same argument to each individual integral on the right side. Therefore, formula (8.84) suggests that any initial disturbance in f and g is carried by $\tilde{f}$ and $\tilde{g}$ by simply shifting these functions to the right and to the left with speed c.

Another consequence of D'Alembert's solution is that it shows that any disturbance in the initial data propagates to the left and to the right with speed c. In fact, and for the sake of argument let us consider the case where $g \equiv 0$, D'Alembert's solution reduces to

$$u(x,t) = \frac{1}{2}(\tilde{f}(x-ct) + \tilde{f}(x+ct)),$$

demonstrating that the initial disturbance $u(x,0)$ splits into two equal parts, half of which travels to the right and the other half to the left. Once these waves reach the boundary they simply reflect and reverse their trajectories toward each other and combine to reconstruct the initial disturbance.

Before leaving this section we point out an interesting application of (8.84) in the context of a tsunami formed in deep water and approaching a coastline. According to our observations, this disturbance will travel with speed $c = \sqrt{gH}$. In waters that are about 5 kilometers deep (so that $H = 5{,}000$ meters), with a standard value of $g = 9.8$ m/s^2, we arrive at $c = 221.36$ m/s, or a speed a little over 700 km/h. When this wave enters shallower waters and approaches a coastline, the speed of the fluid particles that are closer to the shoreline decreases, since H decreases, while the fluid particles are farther away from the coastline are still experiencing the higher speed. Consequently the wave has a tendency to climb on itself, its amplitude rising steadily as it approaches the coastline and bringing an enormous amount of potential energy into the shoreline.

Problems 8.9

1. Show by direct differentiation that $F(x-ct)$ and $G(x+ct)$ are solutions of the linear wave equation (8.53).

2. Show that the D'Alembert solution (8.84) satisfies the initial and boundary conditions in (8.54)–(8.55).

3. Consider the wave equation $u_{tt} - 9u_{xx} = 0$. Find its general solution.

4. Consider the wave equation $u_{tt} - 9u_{xx} = 0$ with initial conditions

$$u(x,0) = \begin{cases} \cos x, & \text{when } -\frac{\pi}{2} < x < \frac{\pi}{2}, \\ 0 & \text{otherwise,} \end{cases}$$

and $g(x) \equiv 0$. Plot the graph of the solution at $t = 0, 0.1, 0.2$ and 0.3.

5. Consider the wave equation $u_{tt} - 4u_{xx} = 0$ on the real line with initial conditions $u(x,0) \equiv 0$ and

$$g(x) = \begin{cases} x, & \text{when } 0 < x < 1, \\ 2 - x & \text{when } 1 < x < 2, \\ 0 & \text{otherwise.} \end{cases}$$

Plot the graph of the solution at $t = 0, 0.1, 0.2$ and 0.3.

8.10 D'Alembert's Solution in MATLAB

The formula in (8.84) can be readily constructed in MATLAB. The main issue is to use appropriate commands within MATLAB to construct $\tilde{f}$ and $\tilde{g}$ properly. We show the code for the same initial-boundary value problem in (8.72). The key in the following program is in the definitions of the `subdomain` statements and the use of MATLAB's `any` command. The purpose of these definitions is to allow MATLAB's `If` statement to work properly with vectors – one could write this program without using `any`, but by using it we are taking advantage of MATLAB's powerful capabilities with vector operations.

As described earlier, $\hat{f}$'s definition, which is defined in `fhat.m` below, starts out in the domain $(0, 3)$ – `subdomain1` and `subdomain2` are used to define $\hat{f}$, which equals f in $(0, 3)$, in this interval. `subdomain3` is the extension of the domain to the symmetric interval $(-3, 0)$, and `subdomain4` and `subdomain5` are needed to define $\hat{f}$ outside of the interval $(-3, 3)$ by its natural periodic extension.

```
% Definition of fhat.m
%
function y = fhat(x),
%
subdomain1 = (x >= 0) & (x < 1);
subdomain2 = (x >= 1) & (x <= 3);
```

```
subdomain3 = (x >= -3) & (x < 0);
subdomain4 = x > 3;
subdomain5 = x < -3;
%
if any(subdomain2)
   x(subdomain2) = -1/2*(x(subdomain2) - 3);
end
%
if any(subdomain3)
   x(subdomain3) = - fhat(-x(subdomain3));
end
%
if any(subdomain4)
   x(subdomain4) = fhat(x(subdomain4)-6);
end
%
if any(subdomain5)
   x(subdomain5) = fhat(x(subdomain5)+6);
end
%
%
y = x;
```

The function `fhat.m` is called upon by `DAlembert.m` listed below:

```
clf;
c=2;
L=3;
x=0:0.01:L;
rows=4;
for i=1:8
   subplot(rows,2,i)
   t=0.3*(i-1);
   plot(x,0.5*(fhat(x-c*t)+fhat(x+c*t)));
   title(['t = ', num2str(t)]);
   axis([0 3 -1, 1])
   hold on
end
```

Figure 8.3 shows the output of this program.

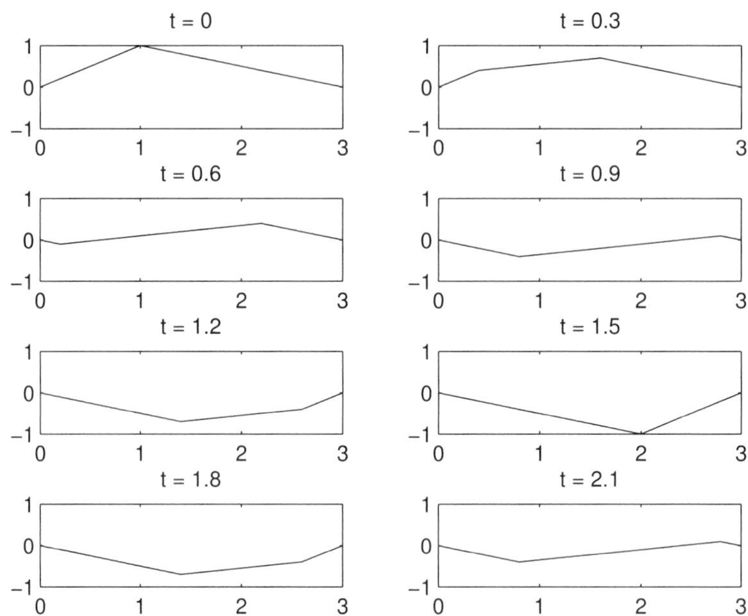

FIGURE 8.3: Using the D'Alembert Method, this figure shows the snapshots of u where u is the solution to the initial-value problem $u_{tt} - 4u_{xx} = 0$, $u(0,t) = u(3,t) = 0$, $u(x,0) = f(x)$, with f given in (8.72), and $g \equiv 0$.

Problems 8.10

1. Consider the IBVP problem

$$u_{tt} = 0.1\, u_{xx},$$

$$u(0,x) = u(1,x) = 0,$$

and

$$u(x,0) = \sin \pi x, \quad u_t(x,0) \equiv 1.$$

Find the solution to this problem in MATLAB and plot its representative snapshots over a period.

2. Consider the IBVP

$$u_{tt} = 2.23\, u_{xx},$$

subject to the boundary conditions

$$u(0,t) = 0, \quad u(1.41,t) = 0,$$

and initial conditions

$$u(x,0) = 0.1f(10(x-0.7)), \quad g(x) \equiv 1,$$

where f is defined by

$$f(x) = \begin{cases} \frac{1}{4}(1+\cos x)^2, & \text{when } -\pi \le x \le \pi, \\ 0 & \text{otherwise.} \end{cases}$$

Find the solution to this problem in MATLAB and plot its representative snapshots over a period.

8.11 Method of Lines and Wave Equation

In the previous sections we applied the separation of variables method and the characteristics method to the wave equation

$$u_{tt} = c^2 u_{xx}, \qquad x \in (0, L), \tag{8.85}$$

subject to boundary conditions

$$u(0,t) = u(L,t) = 0, \tag{8.86}$$

and initial conditions

$$u(x,0) = f(x),\ u_t(x,0) = g(x). \tag{8.87}$$

and obtained solutions to this IBVP. We now introduce another effective method to find a numerical solution to IBVP for (8.85), using a combination of the finite difference method and the method of lines, where we take advantage of the build-in ODE solver `ode45` of MATLAB. Referring to the PDE in (8.85), we discretize the spatial derivative u_{xx} using the centered difference scheme

$$u_{xx}(x_i, t) \approx \frac{1}{h^2}\left[u_{i+1}(t) - 2u_i(t) + u_{i-1}(t)\right] \tag{8.88}$$

and leave the time dependence intact. This semi-discretized approach replaces the PDE $u_{tt} = c^2 u_{xx}$ in (8.85) by the system of ODEs

$$\frac{d^2 u_i}{dt^2} = \frac{c^2}{h^2}(u_{i+1}(t) - 2u_i(t) + u_{i-1}(t)), \qquad i = 1, 2, ..., n, \tag{8.89}$$

where

$$u_i(t) = u(x_i, t)$$

corresponds to the solution values at the interior points $x_i \in (0, L)$. This system is supplemented by the discretized initial data obtained from the information in (8.87). The boundary conditions in (8.86) lead to

$$u_0(t) = u_{n+1}(t) = 0. \tag{8.90}$$

To implement (8.89) in MATLAB we appeal to `ode45`. To prepare this system for `ode45`, which is written to apply to first order systems, we reduce (8.89) to a first order system by defining a new variable v_i as the first derivative of u_i and convert the system of n equations in (8.89) to the following system of $2n$ equations:

$$\frac{du_i}{dt} = v_i, \quad \frac{dv_i}{dt} = \frac{c^2}{h^2}(u_{i+1}(t) - 2u_i(t) + u_{i-1}(t)), \qquad i = 1, 2, ..., n, \tag{8.91}$$

with initial conditions

$$u_i(0) = f(x_i), \quad v_i(0) = g(x_i). \tag{8.92}$$

The following program shows how to implement the method of line for the example in (8.72). The first file, called `waveeqSYS1.m`, introduces the equations in (8.91). The second file, called `waveeqMOLRun.m`, calls on `ode45` and on `waveeqSYS1.m` to solve (8.91)–(8.92). First `waveeqSYS1.m`:

```
function yprime=waveeqSYS1(t,y);
%
global nn h c;
% y represents u and v
u=y(1:nn); v=y(nn+1:2*nn);
term1=c^2/(h^2)*(u(3:nn)-2*u(2:nn-1)+u(1:nn-2));
uprime=v;
vprime=[0; term1; 0];
yprime=[uprime; vprime];
```

Next, we present the program `waveeqMOLRun.m`, which calls `waveeqSYS1.m`:

```
clear all
clf
a1=cputime; % start cpu clock
global nn h c;
c=2;
h=0.01;
%
x=0:h:3;
```

```
nn=length(x);
%
u0=fhat(x);
v0=zeros(nn,1);
%
plot(x,u0);
axis([0 3  -1 1])
drawnow
y0=[u0 v0'];
for i=1:500
    [t,y]=ode45('waveeqSYS1',[0,0.04], y0);
    approximate=y(length(t),1:nn);
    plot(x,approximate)
    axis([0 3 -1 1])
    time=i*t(length(t));
    title(['Wave Equation, Method of Line, time =',
     num2str(time)]);
    drawnow
    y0=y(length(t),:);
end
a2=cputime;  % end cpu clock
a2-a1
```

The M-file `fhat.m` was defined in Section 8.10. Note the use of `drawnow` to animate the graphs of $u(t, x)$ for specific t's (in this case 0.04), as well as the use of `cputime` to monitor how much computing time it takes to solve this run of `waveeqMOLRun.m`. Figure 8.4 shows the snapshot $u(x, 20)$. With the specific stepsize used in this example, `ode45` solves an initial value problem for a system consisting of 602 equations. One of the features to note in this simulation is the small oscillations in the solution when $t = 20$. This lack of accuracy is the result of the initial condition f not being differentiable. By contrast when we apply this program to $f(x) = \sin \frac{\pi x}{3}$ we end up with a more accurate simulation, as seen in Figure 8.5.

The approach we introduced for solving the wave equation can be extended for considerably more complicated problems, including nonlinear PDEs, systems of PDEs, and higher-dimensional problems. We will explore some of these possibilities in the projects at the end of this chapter.

Problems 8.11

1. Use MATLAB and generate the graphs in Figures 8.4 and 8.5.

2. Alter the program `waveeqMOLRun.m` and `waveeqSYS1.m` appropri-

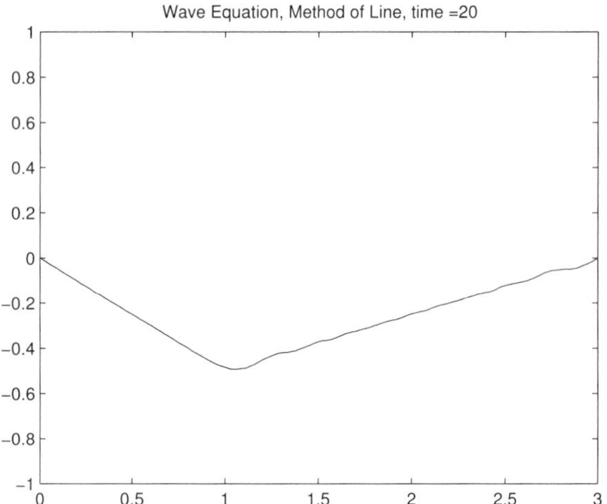

FIGURE 8.4: Using the Method of Line, this figure shows the snapshot of $u(x, 20)$ where u is the solution to the initial-value problem $u_{tt} - 4u_{xx} = 0$, $u(0,t) = u(3,t) = 0$, $u(x,0) = f(x)$, with f given in (8.72), and $g \equiv 0$.

ately to solve the following initial problems for the wave equation (8.53) in the interval $(0, 3)$, with boundary condition $u(0,t) = u(3,t) = 0$ and with initial data

(a) $f(x) = 0$, $g(x) = \sin\frac{\pi x}{3}$.

(b) $f(x) = \begin{cases} x & 0 < x < \frac{3}{2}, \\ 3 - x & \frac{3}{2} \le x < 3 \end{cases}$ and $g(x) = \sin\frac{\pi x}{3}$.

8.12 Project A: Method of Characteristics for General PDEs

In Section 8.9 we motivated the characteristics method by manipulating the formula (8.75), which was obtained based on the separation of variables and the Fourier series method. We now introduce an independent derivation of the D'Alembert formula. Our strategy is to show that under a suitable transformation (ξ, η) of the independent variables (x, t), we are able to convert the wave equation

$$u_{tt} - c^2 u_{xx} = 0, \tag{8.93}$$

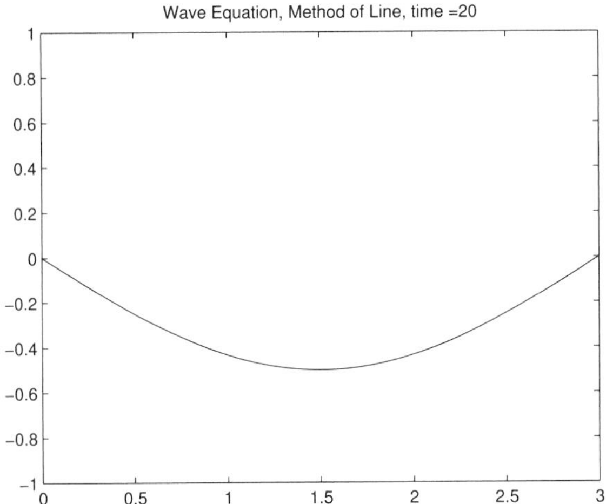

FIGURE 8.5: Using the Method of Line, this figure shows the snapshot of $u(x, 20)$ where u is the solution to the initial-value problem $u_{tt} - 4u_{xx} = 0$, $u(0,t) = u(3,t) = 0$, $u(x,0) = \sin\frac{\pi x}{3}$, and $g \equiv 0$.

to the second–order partial differential equation

$$U_{\xi\eta} = 0. \tag{8.94}$$

which can then be easily solved by simple integration with respect to ξ and η.

1. We start by introducing a general change of variables (ξ, η) by

$$\xi = h(x,t), \qquad \eta = k(x,t). \tag{8.95}$$

We seek h and k in such a way that the transformed (8.93) takes the form (8.94). Let U be related to u through the relation

$$U(\xi, \eta) \equiv u(x,t). \tag{8.96}$$

Use the chain rule of differentiation to show that

(a)

$$\frac{\partial u}{\partial t} = \frac{\partial U}{\partial \xi}\frac{\partial h}{\partial t} + \frac{\partial U}{\partial \eta}\frac{\partial k}{\partial t}. \tag{8.97}$$

and

$$\frac{\partial u}{\partial x} = \frac{\partial U}{\partial \xi}\frac{\partial h}{\partial x} + \frac{\partial U}{\partial \eta}\frac{\partial k}{\partial x}. \tag{8.98}$$

2. By applying the chain rule of differentiation a second time show that (assume that U is smooth enough that $U_{\xi\eta} = U_{\eta\xi}$ holds)

   (a)

$$\frac{\partial^2 u}{\partial t^2} = (\frac{\partial h}{\partial t})^2 \frac{\partial^2 U}{\partial \xi^2} + 2\frac{\partial h}{\partial t}\frac{\partial k}{\partial t}\frac{\partial^2 U}{\partial \xi \partial \eta} +$$
$$(\frac{\partial k}{\partial t})^2 \frac{\partial^2 U}{\partial \eta^2} + \frac{\partial U}{\partial \xi}\frac{\partial^2 h}{\partial t^2} + \frac{\partial U}{\partial \eta}\frac{\partial^2 k}{\partial t^2}, \tag{8.99}$$

   and

   (b)

$$\frac{\partial^2 u}{\partial x^2} = (\frac{\partial h}{\partial x})^2 \frac{\partial^2 U}{\partial \xi^2} + 2\frac{\partial h}{\partial x}\frac{\partial k}{\partial x}\frac{\partial^2 U}{\partial \xi \partial \eta} +$$
$$(\frac{\partial k}{\partial x})^2 \frac{\partial^2 U}{\partial \eta^2} + \frac{\partial U}{\partial \xi}\frac{\partial^2 h}{\partial x^2} + \frac{\partial U}{\partial \eta}\frac{\partial^2 k}{\partial x^2}. \tag{8.100}$$

3. Transform equation (8.93) to (8.94) by substituting (8.99) and (8.100) in (8.93) and rearranging terms to show that

$$u_{tt} - c^2 u_{xx} = 0$$

   is equivalent to

$$A\frac{\partial^2 U}{\partial \xi^2} + 2B\frac{\partial^2 U}{\partial \xi \partial \eta} + C\frac{\partial^2 U}{\partial \eta^2} + D\frac{\partial U}{\partial \xi} + E\frac{\partial U}{\partial \eta} = 0, \tag{8.101}$$

   where

$$A = (\frac{\partial h}{\partial t})^2 - c^2(\frac{\partial h}{\partial x})^2, \tag{8.102}$$

$$B = \frac{\partial h}{\partial t}\frac{\partial k}{\partial t} - \frac{\partial h}{\partial x}\frac{\partial k}{\partial x}, \tag{8.103}$$

$$C = (\frac{\partial k}{\partial t})^2 - c^2(\frac{\partial k}{\partial x})^2, \tag{8.104}$$

$$D = \frac{\partial^2 h}{\partial x^2}, \tag{8.105}$$

$$E = -\frac{\partial^2 k}{\partial t^2}. \tag{8.106}$$

   To arrive at (8.94) choose h and k such that $A = C = 0$. Show that these constraints lead to identical differential equations for h and k:

$$\frac{\partial h}{\partial t} = \pm c\frac{\partial h}{\partial x}, \qquad \frac{\partial k}{\partial t} = \pm c\frac{\partial k}{\partial x}. \tag{8.107}$$

4. Show that
$$h(x,t) = x \pm ct \tag{8.108}$$
are the only solutions of (8.107).

5. Because the function k in (8.107)b satisfies the same equation as h, one ends up with the same set of solutions for k. Hence, we use one of the functions in (8.107), say $x - ct$, for h, and the other for k.

6. Show that with these choices for h and k the coefficient B is not zero while D and E are identically zero. Thus (8.101) reduces to (8.94). The curves
$$\xi = x - ct \qquad \eta = x + ct,$$
are called the characteristic lines of the wave equation.

7. Show that
$$U(\xi, \eta) = F(\xi) + G(\eta), \tag{8.109}$$
is the solution to (8.94), where F and G are arbitrary smooth functions of their arguments. Thus
$$u(x,t) = F(x - ct) + G(x + ct), \tag{8.110}$$
is the general solution to the original wave equation in (8.93).

8.13 Project B: Variations on the Method of Lines

The method of line was introduced for the linear wave equation in Section 8.11. This method is also very effective for more complex problems, a few of which we will explore here.

1. Consider the initial value problem
$$u_{tt} = c^2 u_{xx} = F(x,t), \tag{8.111}$$
subject to the initial and boundary conditions
$$u(0,t) = u(L,t) = 0, \quad u(x,0) = u_t(x,0) \equiv 0.$$
Alter the programs `waveeqSYS1.m` and `waveeqMOLRun.m` appropriately to apply the method of line to solve (8.111). Apply the altered programs to (8.111) if

(a) $F(x,t) \equiv 1$.

(b) $F(x,t) = \sin t$.

(c) $F(x,t) = \frac{x^2 \sin t}{1+x^2}$.

2. Alter the programs `waveeqSYS1.m` and `waveeqMOLRun.m` appropriately to apply the method of line to solve the linear wave equation (8.53) with the initial data in (8.55) but with boundary conditions

$$u(0,t) = b_1(t), \quad u(L,t) = b_2(t). \tag{8.112}$$

Apply the altered programs to find approximate solutions of the following problems (in each case let $f = g = 0$):

(a) $b_1(t) = 1 - \cos t, \quad b_2(t) \equiv 0$.

(b) $b_1(t) = 1 - \cos t, \quad b_2(t) = -\sin t$.

3. Extend the method of line to solve the following initial-boundary value problem:

$$u_{tt} + \alpha u_t + \beta u - c^2 u_{xx} = F(x,t), \tag{8.113}$$

subject to the initial and boundary conditions

$$u(0,t) = u(L,t) = 0, \quad u(x,0) = f(x),\ u_t(x,0) = g(x).$$

With α, β, c, L, F, f and g of your own choosing, run the updated versions of `waveeqSYS1.m` and `waveeqMOLRun.m`.

4. Extend the method of line to solve the following nonlinear initial-boundary value problem:

$$u_{tt} - c^2 u_{xx} = \sin u, \tag{8.114}$$

$$u(0,t) = u(L,t) = 0, \quad u(x,0) = f(x),\ u_t(x,0) = g(x).$$

8.14 Project C: An Inverse Problem

We have already encountered the Initial-Boundary Value Problem

$$u_{tt} = c^2 u_{xx}, \tag{8.115}$$

$$u(0,t) = u(L,t) = 0, \qquad u(x,0) = f(x),\ u_t(x,0) = g(x).$$

Using separation of variables and the Fourier series method, we arrived at the following series solution of this IBVP:

$$u(x,t) = \sum_{n=1}^{\infty} (a_n \cos \frac{n\pi ct}{L} + b_n \sin \frac{n\pi ct}{L}) \sin \frac{n\pi x}{L}, \qquad (8.116)$$

where, with $\phi_n(x) = \sin \frac{n\pi x}{L}$, the coefficients a_n and b_n are determined from the following Fourier formulas:

$$a_n = \frac{(f, \phi_n)}{(\phi_n, \phi_n)}, \qquad b_n = \frac{(f, \phi_n)}{n\pi c(\phi_n, \phi_n)}. \qquad (8.117)$$

The question we would like to consider in this project is the inverse of how the IBVP in (8.115) is posed. Instead of having knowledge of f and g, we would like to investigate how much information we can obtain about $u(x,t)$ if we have some data about u at a fixed point x_0 where $0 < x < L$. For example, suppose that we have been able to collect the following data

$$u(x_0, t) = h(t), \qquad \text{for} \qquad T_0 < t < T_1. \qquad (8.118)$$

Is it possible to recover u at other values of x different from x_0? The goal here is to take advantage of (8.118), knowing that u must satisfy (8.115) for some f and g, and to explore if it is possible to recover the coefficients a_n's and b_n's from our knowledge of h.

We begin the analysis of this problem by observing that (8.118) implies that h and a_n's and b_n's are related by

$$h(t) = \sum_{n=1}^{\infty} (a_n \cos \frac{n\pi ct}{L} + b_n \sin \frac{n\pi ct}{L}) \sin \frac{n\pi x_0}{L}. \qquad (8.119)$$

Our approach here will be to use the *Galerkin* method. Let

$$\psi_{1,n}(t) = \cos \frac{n\pi ct}{L}, \qquad \chi_n(t) = \sin \frac{n\pi ct}{L}, \qquad (8.120)$$

and

$$\bar{a}_n = \sin \frac{n\pi x_0}{L} a_n, \qquad \bar{b}_n = \sin \frac{n\pi x_0}{L} b_n, \qquad (8.121)$$

in terms of which (8.119) becomes

$$h(t) = \sum_{n=1}^{\infty} \bar{a}_n \psi_n(t) + \bar{b}_n \chi_n(t). \qquad (8.122)$$

The Galerkin method suggests taking the inner product of (8.122) with each $\psi_{i,j}(t)$ as i ranges over one and two and j ranges over all positive integers. We end up with the following infinite system of algebraic equations:

$$\begin{array}{rcl}
\sum_{n=1}^{\infty} \bar{a}_n(\psi_n, \psi_1) + \bar{b}_n(\chi_n, \psi_1) & = & (h, \psi_1), \\
\sum_{n=1}^{\infty} \bar{a}_n(\psi_n, \psi_2) + \bar{b}_n(\chi_n, \psi_2) & = & (h, \psi_2), \\
\sum_{n=1}^{\infty} \bar{a}_n(\psi_n, \psi_3) + \bar{b}_n(\chi_n, \psi_3) & = & (h, \psi_3), \\
\cdots & & \cdots \\
\cdots & & \cdots \\
\sum_{n=1}^{\infty} \bar{a}_n(\psi_n, \chi_1) + \bar{b}_n(\chi_n, \chi_1) & = & (h, \chi_1), \\
\sum_{n=1}^{\infty} \bar{a}_n(\psi_n, \chi_2) + \bar{b}_n(\chi_n, \chi_2) & = & (h, \chi_2), \\
\sum_{n=1}^{\infty} \bar{a}_n(\psi_n, \chi_3) + \bar{b}_n(\chi_n, \chi_3) & = & (h, \chi_3), \\
\cdots & & \cdots \\
\cdots & & \cdots
\end{array} \tag{8.123}$$

1. Let $T_0 = 0$ and $T_1 = T$. Show that if $T = \frac{2L}{c}$ then

$$a_n = C_n \int_0^T h(t)\psi_{1,n}(t)\, dt, \quad b_n = C_n \int_0^T h(t)\chi_n(t)\, dt, \tag{8.124}$$

   where

$$C_n = \frac{c}{L \sin \frac{n\pi x_0}{L}},$$

   for $n = 1, 2, \ldots$. Observe that $\sin \frac{n\pi x_0}{L}$ vanishes whenever x_0 is a rational multiple of L. How does this affect the practical computation of a_n's and b_n's? Do we desire T to be large or small in order to avoid the curse of hitting one of the undesirable x_0's?

2. If T is not an integer multiple of $\frac{2L}{c}$, then the formulas in (8.124) are no longer valid. One way to obtain approximate values for a_n's and b_n's is to truncate the infinite series in (8.123) and convert the information there to understanding the behavior of a system of simultaneous linear equations. With that in mind, replace (8.123)

by

$$\begin{aligned}
\textstyle\sum_{n=1}^{M} \bar{a}_n(\psi_n, \psi_1) + \bar{b}_n(\chi_n, \psi_1) &= (h, \psi_1), \\
\textstyle\sum_{n=1}^{M} \bar{a}_n(\psi_n, \psi_2) + \bar{b}_n(\chi_n, \psi_2) &= (h, \psi_2), \\
\textstyle\sum_{n=1}^{M} \bar{a}_n(\psi_n, \psi_3) + \bar{b}_n(\chi_n, \psi_3) &= (h, \psi_3), \\
\ldots & \quad \ldots \\
\ldots & \quad \ldots \\
\textstyle\sum_{n=1}^{M} \bar{a}_n(\psi_n, \chi_1) + \bar{b}_n(\chi_n, \chi_1) &= (h, \chi_1), \\
\textstyle\sum_{n=1}^{M} \bar{a}_n(\psi_n, \chi_2) + \bar{b}_n(\chi_n, \chi_2) &= (h, \chi_2), \\
\textstyle\sum_{n=1}^{M} \bar{a}_n(\psi_n, \chi_3) + \bar{b}_n(\chi_n, \chi_3) &= (h, \chi_3), \\
\ldots & \quad \ldots \\
\ldots & \quad \ldots
\end{aligned} \tag{8.125}$$

Show that this system is equivalent to $A\mathbf{x} = \mathbf{b}$ where

$$\mathbf{x} = \begin{bmatrix} \bar{a}_1 & \bar{a}_2 & \ldots & \bar{a}_n & \bar{b}_1 & \bar{b}_2 & \ldots & \bar{b}_n \end{bmatrix}^{\mathrm{T}} \tag{8.126}$$

and

$$\mathbf{b} = \begin{bmatrix} (h, \phi_1) & (h, \phi_2) & \ldots & (h, \chi_1) & (h, \chi_2) & \ldots \end{bmatrix}^{\mathrm{T}}.$$

Determine the matrix A. Let $D(n, T)$ denote the determinant of A. With $c = 3$, $L = 2$, plot the graph of this function as a function of n and T.

8.15 Project D: Exact Solutions of the Rotating Shallow Water Equations

This project is motivated by the work presented in the 1987 paper of Benoit Cushman-Roisin [10] where an algorithm is proposed for obtaining exact solutions of RSWE in (8.30). As pointed out in [10], the search for finding such solutions goes back to at least the 1930s, in the work of G. R. Goldsborough [11], and later pursued by F. K. Ball [12], [13], and by W. C. Thacker [14]. The paper [10] has an excellent introduction and summary of early research on this topic.

The goal in this project is to develop a MATLAB program, primarily based on ode45, to seek special solutions of the system of the PDEs in (8.30), which will be nearly exact solutions of this system, the flaw being only in the limitations associated with numerical approximations that result from solving a system of ODEs. These solutions will have simple spatial structures, where the x and y dependence will be at most quadratic, but despite their simplicity, these solutions are quite useful when investigating some of the general features of large-scale circulation in the open oceans, as well as in monitoring the accuracy of numerical schemes for the PDEs in (8.30).

We begin our development with the template described in (8.46).

1. Substitute the expressions (8.46) into the PDEs in (8.30) to obtain twelve nonlinear differential equations for the twelve unknown functions of t

$$A(t), B(t), C(t), D(t), E(t), F(t),$$

$$U_0(t), U_1(t), U_2(t),$$

$$V_0(t), V_1(t), V_2(t).$$

   Compare the system you obtain to the one on page 236 of [10]. If you have access to *Mathematica* or MATLAB's symbolic manipulator capability, employ them to obtain the twelfth order system. You will notice small discrepancies between the equations you obtain and the ones listed in [8.46].

2. As is the case with solving any IVP using ode45, we need to first define an M-file that contains the equations. One way to construct this M-file, which we call EllipticVortexEqns.m, is by simply enumerating the right side of the twelve ODEs on page 236 of [10] as follows:

```
function zprime=EllipticVortexEqns(t,z)
global f g
%
A=z(1);B=z(2);C=z(3);D=z(4);E=z(5);F=z(6);
U0=z(7);U1=z(8);U2=z(9);
V0=z(10);V1=z(11);V2=z(12);
eqn1=-(3*U1+V2)*A-2*V1*B;
eqn2=-U2*A-2*(U1+V2)*B-V1*C;
eqn3=-2*U2*B-(U1+3*V2)*C;
eqn4=-U0*A-V0*B-(2*U1+V2)*D-V1*E;
eqn5=-U0*B-V0*C-U2*D-(U1+2*V2)*E;
eqn6=-2*U0*D-V0*E-(U1+V2)*F;
```

```
eqn7=-U0*V1-U2*V0+f*V0-2*g*D;
eqn8=U1^2-U2*V1+f*V1-2*g*A;
eqn9=-U1*U2-U2*V2+f*V2-2*g*B;
eqn10=-U0*V1-V0*V2-f*U0-2*g*E;
eqn11=-U1*V1-V1*V2-f*U1-2*g*B;
eqn12=-U2*V1-V2^2-f*U2-2*g*C;
zprime=[eqn1;eqn2;eqn3;eqn4;eqn5;eqn6;...
     eqn7;eqn8;eqn9;eqn10;eqn11;eqn12];
```

A typical run of `ode45` requires values of f and g, and initial values for the vector `z`. The following is an example of such a run; we call the M-file `RunEllipticVortex.m`, which plots the graph of $h(1,1,t)$ for $t \in (0, \frac{2\pi}{f})$:

```
clf
clear all
global f g
f=1e-6;
g=10;
A=4;C=1;B=1;
D=0;E=0;F=1;
initial=[A B C D E F 0 0 0 0 0 0];
options=odeset('RelTol',1e-6,'AbsTol',...
     1e-10*ones(12,1));
[t,z]=ode45(@EllipticVortexEqns,[0 2*pi/f],...
           initial,options);
X=1;Y=1;
for i=1:length(t)
    hh(i)=z(i,1)*X.^2+2*z(i,2)*X.*Y+z(i,3)*Y.^2+...
          2*z(i,4)*X+2*z(i,5)*Y+z(i,6)*ones(size(X));
end
plot(t,hh)
```

Execute `RunEllipticVortex.m` and report on MATLAB's response.

3. MATLAB has several ODE solvers, `ode45` being one of them. Experiment with the list of possible ODE solvers (for example `ode23`, `ode23s`, etc.) and replace `ode45` in `RunEllipticVortex.m` and report on whether the output is influenced by the choice of the ODE solver.

4. The ODEs in this system have the special feature that the right sides are generally quadratic or cubic in the variable `z`. These equations typically have the blowup-in-finite-time feature, in that, for

certain initial conditions, the solution becomes infinite in finite time. To see a simple example, consider the *Riccati equation*

$$\frac{dx}{dt} = x^2,\ x(0) = x_0. \tag{8.127}$$

Show, by direct differentiation, that $x(t) = \frac{x_0}{1-x_0 t}$ is the exact solution of the above IVP. Note that if x_0 is negative, then $x(t)$ exists for all $t > 0$, but if x_0 is positive, then x becomes unbounded by the time $t = \frac{1}{x_0}$. This blow-up phenomenon is also inherent in the twelfth-order system.

(a) Let $f = 1$ and $g = 1$. Select and keep fixed the `RelTol` and `AbsTol` in the `ode45`'s `options`. Select the initial data for `z` at random from the interval $(-1, 1)$. Execute `RunEllipticVortex.m` for $t \in (0, 2\pi)$ a number of times, but no less than 20 times. Report on the behavior of `ode45`. How often does it seem that the twelfth-order system "blows up in finite time"?

(b) Let $f = 0.0001$ and $g = 10$, where now these values are closer to the physical values we expect. Select an initial condition `initial` at random from the interval $(-1, 1)$. Execute `RunEllipticVortex.m` for $t \in (0, \frac{2\pi}{f})$ first for `initial`, and next for

```
10^{-i}*initial
```

with i ranging from 1 to 10. Report on the behavior of `ode45` as the initial condition gets smaller.

8.16 Project E: Courant–Friedrichs–Lewy Condition

The topic of the numerical analysis of ODEs and PDEs has a rich history in applied mathematics and has been treated in many excellent and highly recommended texts, including the texts by Morton and Mayers [17], and Gottlieb and Orszag [19]. In Chapter 5 we touched on the mathematical issue of the numerical stability of ODEs, specifically in the context of the forward and backward finite difference methods. In this project we explore the concept of numerical stability for the wave equation and ask the reader to experiment with the Courant-Friedrich-Lewy (CFL) condition, defined below, to appreciate its impact on the stability of finite difference methods. The approach here is computational and experimental, and the reader is invited and strongly encouraged to investigate further the analysis that leads to the CFL condition, in [17] and [19], or by visiting the many web sites that are dedicated to this topic.

1. Consider the advection equation

$$u_t + cu_x = 0, \quad u(x,0) = f(x), \quad u(0,t) = u(2\pi,t). \qquad (8.128)$$

   (a) Let $u_{i,j}$ stand for $u(x_i, t_j)$, where $x_i = i\Delta x$, $i = 1, ..., N$, and $t_j = j\Delta t$, $j = 0, 1, ...$

   (b) Begin with the Forward Euler Method and replace u_t and u_x in (8.128) by $\frac{1}{\Delta t}(u_{i,j+1} - u_{i,j})$ and $\frac{1}{\Delta x}(u_{i+1,j} - u_{i,j})$, respectively, to arrive at

$$u_{i,j+1} = u_{i,j} - c\lambda(u_{i+1,j} - u_{i,j}), \qquad (8.129)$$

   where $\lambda = \frac{\Delta t}{\Delta x}$.

   (c) Let $f(x) = \sin x$. Note that the exact solution of the IBVP (8.128) is

$$u(x,t) = \sin(x - ct).$$

   Write a MATLAB program to compute $u_{i,j}$, with c and λ as parameters. The main goal here is to arrive at the CFL condition

$$c\lambda \leq 1 \qquad (8.130)$$

   as a condition that suggests that the numerical computations are stable and bounded. To that end, fix c and N (and hence Δx, the number of partitions of the domain $(0, 2\pi)$). Compute the approximation to $u(\pi, 1)$ for various values of Δt. Report on the accuracy of the computation in relation to Δt and (8.130).

2. Consider the following IBVP for the wave equation:

$$u_{tt} = c^2 u_{xx}, \qquad x \in (0, L), \qquad (8.131)$$

$$u(0,t) = u(L,t) = 0, \quad u(x,0) = f(x), \; u_t(x,0) = g(x). \qquad (8.132)$$

   (a) Convert the wave equation to a first order system of PDEs by introducing the two dependent variables v and w defined by

$$v = u_t, \quad w = u_x. \qquad (8.133)$$

   Show that (8.131)–(8.132) is equivalent to

$$w_t = v_x, \quad v_t = c^2 w_x, \qquad (8.134)$$

subject to

$$v(0,t) = v(L,t) = 0, \tag{8.135}$$

and

$$v(x,0) = g(x),\ w(x,0) = f'(x). \tag{8.136}$$

(b) Following the strategy of the previous problem, discretize (8.133)–(8.136) by using the centered difference method in x and the Forward Euler Method in t. Write down the finite difference equations that $v_{i,j}$ and $w_{i,j}$ satisfy.

(c) Let $L = 2\pi$, $f(x) = \sin x$, $g(x) = x(2\pi - x)$. Fix N and c. Write a MATLAB program to compute $(v(\pi,1), w(\pi,1))$. Compare the approximate solution to the one you obtain by either using the Fourier series method or by the D'Alembert method. Report on the accuracy of the computation in relation to the wave equation's CFL condition

$$c^2\lambda^2 \leq 1, \tag{8.137}$$

where λ is, as before, the ratio of Δt and Δx.

8.17 References

1. Batchelor, G., *An Introduction to Fluid Dynamics*, Cambridge University Press, 2000.

2. Stoker, J., *Water Waves: The Mathematical Theory with Applications*, Wiley Classics Library Edition, 1992.

3. Gerbeau, J.-F., Perthame, B., "Derivation of viscous Saint-Venant system for laminar shallow water; numerical validation," *Discrete and Dynamical Systems*, Series B, Vol 1, no. 1, 2001, pp. 89–102.

4. LeVeque, R., *Finite Difference Methods for Ordinary and Partial Differential Equations*, SIAM 2007.

5. George, D. L., "Finite volume methods and adaptive refinement for tsunami propagation and inundation," Dissertation, University of Washington, 2006.

6. Cheng, B., Tadmor, E., "Approximate periodic solutions for the rapidly rotating shallow-water and related equations," in *Water waves, Theory and Experiment*, Proceedings of the Conference held in Howard University, May 2008 (M. F. Mahmood, D. Henderson, H. Segur, eds), World Scientific (2010), pp. 69–78.

7. Kim, J., LeVeque, R. J., "Two-layer shallow water system and its applications," Proceedings of the Twelfth International Conference on Hyperbolic Problems, 2008. *Proc. Symp. Appl. Math.*, 67, 2009, pp. 737–743.

8. Greenberg, J. M., Hartig, D., "A new Lagrangian shallow water circulation model", *Far East Journal of Ocean Research*, 2009, pp. 93–156.

9. Pedlosky, J., *Geophysical Fluid Dynamics*, 2nd Edition, Springer Verlag, New York, 1986.

10. Cushman-Roisin, B., "Exact analytical solutions for elliptical vortices of the shallow-water equations", *Tellus*, 39A, 1987, pp. 235–244.

11. Goldsborough, G., "The tidal oscillations in an elliptic basin of variable depth", *Proceedings of the Royal Society A*, Vol 130, 1930, pp. 157–167.

12. Ball, F., "Some general theorems concerning the fluid motion of a shallow liquid lying on a paraboloid," *Journal of Fluid Mechanics*, Vol 17, 1963, pp. 240–256.

13. Ball, F., "The effect of rotation on the simpler modes of motion of a liquid in an elliptic paraboloid," *Journal of Fluid Mechanics*, Vol 22, 1965, pp. 529–545.

14. Thacker, W., "Some exact solutions to the nonlinear shallow-water wave equation," *Journal of Fluid Mechanics*, Vol 107, 1981, pp. 499–508.

15. Malek-Madani, R., *Advanced Engineering Mathematics with Mathematica® and MATLAB®*, Addison-Wesley, 1998.

16. Miller, R., *Numerical Modeling of Ocean Circulation*, Cambridge University Press, 2007.

17. Morton, K., Mayers, D., *Numerical Solution of Partial Differential Equations*, Cambridge University Press, Second edition, 2005.

18. Stanoyevitch, A., *Introduction to Numerical Ordinary and Partial Differential Equations Using MATLAB*, John Wiley & Sons, 2005.

19. Gottlieb, D., Orszag, S., *Numerical Analysis of Spectral Methods: Theory and Applications*, SIAM, 1977.

Chapter 9

Wind-Driven Ocean Circulation: Stommel and Munk Models

9.1 Introduction

In 1948 Henry Stommel, in the seminal paper entitled "The Western Intensification of Wind-driven Ocean Currents," see [1], proposed a simple model for the Gulf Stream based on the fundamental equations of geophysical fluid dynamics. In that paper Stommel concentrated on isolating the parameters that lead to the generation of boundary layers on the western boundaries of large basins in the northern hemisphere, reminiscent of the Gulf and Kuroshio Streams. Stommel showed that the variations of the Coriolis parameter f with latitude are primarily responsible for the formation of boundary layers in wind-driven circulations. Specifically, in the Stommel model the only forces present are the ones due to Coriolis, the wind stress, and an additional frictional force whose presence, albeit somewhat artificial, is intended to help with writing down a well-posed boundary value problem for a second order PDE for the stream function of the flow. A relatively careful derivation of this model will be one of the main features of this chapter.

Almost concurrently with the appearance of Stommel's paper, Walter Munk, in the paper entitled "On the Wind-driven Ocean Circulation," see [2], introduced a slightly different model of circulation in the North Atlantic. The key departure of Munk's model from Stommel's is in the way dissipation enters into the model. Munk, taking into account the turbulent nature of the flow, introduces an internal dissipation mechanism through the viscous stresses. This model then leads to a fourth order PDE for the underlying stream function.

In this chapter we will develop the basic tenets of the Stommel and Munk models. We will derive the underlying PDEs and the boundary-value problem that govern the behavior of the stream function in each model, use separation of variables and find the exact solution when possible, and then use MATLAB to generate graphs of the typical streamlines.

We then apply the finite difference methodology we have developed so far, and find approximate solutions for each boundary value problem.

Before describing the Stommel and Munk models in detail, we present two simple boundary value problems, one that mimics flows in an unbounded rectangular bay, where the flow is incompressible and irrotational, and the other a rotational flow in a bounded rectangular region. These two flows lead to the familiar eigenvalues and eigenfunctions or the normal modes of the Laplace operator, similar to the ones we have already encountered in the previous chapters, and point to fundamental structures that also appear in the Stommel and Munk models.

9.2 Flow in a Rectangular Bay — Normal Modes

We now take up the study of deriving the set of all possible irrotational flows of an incompressible fluid in the region Ω given by

$$\{(x,y)|\, x<0, 0<y<h\}. \tag{9.1}$$

We will think of the region Ω as the horizontal cross section of a semi-infinite bay, where the fluid flows we consider do not change with depth; we therefore assume that these flows are two-dimensional. We further consider steady-state flows, that is

$$\mathbf{v} = \langle u(x,y), v(x,y), 0\rangle. \tag{9.2}$$

Since the fluid is incompressible, $\mathbf{v}$ is associated with a stream function $\psi(x,y)$ through the relations

$$u = \frac{\partial \psi}{\partial y}, \quad v = -\frac{\partial \psi}{\partial x}, \tag{9.3}$$

and since the flow is irrotational, $\mathbf{v}$ is curl-free, that is $\frac{\partial v}{\partial x} - \frac{\partial u}{\partial y}$ must vanish, which in turn requires that ψ satisfy *Laplace's Equation*

$$\Delta\psi = \frac{\partial^2 \psi}{\partial x^2} + \frac{\partial^2 \psi}{\partial y^2} = 0. \tag{9.4}$$

Equation (9.4) is supplemented with the boundary conditions at the three boundaries $y=0$, $y=h$ and $x=0$, which we assume are impenetrable seawalls, so that

$$v(x,0) = v(x,h) = 0 \qquad \text{where } x<0, \tag{9.5}$$

and

$$u(0,y) = 0, \qquad \text{where } 0 < y < h. \tag{9.6}$$

Note that (9.5) and (9.6) do not allow fluid particles to cross the seawalls but do allow particles that are located on these boundaries to move along them.

In terms of the stream function ψ the relations (9.5) and (9.6) take the form

$$\frac{\partial \psi(x,0)}{\partial x} = \frac{\partial \psi(x,h)}{\partial x} = \frac{\partial \psi(0,y)}{\partial y} = 0. \tag{9.7}$$

After an integration each relation in (9.7) reduces to

$$\psi(x,0) = \text{const}, \qquad \psi(x,h) = \text{const}, \qquad \psi(0,y) = \text{const}. \tag{9.8}$$

We anticipate that the seawall boundaries constitute a single contour of the stream function ψ and therefore set the three constants in (9.8) equal to the same constant. Since any stream function is known only up to a constant, without loss of generality we set this constant to zero. Hence the three boundary conditions in (9.6) and (9.7) reduce to $\psi(x,y) = 0$, which we write as the *Dirichlet* boundary condition

$$\psi|_{\partial\Omega} = 0. \tag{9.9}$$

In summary, the set of fluid flows we are seeking in the bay Ω must satisfy Laplace's equation subject to the zero boundary condition (9.9).

As in the case of the wave equation, we seek solutions of (9.4) in separated form:

$$\psi(x,y) = F(x)G(y). \tag{9.10}$$

Substituting (9.10) in Laplace's equation and dividing by FG leads to

$$\frac{F''}{F} + \frac{G''}{G} = 0, \qquad \text{for all } x \text{ and } y, \tag{9.11}$$

from which we conclude that each fraction in (9.11) must be a constant. We assume the constant $\frac{F''}{F}$ is positive (see the problems at the end of the section for all other cases) and denote it by λ^2. Hence

$$\frac{F''}{F} = \lambda^2, \qquad \frac{G''}{G} = -\lambda^2, \tag{9.12}$$

or

$$F'' - \lambda^2 F = 0 \qquad \text{and} \qquad G'' + \lambda^2 G = 0. \tag{9.13}$$

The general solutions of these equations are $F(x) = c_1 e^{\lambda x} + c_2 e^{-\lambda x}$ and $G(y) = c_3 \cos \lambda y + c_4 \sin \lambda y$. The stream function ψ, as expressed in (9.10) in terms of F and G, now takes the form

$$\psi(x,y) = (c_1 e^{\lambda x} + c_2 e^{-\lambda x})(c_3 \cos \lambda y + c_4 \sin \lambda y). \tag{9.14}$$

The constants c_1, c_2, c_3, c_4 and λ must now be determined so that the stream function ψ complies with the boundary conditions (9.9). The first boundary condition, that $\psi(x, 0)$ vanishes for all negative values of x, when applied to (9.14), leads to

$$0 = (c_1 e^{\lambda x} + c_2 e^{-\lambda x}) c_3 \tag{9.15}$$

for all $x < 0$, which leads to the conclusion that c_3 must vanish. With this information in hand, the function ψ in (9.14) reduces to

$$\psi(x, y) = (Ae^{\lambda x} + Be^{-\lambda x}) \sin \lambda y, \tag{9.16}$$

where A and B stand for the two products $c_4 c_1$ and $c_4 c_2$. The second part of the boundary conditions, that $\psi(0, y)$ must vanish, when applied to (9.16), requires that

$$(A + B) \sin \lambda y = 0 \tag{9.17}$$

for all y with $0 < y < h$. Hence $A + B$ equals zero or $B = -A$. This expression now reduces (9.16) to

$$\psi(x, y) = C \sinh \lambda x \sin \lambda y, \tag{9.18}$$

where we have used the identity $\sinh z = \frac{1}{2}(e^z - e^{-z})$ and have defined C to stand for the remaining constants. The final boundary condition, that $\psi(x, h) = 0$, requires that C and λ satisfy the expression

$$C \sinh \lambda x \sin \lambda h = 0 \tag{9.19}$$

for all negative x. The above expression leads to $\sin \lambda h = 0$ since allowing C to vanish would lead to $\psi(x, y)$ vanishing identically. The latter equation in λ is met if λ is selected so that λh is an integer multiple of π. Hence, we have discovered that there are infinitely many candidates for λ, the constant of separation of variables in (9.12), of the form

$$\lambda_n = \frac{n\pi}{h}, \qquad n = 1, 2, 3, ... \tag{9.20}$$

The associated stream functions ψ_n are

$$\psi_n(x, y) = C_n \sinh \frac{n\pi x}{h} \sin \frac{n\pi y}{h}. \tag{9.21}$$

We summarize these findings in the following theorem.

Theorem 9.2.1 (Flow in the Bay)
The boundary value problem (9.4) and (9.9) has infinitely many solutions

given by (9.21). These solutions are linearly independent in the domain $(-\infty, 0) \times (0, h)$.

Putting it slightly differently, and recalling the definition of an eigenvalue μ and its attendant eigenfunction u of an operator L (that $L[u] = \mu u$, subject to boundary conditions), we have shown that the functions ψ_n in (9.21) are all eigenfunctions of the Laplace operator Δ corresponding to eigenvalue $\mu = 0$. We will study the eigenvalue-eigenfunction problem for the Laplacian in more detail in the next section in the context of bounded domains Ω, and will arrive at the conclusion that there are generally only a finite number of eigenfunctions corresponding to an eigenvalue μ when Ω is bounded . The fact that we have obtained infinitely many eigenfunctions corresponding to $\mu = 0$ for the domain Ω defined by (9.1) is due to the unboundedness of this domain.

The eigenfunctions or normal modes (9.21) are special solutions of the BVP (9.4), (9.9), each having a single parameter, C_n, at our disposal to manipulate to construct further solutions of the BVP. Each normal mode is characterized by its structure in the y-direction, each displaying a flow that has several invariant regions, regions that do not communicate or mix. Figure 9.1 shows the first four normal modes or eigenfunctions (9.21). The n-th normal mode consists of n invariant regions where fluid particles are confined to stay. The flow direction in any invariant region is in the opposite direction to the flow direction in its neighboring invariant regions. Two adjacent invariant regions are separated by an invariant curve $y = \text{const}$, where, much like the seawall boundaries $x = 0$, $y = 0$ and $y = h$, these curves act as artificial seawalls or barriers, impenetrable by fluid particles. These internal "boundaries" are not picked up by the `contour` command of MATLAB because they are only continuous and not differentiable. In fact, the set of points corresponding to the seawall boundaries and these internal invariant curves correspond to the single contour $\psi_n(x, y) = 0$. Figure 9.2 shows the graph of $z = \psi_3(x, y)$, where it is relatively clear that the curves $y = k$, where $\sin 3k = 0$ (here $h = \pi$), are zeros of ψ_3. A more illuminating way of computing the zero-contour level of any normal mode is to use the special option within `contour` to plot a single contour. Here is how one proceeds:

```
[X,Y]=meshgrid(-3:0.01:0.1,-0.1:0.01:pi+0.1);
contour(X,Y,FlowBay(0.1,pi,3,X,Y),[0 0])
```

where `FlowBay.m` is the M-file

```
function z=FlowBay(A,h,n,x,y);
%
z=A*sinh(n*pi*x/h).*sin(n*pi*y/h);
```

The output is shown in Figure 9.3, which clearly shows the seawalls and the internal barriers. This figure is obtained by

```
contour(x,y,psi(x,y),[0 0])
```

after `psi`, `x`, and `y` have been properly defined.

As we saw in the discussion of the wave equation in Chapter 8, the set of functions $\{\sin\frac{n\pi y}{h}\}$ form a basis for the set of functions f for which formulas such as (8.66) and (8.67) make sense. [1] In the case of the wave equation, we applied the Fourier series approach to construct a solution that satisfies the appropriate initial data (again, see (8.66)). It turns out that this approach can also be useful in the context of the boundary value problem for the flow in a bay, which we elaborate on now.

We seek solutions of the boundary value problem (9.4) and

$$\psi(x,0) = \psi(x,h) = \psi(0,y) = 0, \tag{9.22}$$

with the *additional* boundary condition

$$\psi(a,y) = f(y), \tag{9.23}$$

where f is a given function of $y \in (0,h)$. We can obtain the solution to Laplace's equation (9.4), subject to the boundary conditions (9.22)–(9.23), by considering the linear combination of the normal modes (9.21), namely,

$$\psi(x,y) = \sum_{n=1}^{\infty} C_n \sinh\frac{n\pi x}{h}\sin\frac{n\pi y}{h}, \tag{9.24}$$

since each normal mode in (9.21) satisfies Laplace's equation as well as the three boundary conditions in (9.22). We note that the fourth boundary condition (9.23) would be satisfied if we could select the coefficients C_n's in (9.24) so that

$$f(y) = \sum_{n=1}^{\infty} C_n \sinh\frac{n\pi a}{h}\sin\frac{n\pi y}{h}, \tag{9.25}$$

that is, select these coefficients so that the expression $C_n \sinh\frac{n\pi a}{h}$ is the

[1] Here we use the term "basis" informally and rather loosely, to mean that a large collection of functions f, those that satisfy the boundary conditions $f(0) = f(h) = 0$ and are square-integrable in the interval $(0,h)$, can be approximated by the set of functions $\{\sin\frac{n\pi y}{h}\}$. We do not take up in this text the proper definitions related to the *completeness* of this basis and the appropriate topology in which the partial sums converge to f, and instead refer the interested reader to classic texts in Advanced Calculus and Real Analysis for this subject.

Fourier sine coefficient of the function f when this function is expanded in terms of $\sin \frac{n\pi y}{h}$, i.e.,

$$C_n \sinh \frac{n\pi a}{h} = \frac{2}{a} \int_0^h f(y) \sin \frac{n\pi y}{h}\, dy,$$

which results in

$$C_n = \frac{2}{a \sinh \frac{n\pi a}{h}} \int_0^h f(y) \sin \frac{n\pi y}{h}\, dy. \tag{9.26}$$

We summarize the above discussion in the following theorem:

Theorem 9.2.2 (A BVP for Laplace's Equation)
The Boundary Value Problem (9.4) subject to the four boundary conditions stated in (9.22)–(9.23) can be solved by means of the Fourier series solution (9.24) whose coefficients C_n are given in (9.26).

Some of the exercises at the end of this section address this approach with concrete examples of f. We end this section by appealing to MATLAB to demonstrate that the invariance structure of the individual normal modes can be broken if we consider arbitrary linear combination of normal modes. For example, consider the stream function

$$\psi(x, y) = \psi_1(x, y) + \psi_2(x, y) + \psi_3(x, y) \tag{9.27}$$

where $h = \pi$ and $C_1 = 1$, $C_2 = -0.01$ and $C_3 = -0.0015$. The streamlines of this flow are shown in Figure 9.4. The figure shows the nontrivial way that the three normal modes of (9.21) interact and how the invariance regions of an individual normal mode affects modes.

Problems 9.2

1. Show that the stream function ψ of a two-dimensional irrotational and incompressible flow must satisfy the Laplace equation.

2. Show by direct differentiation that the expression (9.21) is a solution of the BVP (9.9).

3. While applying the separation of variables method to the Laplace Equation we concluded that the fraction $\frac{F''}{F}$ must be constant, but then made the assumption in (9.12) that this constant is positive and proceeded to arrive at the normal modes (9.21). Assume now that this constant is negative and proceed to determine the resulting normal modes.

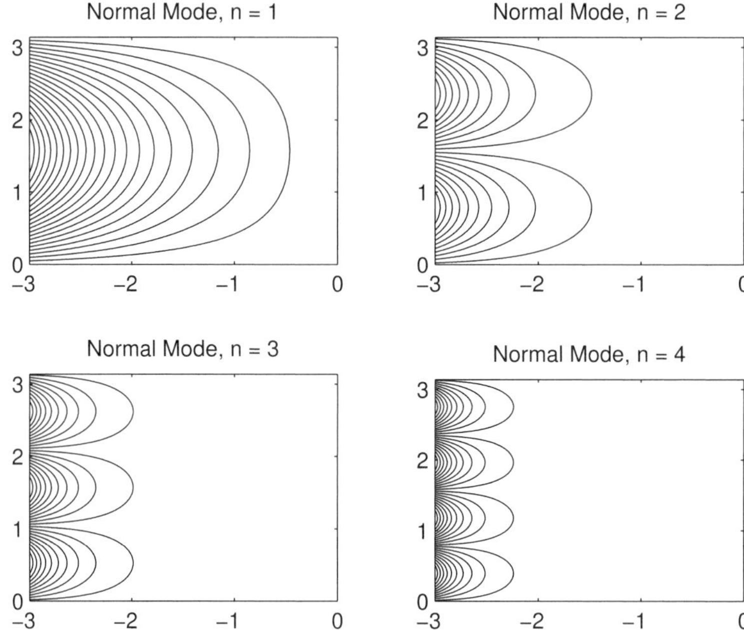

FIGURE 9.1: Contours of the first four eigenfunctions or normal modes (9.21) of the BVP (9.4), (9.9). Here $C_n = 10^{-n}$, n ranging from 1 through 4. Note that the n-th mode has n invariant regions, where fluid particles don't mix. The direction of the flow in each invariant region is opposite of the direction of the flow in its neighboring invariant region.

4. Referring to the discussion in the previous problem, assume that the ratio $\frac{F''}{F}$ is zero and proceed to obtain the resulting normal modes.

5. Consider the first normal mode ψ_1 in (9.21), with $h = \pi$ and $C_1 = 1$. Find the associated velocity and acceleration fields. Plot the graphs of the streamlines and the velocity and acceleration fields on the same screen.

6. Consider the second normal mode ψ_2 with $h = \pi$ and $C_2 = 1$. Use MATLAB's `ode45` and plot the graphs of the particle paths whose initial positions are located at $(-1, i)$ with i ranging from 0 to π at an increment of 0.1. Plot on one screen the graphs for one unit of time in the future and one unit of time in the past. Color the future trajectories blue and the past trajectories red.

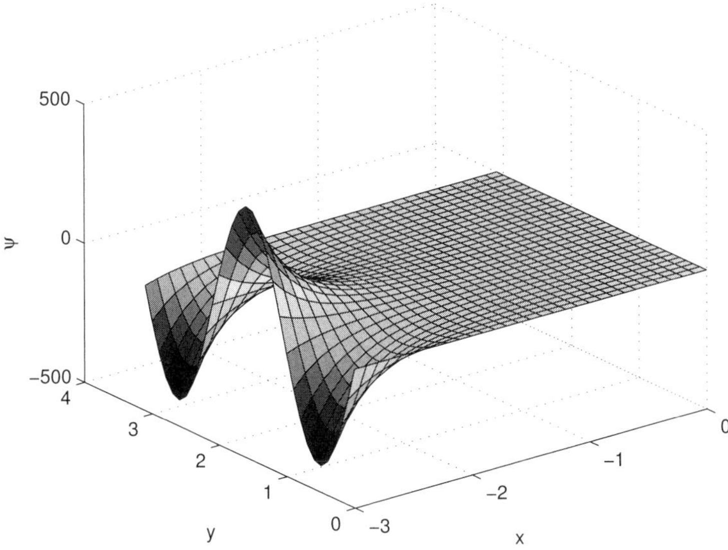

FIGURE 9.2: The surface plot of $\psi_3(x, y) = 0.1 \sinh 3x \sin 3y$.

7. Returning to the BVP in Theorem 9.2.2, consider the case where f is given by

$$f(y) = y(h - y).$$

   (a) Compute the Fourier coefficients C_n in (9.26) as a function of n and h.

   (b) Let $h = 1.3$. Write a MATLAB program to compute $S_N(x)$, the N-th partial sum of the series solution of the BVP, where S_N is

$$S_N(x) = \sum_{n=1}^{N} C_n \sinh \frac{n\pi x}{h} \sin \frac{n\pi y}{h}.$$

   Your program should define the function S_N, which may use the `global` command to pass parameter h to the computation of S_N. Experiment with `quad`, `quadl` and `quadv` to decide which one leads to the best solution for computing the Fourier coefficients. Apply this program to the cases where $N = 8$, $N = 16$ and $N = 32$ and plot the graph of the resulting stream function in each case.

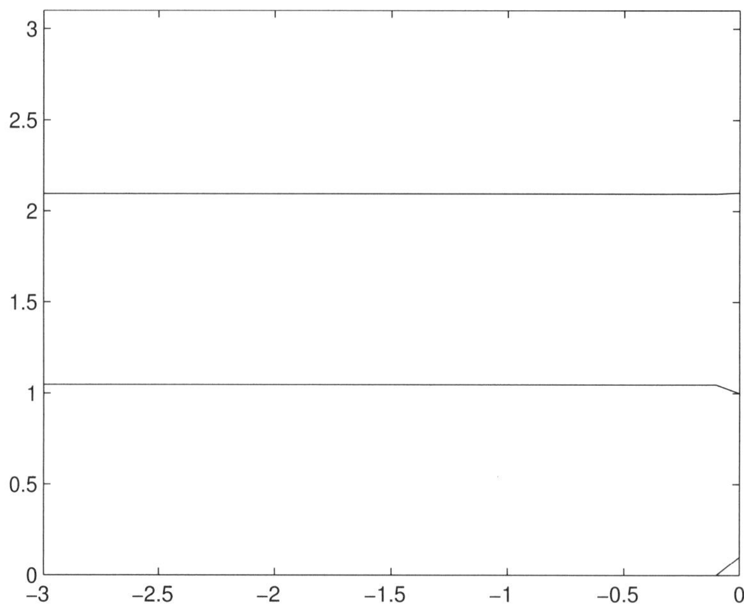

FIGURE 9.3: The plot of the zero contour of $\psi_3(x, y) = 0.1 \sinh 3x \sin 3y$, showing the seawall boundaries and the two internal barriers $y = \frac{\pi}{3}$ and $y = \frac{2\pi}{3}$.

9.3 Eigenfunctions of the Laplace Operator

In the previous section we studied a collection of irrotational fluid flows that are related to the Laplace operator. We saw that these flows are normal modes or eigenfunctions of the Laplace operator corresponding to this operator's zero eigenvalue, i.e., the stream function of these flows satisfy the Laplace equation given in (9.4). We now take up the study of the structure of rotational fluid flows by first concentrating on constructing eigenfunctions of nonzero eigenvalues of the Laplace operator, that is, determining the pair (λ, ψ) such that

$$\Delta\psi = -\lambda^2\psi, \quad \psi_{\partial\Omega} = 0, \tag{9.28}$$

and in the subsequent section solve the *Poisson* equation

$$\Delta\psi = f(x, y), \quad \psi_{\partial\Omega} = 0.$$

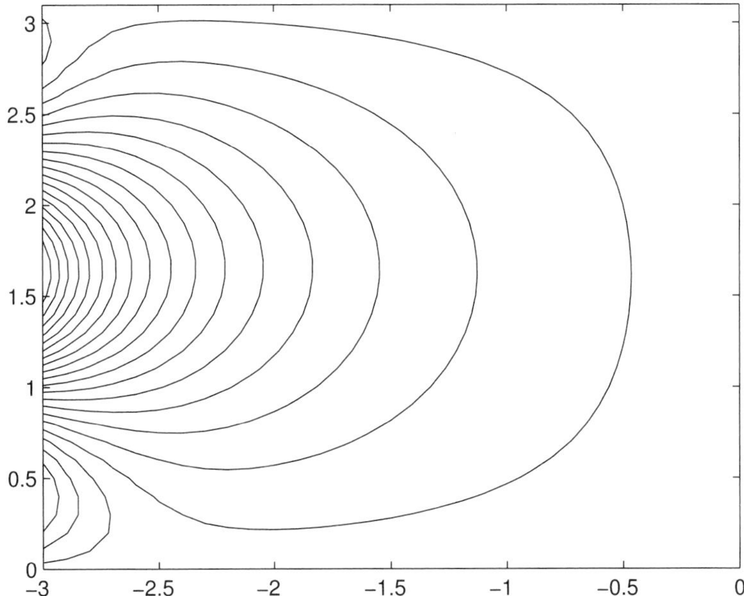

FIGURE 9.4: The streamlines of a linear combination of the normal modes (9.27) where the invariance structure of individual normal modes are altered and deformed.

We now consider the domain Ω as the rectangle

$$\Omega = \{(x, y) |\, 0 < x < a,\, 0 < y < b\}, \tag{9.29}$$

which is bounded, in contrast to the domain in the "flow in the bay" we considered in the previous section.

Returning to (9.28), we begin the process of constructing a solution as we did with the Laplace equation, by applying the method of separation of variables. We seek solutions ψ of the form

$$\psi(x, y) = F(x)G(y). \tag{9.30}$$

After substituting this template into (9.28), we arrive at the expression

$$\frac{F''}{F} + \frac{G''}{G} = -\lambda^2$$

which gives the two sets of ODEs

$$\frac{F''}{F} = -\mu^2, \quad \frac{G''}{G} + (\lambda^2 - \mu^2) = 0,$$

or equivalently

$$F'' + \mu^2 F = 0, \quad G'' + (\lambda^2 - \mu^2)G = 0. \tag{9.31}$$

Here μ is the constant of separation of variables. At this point both λ and μ are unknown, but we assume, anticipating oscillating solutions in (9.31), that $\lambda > \mu$. With this assumption, the solutions to (9.31) are

$$F(x) = c_1 \sin \mu x + c_2 \cos \mu x, \quad G(y) = c_3 \sin \gamma y + c_4 \cos \gamma y, \tag{9.32}$$

where

$$\gamma = \sqrt{\lambda^2 - \mu^2}. \tag{9.33}$$

With this expression for γ, the eigenfunction ψ takes the form

$$\psi(x, y) = (c_1 \sin \mu x + c_2 \cos \mu x)(c_3 \sin \gamma y + c_4 \cos \gamma y). \tag{9.34}$$

We now begin to apply the Dirichlet boundary condition $\psi|_{\partial\Omega} = 0$ at the four segments of the boundary $x = y = 0$ and $x = a$ and $y = b$. When $x = 0$, the expression (9.34) reduces to

$$\psi(0, y) = 0 = c_2(c_3 \sin \gamma y + c_4 \cos \gamma y), \quad \text{for all } 0 < y < b,$$

which results in $c_2 = 0$. The stream function ψ in (9.34) then reduces to

$$\psi(x, y) = (A \sin \gamma y + B \cos \gamma y) \sin \mu x, \tag{9.35}$$

where A and B are the appropriate combinations of c_1, c_3 and c_4. Next, if we apply the boundary condition at $y = 0$, we see that ψ in (9.35) reduces to

$$\psi(x, y) = A \sin \mu x \sin \gamma y. \tag{9.36}$$

The remaining two boundary conditions at $x = a$ and $y = b$ will now determine μ and γ. First, the expression $0 = \psi(a, y) = A \sin \mu a \sin \gamma y$ is satisfied if we choose μ so that $\sin \mu a = 0$, or if

$$\mu_n = \frac{n\pi}{a}, \quad n = 1, 2, ... \tag{9.37}$$

Similarly, the boundary condition $\psi(x, b) = 0$ dictates that γ satisfy the relation $\sin \gamma b = 0$ or

$$\gamma_m = \frac{m\pi}{b}. \tag{9.38}$$

Recalling the connection between λ, μ and γ from (9.33), we have

$$\lambda_{nm}^2 = (\frac{n^2}{a^2} + \frac{m^2}{b^2})\pi^2. \tag{9.39}$$

with the corresponding eigenfunctions

$$\psi_{nm}(x, y) = A_{nm} \sin \frac{n\pi x}{a} \sin \frac{m\pi y}{b}. \tag{9.40}$$

We summarize these conclusions in the following theorem.

Theorem 9.3.1 (Normal Modes of the Laplace Operator) *The eigenvalue-eigenfunctions pair (λ, ψ) of the Laplace operator are given by the expressions (9.39) and (9.40).*

Several of the exercises explore the structure of these solutions.

Problems 9.3

1. Show that the functions $\{\sin \frac{n\pi x}{a} \sin \frac{m\pi y}{b}\}$ are linearly independent in the domain Ω.

2. Show by direct differentiation that the expression (9.34) is a solution of (9.28).

3. Show by direct differentiation that the expression (9.40) is a solution of (9.28) with eigenvalues defined by (9.39).

4. Determine whether the BVP

$$\Delta\psi = \lambda^2\psi, \quad \psi|_{\partial\Omega} = 0$$

   has a nontrivial solution pair (λ, ψ) with λ a real number.

5. Consider the (m, n)-th normal mode in (9.40). Determine the velocity, acceleration and the vorticity of the fluid flow associated with this mode.

6. Let ψ_{nm} be the (n, m)-th eigenfunction defined in (9.40). Plot the contours of this normal mode and the induced velocity field when

   (a) $n = m = 1$, $A_{nm} = 1$, $a = 3$ and $b = 2$.

   (b) $n = 2$, $m = 3$, $A_{nm} = 1$, $a = 5$ and $b = 3$.

7. Consider the special case $a = b = L$, i.e., when the domain Ω is a square of length L. In this setting the eigenvalues λ_{nm}^2 become

$$\lambda_{nm}^2 = \frac{\pi^2(n^2 + m^2)}{L^2}.$$

   Since $n^2 + m^2 = m^2 + n^2$ all eigenfunctions corresponding to λ_{nm}^2 have at least multiplicity two because ψ_{nm} is linearly independent of ψ_{mn} (prove this). Show that a consequence of this fact is that any linear combination

$$c_1\psi_{nm} + c_2\psi_{mn}$$

is an eigenfunction corresponding to the same eigenvalue λ^2_{nm}. Show that there are cases of distinct n and m for which multiple eigenfunctions ψ_{nm} correspond to the same eigenvalue λ^2_{nm}. For example, with $n = 3$ and $m = 4$, or $n = 4$ and $m = 3$ both ψ_{34} and ψ_{43} are eigenfunctions corresponding to the eigenvalue $\frac{25\pi^2}{L^2}$. Let $L = \pi$.

(a) Show that $c\psi_{34} + d\psi_{43}$ is an eigenfunction of the Laplacian with eigenvalue 25. Plot the contours of this normal mode with $c = 1$ and $d = 3$.

(b) Show that $\psi_{5,12}$ and $\psi_{12,5}$ are both eigenfunctions of the Laplacian. What is the corresponding eigenvalue?

8. Let $a = b = \pi$. Consider the case of $\lambda^2 = 50$. Show that there are three eigenfunctions corresponding to this eigenvalue, $\psi_{1,7}$, $\psi_{7,1}$ and $\psi_{5,5}$. Let

$$\psi = \psi_{1,7} + \psi_{7,1} + \psi_{5,5}$$

where the coefficients A_{nm} in (9.40) are chosen at random from the interval $(0, 1)$. Plot the contours of this stream function.

9. Let $a = b = \pi$. Consider the case of $\lambda^2 = 65$. How many linearly independent eigenfunctions does this eigenvalue have? Let

$$\psi = \sum_{i=1}^{p} \psi_i$$

where $\{\psi_i\}$'s are the eigenfuctions. Choose the coefficients A_{nm} in (9.40) at random from the interval $(0, 1)$. Plot the contours of this stream function.

10. Consider the collection of integers $m^2 + n^2$ where m and n range from 1 to 25. Use MATLAB and investigate these numbers in the context of the eigenvalues of the Laplace operator when $a = b = \pi$. List all the eigenvalues that have one, two, three, four, five and six eigenfunctions, if there are any.

9.4 Poisson Equation

The eigenfunctions of the Laplacian may be used quite effectively to study solutions of boundary value problems for the Poisson equation

$$-\Delta\psi = f(x,y), \qquad \psi|_{\partial\Omega} = 0, \tag{9.41}$$

where Ω is the rectangular domain

$$\Omega = \{(x,y)|\, 0 < x < a,\, 0 < y < b\}$$

defined in (9.29). In the context of rotational incompressible flows, the stream function ψ defines a flow whose vorticity is characterized by f.

One approach to computing a solution to the BVP in (9.41), (9.29) is to take advantage of the special properties of the eigenfunctions of the Laplacian and seek a solution to the BVP as an expansion in terms of these normal modes. We have encountered this approach already, in the context of the flow in the bay, but the method is so powerful in applied mathematics that it is worth revisiting its development. The eigenfunction method applies to a wide class of initial and boundary value problems, including all problems that involve linear operators. In certain cases it can even be extended to gain insight into the behavior of solutions of nonlinear problems. It is also a proper tool to use with the method of lines and the Galerkin method, which were introduced earlier.

The main idea is as follows: The action of any operator L on any function ϕ is simply the function ψ given by

$$\psi = L(\phi).$$

Typically the functions ϕ and ψ hardly look alike. For example, when $L = \frac{d}{dx}$ and $\phi(x) = \ln x$, then $\psi(x) = L(\phi) = \frac{1}{x}$, which does not resemble ϕ, and in particular, ψ is not a constant multiple of ϕ. However, when ϕ is an eigenfunction of L, the situation is quite different, because now ψ simply ends up being a constant multiple of ϕ, i.e., we have the critical relation

$$L(\phi) = \mu\,\phi. \tag{9.42}$$

To rephrase, eigenfunctions of any operator L, including the Laplace operator, are special, in that the operator L's action on any of its eigenfunctions essentially leaves that function intact, except for the scaling factor μ. We remark in passing that we have already encountered this concept in the context of matrices and have explored its consequences.

The main reason the concept of eigenvalue-eigenfunction plays a critical role in seeking solutions of PDEs is because of the ease by which the operator L acts on its eigenfunctions. To see this in an abstract setting, consider a PDE that is expressed as

$$L(\psi) = f. \tag{9.43}$$

In the special case of our Poisson BVP, $L = -\Delta$. Let ψ_n bet the eigenfunction of L with the associated eigenvalues μ_n, that is,

$$L(\psi_n) = \mu_n \psi_n. \tag{9.44}$$

An important property of eigenfunctions $\{\psi_n\}$is that they often form a *basis* for the collection of functions of interest to us, meaning that a large class of functions f can be expanded in terms of ψ_n:

$$f(x) = \sum_{n=1}^{\infty} a_n \psi_n(x), \tag{9.45}$$

where a_n's are constant. We will refer to the coefficients a_n's as the (generalized) *Fourier* coefficients of f in terms of the basis functions or the eigenfunctions $\{\psi_n\}$. We have already encountered examples of expansions of the form (9.45), specifically when discussing the wave equation where $\psi_n(x) = \sin \frac{n\pi x}{L}$, and in the case of the Flow in the Bay, where $\psi_n(y) = \sin \frac{n\pi y}{h}$. As was pointed out earlier, we are not developing here the rigorous mathematics that describes the sense in which the partial sums of the infinite series in (9.45) converge to f, and refer the interested reader to [3], among many other texts on PDEs, for details. Instead, here we concentrate on how to implement this approach, and combine it with MATLAB's capabilities, to obtain approximate solution to the Poisson equation.

Returning to (9.43), we note that we are dealing with two functions, f and ψ, in this PDE, where f is the known forcing term, and ψ is the desired solution to this equation. The first step of our strategy is to expand f in terms of the eigenfunctions of L and obtain (9.45). The coefficients a_n's are determined from the application of a typical Fourier coefficient formula. The second step is to seek the solution ψ also as an expansion in terms of the eigenfunctions of L:

$$\psi(x) = \sum_{n=1}^{\infty} b_n \psi_n(x). \tag{9.46}$$

Unlike the case of f, where f is a known function so we have access to formulas to compute its Fourier coefficients, the function ψ is unknown so we need to finds b_n's, its Fourier coefficients, indirectly. The piece of information that comes to the rescue of this approach is the original PDE in (9.43). Substituting the two Fourier series (9.45) and (9.46) into (9.43) yields

$$L(\sum_{n=1}^{\infty} b_n \psi_n(x)) = \sum_{n=1}^{\infty} a_n \psi_n(x). \tag{9.47}$$

The operator L is linear, so we can rewrite (9.47), at least formally, as follows:

$$\sum_{n=1}^{\infty} b_n L(\psi_n(x)) = \sum_{n=1}^{\infty} a_n \psi_n(x). \tag{9.48}$$

Next we appeal to (9.44) to replace $L(\psi_n)$ with $\mu_n \psi_n$ and rewrite (9.48) as

$$\sum_{n=1}^{\infty} b_n \mu_n \psi_n(x) = \sum_{n=1}^{\infty} a_n \psi_n(x). \tag{9.49}$$

Since the functions $\{\psi_n\}$ form a basis, they are linearly independent. Hence the coefficients $b_n \mu_n$ and a_n in the two series must be the same, that is,

$$b_n = \frac{a_n}{\mu_n}. \tag{9.50}$$

We summarize this discussion in the following theorem

Theorem 9.4.1
The solution ψ to the PDE defined by (9.43), subject to boundary conditions, can be obtained as an infinite series (9.45), where the coefficients b_n are given by (9.50).

We now return to the Poisson BVP (9.41) and apply the findings in Theorem 9.4.1 to obtain its solution. Here $L = -\Delta$ whose eigenfunctions are

$$\psi_{nm} = \sin \frac{n\pi x}{a} \sin \frac{m\pi y}{b},$$

when Ω is the rectangular domain in (9.29). As we have seen earlier

$$L(\psi_{nm}) = -\Delta\psi_{nm} = \lambda_{nm}^2 \psi_{nm},$$

where λ_{nm} are defined in (9.39). Following the strategy that led to Theorem 9.4.1, we expand f and ψ in terms of the eigenfunctions ψ_{nm}:

$$f(x, y) = \sum_{n,m}^{\infty} a_{nm} \sin \frac{n\pi x}{a} \sin \frac{m\pi y}{b}, \tag{9.51}$$

where a_{nm} are determined as before, that is,

$$a_{nm} = \frac{4}{ab} \int_0^a \int_0^b f(x,y) \sin \frac{n\pi x}{a} \sin \frac{m\pi y}{b} \, dxdy. \tag{9.52}$$

We now seek the solution ψ to the BVP in the form

$$\psi(x,y) = \sum_{n,m}^{\infty} b_{nm} \sin \frac{n\pi x}{a} \sin \frac{m\pi y}{b}, \tag{9.53}$$

where the scalars ψ_{nm} are to be determined. Since the function ψ is a solution of the Poisson equation in (9.41), we have

$$-\Delta\psi = f,$$

or

$$-\Delta(\sum_{n,m}^{\infty} b_{nm} \sin \frac{n\pi x}{a} \sin \frac{m\pi y}{b}) = \sum_{n,m}^{\infty} a_{nm} \sin \frac{n\pi x}{a} \sin \frac{m\pi y}{b}. \tag{9.54}$$

The Laplacian is a linear operator, that is $L(f+g) = L(f) + L(g)$, as you can verify easily by noticing that

$$-\Delta(f+g) = -\Delta f - \Delta g. \tag{9.55}$$

When we apply this feature of the Laplacian to the left side of (9.54), we end up with

$$-\Delta(\sum_{n,m}^{\infty} b_{nm} \sin \frac{n\pi x}{a} \sin \frac{m\pi y}{b}) = -\sum_{n,m}^{\infty} b_{nm} \Delta(\sin \frac{n\pi x}{a} \sin \frac{m\pi y}{b}) =$$

$$\sum_{n,m}^{\infty} \lambda_{nm}^2 b_{nm} \sin \frac{n\pi x}{a} \sin \frac{m\pi y}{b}. \tag{9.56}$$

Hence (9.54) reduces to

$$\sum_{n,m}^{\infty} \lambda_{nm}^2 b_{nm} \sin \frac{n\pi x}{a} \sin \frac{m\pi y}{b} = \sum_{n,m}^{\infty} a_{nm} \sin \frac{n\pi x}{a} \sin \frac{m\pi y}{b}. \tag{9.57}$$

Since the eigenfunctions ψ_{nm} are linearly independent, the coefficients of the two series in (9.57) must be equal, from which we deduce that b_{nm} are

$$b_{nm} = \frac{a_{nm}}{\lambda_{nm}^2} \tag{9.58}$$

as predicted by Theorem 9.4.1. We summarize the above discussion in the following theorem.

Theorem 9.4.1 (Solution to the Poisson BVP)
The solution to the BVP in (9.41) is given by the series (9.53), where the coefficients b_{nm} are given by

$$b_{nm} = \frac{4}{ab\lambda_{nm}^2}\int_0^a\int_0^b f(x,y)\sin\frac{n\pi x}{a}\sin\frac{m\pi y}{b}\,dxdy. \tag{9.59}$$

To illustrate the utility of this theorem, consider the Poisson BVP with the forcing term f given by the function

$$f(x,y) = -10\,x^2(3-x)\sin\frac{\pi y}{2} - 50\,y(2-y)^2\sin\frac{\pi x}{3}, \tag{9.60}$$

defined in the domain $\Omega = \{(x,y)|, 0 < x < 3, 0 < y < 2\}$. The following MATLAB code computes the coefficients a_{nm} and b_{nm} and plots the contours of f and ψ as shown in Figure 9.5, taking into account only the 10-th partial sum in each of the series expansions (9.51) and (9.53). Here the notation f_{10} and ψ_{10} stand for the 10-th partial sums of each respective function:

$$f_{10} = \sum_{m=1}^{10}\sum_{n=1}^{10} a_{nm}\sin\frac{n\pi x}{3}\sin\frac{m\pi y}{2},$$

and

$$\psi_{10} = \sum_{m=1}^{10}\sum_{n=1}^{10} b_{nm}\sin\frac{n\pi x}{3}\sin\frac{m\pi y}{2}.$$

Figure 9.5 shows one of the important features of the Laplacian operator, that this operation has the tendency to symmetrize a perturbation that is asymmetric, a feature that we will come back to in the next section.

Here is how Figure 9.5 was obtained in MATLAB:

```
a=3; b=2;
% Define f;  Combine the two lines below on
% a single line when executing in MATLAB
%
f=inline('-10*x.^2.*(3-x).*sin(pi*y/2)+...
    -50*y.*(2-y).^2.*sin(pi*x/3)','x','y');
%
% Define the domain and plot the contours of f
%
[X,Y]=meshgrid(0:0.01:a,0:0.01:b);
```

```
subplot(2,1,1)
contour(X,Y,f(X,Y),'black')
hold on
%
% Compute the Fourier Coefficients of f
%
for m=1:10
    for n=1:10
% Combine the two lines below in a single line
% before executing in MATLAB
        A(n,m)=4/(a*b)*quadv(@(y) quadv(@(x) f(x,y).*....
              sin(n*pi*x/a).*sin(m*pi*y/b),0,a),0,b);

    end
end
%
% Compute the solution psi
%
S=0;
for m=1:10
    for n=1:10
        eigenvalues=-pi^2*(n^2/a^2 + m^2/b^2);
        B(n,m)=A(n,m)/eigenvalues;
        S=S+B(n,m)*sin(n*pi*X/a).*sin(m*pi*Y/b);
    end
end
contour(X,Y,S,'black')
subplot(2,1,2)
```

Note the multiple use of quadv in the above code:

```
quadv(@(y) quadv(@(x) f(x,y).*...
     sin(n*pi*x/a).*sin(m*pi*y/b),0,a),0,b);
```

which computes the iterated integral in (9.59). Alternatively, we could compute the double integration in (9.59) by applying the MATLAB command

```
dblquad
```

which, when applied to (9.59) as

```
dblquad(@(x,y) f(x,y).*sin(n*pi*x/a).*sin(m*pi*y/b),0,a,0,b);
```

leads to a considerably more efficient computation of the Fourier coefficients b_{nm}. In the exercises we will apply the `tic ... toc` command of MATLAB to quantify how much faster `dblquad` computes a Fourier coefficient relative to a double application of `quadv`.

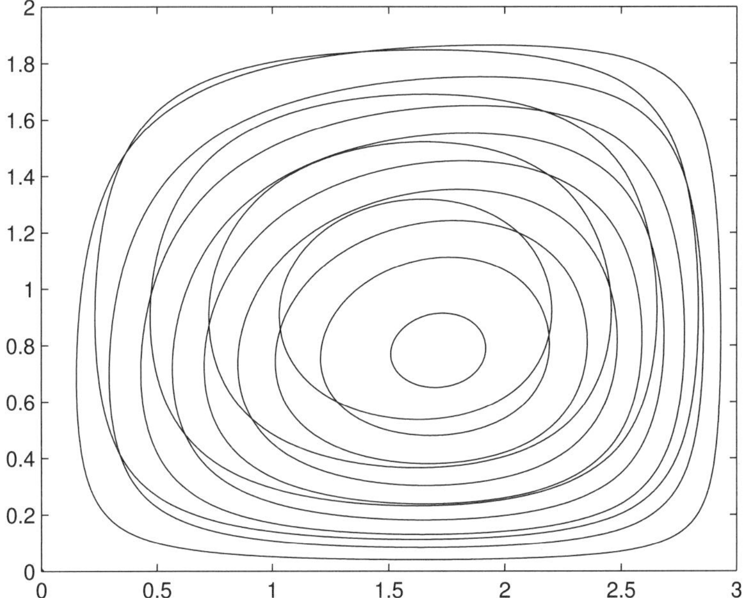

FIGURE 9.5: Contours of 10-th partial sums ψ_{10} and f_{10} of the stream function and the forcing term defined in (9.60), respectively. Contours of ψ_{10} are more or less symmetric in the domain while the contours of f_{10} are asymmetric and bunched up relatively closely to the southern boundary of the domain.

9.4.1 Poisson Equation with Localized Vorticity

One of the curious aspects of the previous example is the relative symmetry of the stream contours even though the forcing term f is asymmetric. To explore this phenomenon more closely, we now consider the Poisson BVP with forcing terms f whose support is in a small disk in the domain. Specifically, we consider functions f of the following form:

$$f(x,y) = \begin{cases} \frac{A}{2\delta}(1+\cos\frac{\pi}{\delta}||\mathbf{r}-\mathbf{P}||), & \text{when } ||\mathbf{r}-\mathbf{P}|| < \delta, \\ 0 & \text{otherwise.} \end{cases} \tag{9.61}$$

Here $\mathbf{P} = \langle x_0, y_0 \rangle$ is a point in the domain Ω and is the center of the disk in which the support of f is located, δ is the radius of the support, chosen small enough so that the disk of radius δ about $P = (x_0, y_0)$ is located entirely in Ω. The amplitude $\frac{A}{2\delta}$ is selected so that the integral of f over the entire domain Ω in (9.28) is A.

The following MATLAB code shows how to plot the graph of f in the special case where the domain Ω is the rectangle $(0, a) \times (0, b)$ with $a = 3$, $b = 2$, $x_0 = 2.5$, $y_0 = 0.5$ and $A = 10$. For δ we select

$$\delta = \frac{1}{2} \min(x_0, y_0, a - x_0, b - y_0)$$

to ensure that the disk of radius δ about P lies entirely in Ω.

```
global a b x0 y0 delta
%
a=3; b=2;
x0=2; y0=0.5;
delta=1/2*min([x0,y0,a-x0,b-y0]);
amp=10;
[X,Y]=meshgrid(0:0.01:a,0:0.01:b);
Z=LocalVorticity(amp,X,Y);
contour(X,Y,Z)
```

which calls on the M-file `LocalVorticity.m`:

```
function z=LocalVorticity(amp,x,y);
%
global a b x0 y0 delta
%
distance=sqrt((x-x0).^2+(y-y0).^2);
z=amp/(2*delta)*(1+cos(pi/delta*distance)).*(distance<delta);
```

Note the use of the boolean command `(distance<delta)`, which is used to reduce f, which is called `LocalVorticity` in the code, to zero outside of the support. When we combine this code with the one developed earlier that led to computing the approximate solution to the Poisson BVP in (9.41), we end up with the figures in Figure 9.6. The figures show the 30-partial sum approximation of each function, the forcing term f and its associated stream function ψ. Note that the stream function ψ_{30} displays a weakened localization relative to the one shown by the forcing term f_{30}.

Problems 9.4

1. Solve the following the Poisson boundary value problem with the following data. In each case plot the graph of the streamlines of the 10-th partial sum ψ_{10} as well as the contours of f on the same screen.

    (a) $a = 4$, $b = 5$ and $f(x, y) = xy(4 - x)(5 - y)$.

    (b) $a = 1$, $b = 1$ and $f(x, y) = 1$. Plot the graph of $\psi_{10}(x, \frac{1}{2})$ and report on the behavior of the function as x approaches the boundary values $x = 0$ and $x = a$.

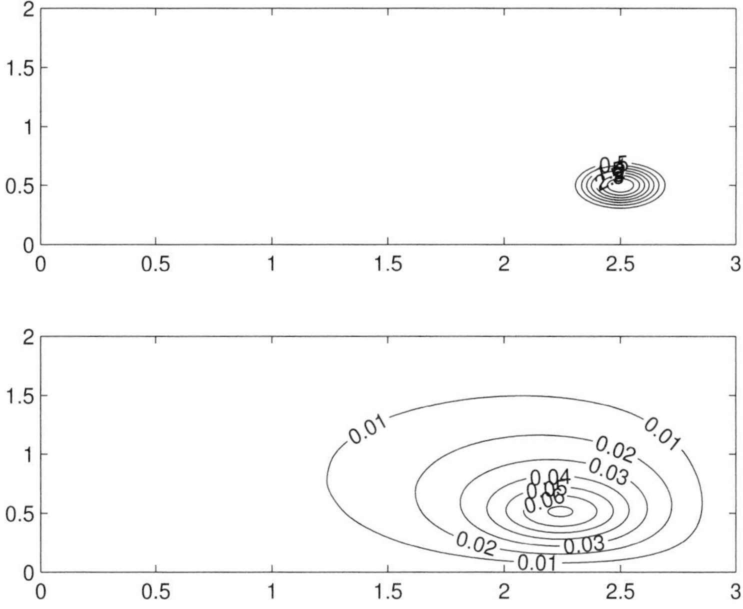

FIGURE 9.6: Contours of the partial sums ψ_{30} and f_{30} of the stream function and its associate forcing term defined in (9.61), respectively.

(c) $a = 1$, $b = 1$ and $f(x, y) = x(1 - x)\sin 2\pi x \sin \pi y - y \sin \pi x \sin 2\pi y$.

2. Use `help` with `tic` and familiarize yourself with the usage of this MATLAB utility. Apply `tic ... toc` to `quadv` as well as to `dblquad` in the computation of the Fourier coefficients of f, where f is define in (9.60), and report the quantitative difference in using these two quadrature utilities.

3. Consider the Poisson boundary value problem with

$$f(x, y) = \begin{cases} 1 & \text{when } (x - \frac{1}{2})^2 + (y - \frac{1}{2})^2 < \frac{1}{16}, \\ 0 & \text{otherwise.} \end{cases}$$

and $a = b = 1$. Display the contours of the stream function ψ when the domain is defined by

(a) $a = b = 1$

(b) $a = b = 10$.

4. The purpose of this problem is to write a MATLAB code based on

the local vorticity function f defined in (9.61). To emphasize that the value of $f(x, y)$ depends on the amplitude A and f's support center $\mathbf{P}$, let

$$f_{A,\mathbf{P}}$$

denote the function f defined in (9.61). Let

$$f = \sum_{i=1}^{N} f_{A_i,\mathbf{P}_i}$$

denote the forcing function associated with N local vortices of amplitude A_i centered at positions $\mathbf{P}_i$'s. Write a MATLAB program that accepts N vortices of different amplitudes and produces the approximation of the stream function with a desired accuracy specified by the user. Apply your program to plot the contours of ψ_{30} if $a = b = 1$ and

(a)

$$A_1 = 3,\ \mathbf{P}_1 = \langle \frac{a}{3}, \frac{b}{3} \rangle,\ A_2 = -2,\ \mathbf{P}_1 = \langle \frac{2a}{3}, \frac{b}{3} \rangle.$$

(b)

$$f = \sum_{i=1}^{2} f_{A_i,\mathbf{P}_i}$$

where A_i is a random number in the interval $(-10, 10)$, and $\mathbf{P}_i$ are chosen at random in the domain $(0, a) \times (0, b)$.

9.5 Stommel Model

The Stommel model is characterized by the desire to conceive of a simple model of circulation in a basin such as the North Atlantic by taking into account only the most prominent forces that affect the flow of the fluid particles, those forces that result in generating the kind of a current that resembles the Gulf Stream, not only geometrically but physically, in that most of the salient parameters that we know should characterize the Gulf Stream are present in the model.

As stated earlier in this chapter, one of the first models for a Gulf Stream like current proposed appeared in [1]. The remainder of the chapter will be dedicated to deriving the governing PDE, the necessary boundary conditions, and finally obtaining the solution to this BVP, analytically and numerically.

The geometry is motivated by considering a basin that has dimensions similar to North Atlantic (about 10,000 kilometers in each horizontal directions, and a depth of a few kilometers). In order to simplify the mathematical complexity of this problem we confine the domain to a cube. Following Stommel's approach, we choose to take into account only the impact of the acceleration due to Coriolis, the effect of the pressure gradient in the flow, which is primarily created due to the wind forcing at the surface, and, in order to have a reasonable steady-state solution, a simple model of frictional forces, taken proportional to velocity. The goal is solve the underlying time-independent boundary value problem for the stream function and to demonstrate the boundary-layer character of the solution on the western boundary of the region as a manifestation of the Gulf Stream. The model derivation we present is somewhat ad hoc but it hopefully has enough content to build the reader's intuition to appreciate which forces have been balanced, while identifying the forces that have been neglected to arrive at the reduced Stommel model.

We will then present the results of the paper [4], by Joe Pedlosky, where we start with the full nonlinear time-dependent equations of geophysical fluid dynamics and show how Stommel's model can be derived from the full model after we are identify certain small non-dimensional parameters.

9.5.1 Governing PDE

We begin by considering a rectangular basin occupied by a homogeneous and incompressible fluid of density ρ_0. When the fluid is stationary its domain B is defined by

$$B = \{(x, y, z) |\, 0 < x < \lambda,\, 0 < y < b,\, 0 < z < d\}$$

When this basin is perturbed its new shape is defined by

$$\Omega = \{(x, y, z) |\, 0 < x < \lambda,\, 0 < y < b,\, 0 < z < d + \eta(x, y, t)\}$$

where $z = \eta(x, y, t)$ represents the air-sea interface.

We assume that the motion is steady and two dimensional so that

$$\mathbf{v} = \langle u(x, y), v(x, y), 0 \rangle.$$

We further assume that the nonlinear convective term of the acceleration, $\mathbf{v} \cdot \nabla \mathbf{v}$, is small compared with the linear Coriolis term, the pressure gradient term, the wind forcing and the frictional forcing, and neglect it in deference to the linear terms in the balance of linear momentum. These assumptions lead to a substantial reduction of the continuity equation and the balance of linear momentum. In particular the continuity equation becomes

$$\frac{\partial u}{\partial x} + \frac{\partial v}{\partial y} = 0, \tag{9.62}$$

and the balance of linear momentum takes the form

$$-\rho_0 f v = -\frac{\partial p}{\partial x} + \rho_0 F_1, \quad \rho_0 f u = -\frac{\partial p}{\partial y} + \rho_0 F_2, \tag{9.63}$$

$$0 = -\frac{\partial p}{\partial z} - \rho_0 g. \tag{9.64}$$

Here $f = 2\Omega \sin\phi$ is the Coriolis parameter, and in the third equation the hydrostatic approximation is assumed. The terms F_1 and F_2 are components of the forcing terms, representing wind forcing and bottom friction. They will be constructed specifically to ensure that (9.63) and (9.64) match the equivalent equations of Stommel.

Since ρ_0 and g are independent of x, y and z, we conclude from (9.64) that $\frac{\partial p}{\partial z}$ is independent of x and y, implying that $\frac{\partial^2 p}{\partial x \partial z} = \frac{\partial^2 p}{\partial y \partial z} = 0$. Keeping this in mind, we return to (9.63) and differentiate them with respect to z to get

$$\frac{\partial(-\rho_0 f v)}{\partial z} = \rho_0 \frac{\partial F_1}{\partial z}, \quad \frac{\partial(\rho_0 f u)}{\partial z} = \rho_0 \frac{\partial F_2}{\partial z}.$$

We further assume that F_1 and F_2 are independent of z. The above equations then imply that both fu and fv are independent of z, and since f is independent of z, we conclude that u and v are functions of x and y only. Although the statement about the independence of u and v on z was assumed from the beginning, it is instructive to note that this conclusion could have been reached by simply assuming that the external forces acting on B are independent of z.

Next we integrate (9.64) with respect to z, from a typical z to $z = d + \eta(x, y, t)$, where, as stated earlier, $d + \eta$ defines the air-sea interface. We get

$$p(x, y, z, t) = p_0 + \rho_0 g(d + \eta(x, y, t) - z), \tag{9.65}$$

where p_0 is the air pressure, assumed constant for simplicity. Replacing p from (9.65) in (9.63) results in eliminating p from the equations of balance of linear momentum:

$$-fv = -g\frac{\partial \eta}{\partial x} + F_1, \quad fu = -g\frac{\partial \eta}{\partial y} + F_2.$$

Next, we integrate both equations with respect to z, from $z = 0$ to $z = d + \eta$:

$$-fv(d + \eta) = -g\frac{\partial \eta}{\partial x}(d + \eta) + F_1(d + \eta),$$

$$fu(d+\eta) = -g\frac{\partial \eta}{\partial y}(d+\eta) + F_2(d+\eta).$$

We assume that η is small and negligible relative to d. We, therefore, arrive at the linearized equations

$$-fvd = -gd\frac{\partial \eta}{\partial x} + dF_1, \quad fud = -gd\frac{\partial \eta}{\partial y} + dF_2. \tag{9.66}$$

With the goal of obtaining the equations in Stommel's paper, we now define F_1 and F_2 in the following special way:

$$dF_1 = -Ru - \gamma \cos\frac{\pi y}{b}, \quad dF_2 = -Rv, \tag{9.67}$$

where γ and R are positive constants. Thus $d\langle F_1, F_2\rangle$ is the sum of two vectors $\mathbf{R} + \mathbf{T}$ where

$$\mathbf{R} \equiv -R\mathbf{v} = \langle -Ru, -Rv\rangle$$

is intended to model the overall impact of the bottom and lateral frictional damping (molecular or eddy viscosity terms are not included in this model). And the traction vector $\mathbf{T}$ defined by

$$\mathbf{T}(y) = \langle -\gamma \cos\frac{\pi y}{b}, 0\rangle,$$

which represents the wind stress/wind forcing on the air-sea interface. Note that $\mathbf{T}(0) = \langle -\gamma, 0\rangle$ so that the wind stress in the lower part of the basin is applying a force in the negative x-direction, while $\mathbf{T}(b) = \langle \gamma, 0\rangle$, signifying the wind in the positive x-direction in the upper part of the basin; $\mathbf{T}$ thus represents a shearing force exerted on the surface of B, very similar to the trade and prevailing winds that have been observed in mid-latitudes in the North Atlantic.

Substituting (9.67) into (9.66) results in the following set of equations for u, v and η:

$$-fv = -g\frac{\partial \eta}{\partial x} - \frac{R}{d}u - \frac{\gamma}{d}\cos\frac{\pi}{b}, \quad fu = -g\frac{\partial \eta}{\partial y} - \frac{R}{d}v. \tag{9.68}$$

Next we eliminate η from (9.68) by cross-differentiating the above equations: Differentiate (9.68a) with respect to y and (9.68b) with respect to x, and subtract to get

$$-f'(y)v - f\frac{\partial v}{\partial y} - f\frac{\partial u}{\partial x} = -\frac{R}{d}\frac{\partial u}{\partial y} + \frac{\gamma\pi}{db}\sin\frac{\pi y}{b} + \frac{R}{d}\frac{\partial v}{\partial x}, \tag{9.69}$$

where we have made use of (9.62) to simplify. Speaking of equation

(9.62), we note that this equation implies that u and v are related to a stream function ψ through the relations

$$u = \frac{\partial \psi}{\partial y}, \qquad v = -\frac{\partial \psi}{\partial x}, \tag{9.70}$$

which, when applied to (9.69), results in the following partial differential equation

$$\Delta \psi + \alpha \psi_x = A \sin \frac{\pi y}{b}, \tag{9.71}$$

where

$$\alpha = \frac{f'(y)d}{R}, \qquad A = \frac{\gamma \pi}{Rb}. \tag{9.72}$$

We complement the PDE in (9.71) with zero boundary conditions, i.e.,

$$\psi|_{\partial B} = 0. \tag{9.73}$$

Our task in the next section is to solve the boundary-value problem (9.71)–(9.73) using separation of variables. But before completing this section we comment that the typical values of the various constants in (9.72) are (see [1])

$$d = 200 \text{ m}, \quad \lambda = 10,000 \text{ km}, \quad b = 2\pi \times 1000 \text{ km},$$

$$\gamma = 1 \text{ dyne/cm}^2, \quad R = 0.02. \tag{9.74}$$

Among these values, the magnitude of $f'(y)$ is worth noting; Since $y = a\phi$ where a is the radius of the planet and ϕ is the latitude, we note that $f(y) = 2\Omega \sin \phi = 2\Omega \sin \frac{y}{a}$. Hence, $f'(y) = \frac{2\Omega}{a} \cos \frac{y}{a}$. In this problem's setting, with $a = 6240$ km, $\Omega = \frac{\pi}{43200}$ and $0 < y < b$, we have

$$1.24575 \text{x} 10^{-13} < f'(y) < 2.33084 \text{x} 10^{-13}.$$

Following Stommel's footsteps, in what follows we assume the value $f'(y) \approx 2 \times 10^{-13}$.

Problems 9.5

1. Review the derivation of the Stommel model in this section and write down all of the assumptions made that enabled us to reduce the fully nonlinear equations of geophysical fluid dynamics to the simple linear PDE (9.71).

2. Show that the PDE in (9.71) is equivalent to the following PDE:
$$e^{-\alpha x}(\nabla \cdot e^{\alpha x}\nabla\psi) = A\sin\frac{\pi y}{b}. \tag{9.75}$$

3. Define the operator L by
$$L(\phi) = e^{-\alpha x}(\nabla \cdot e^{\alpha x}\nabla\phi). \tag{9.76}$$
Show that the functions $\phi_{m,n}(x,y)$ defined by
$$\phi_{m,n}(x,y) = e^{-\frac{\alpha x}{2}}\sin n\pi x\,\sin\frac{m\pi y}{\tau} \tag{9.77}$$
are the eigenfunctions of the L in the domain $(0,1)\times(0,\tau)$ with Dirichlet boundary conditions, i.e., $\phi_{m,n}$ satisfy the relations
$$L(\phi_{m,n}) = -\mu_{m,n}^2\phi_{m,n},$$
where $\mu_{m,n}^2$ is the eigenvalue. Write down the eigenvalues.

4. Show that the eigenfunctions (9.77) are mutually orthogonal with respect to the weight $w(x) = e^{\alpha x}$, that is
$$\int_0^1\int_0^\tau \phi_{m',n'}\phi_{m,n}w(x)\,dy\,dx = 0, \quad \text{if } m \neq m', n \neq n'. \tag{9.78}$$

5. Let $\tau = 0.8$.
   (a) Let $\alpha = 20$. Plot the contours of the first three eigenfunctions $(\phi_{1,1}, \phi_{2,1}, \phi_{1,2})$ and the eigenfunction $\phi_{3,5}$. See Figure 9.7. Next, to appreciate the difference between $\phi_{1,1}$ and $\phi_{2,1}$, plot the difference between these two eigenfunctions at $x = 0.1$ and $x = 0.7$ for $y \in (0,\tau)$. See Figure 9.8.
   (b) Let $\alpha = 200$. Plot the graphs of the eigenfunctions in the previous problem for this value of α. Describe the impact of the size of α on the size of the boundary layer in the first eigenfunction.
   (c) Compute the velocity at the point $P = (x,y) = (0.1, \frac{\tau}{2})$ for the two velocity fields derived from the eigenfunction $\phi_{1,1}$, first when $\alpha = 20$ and next when $\alpha = 200$, and report on the difference between the speed of the fluid located at P for each vector field.

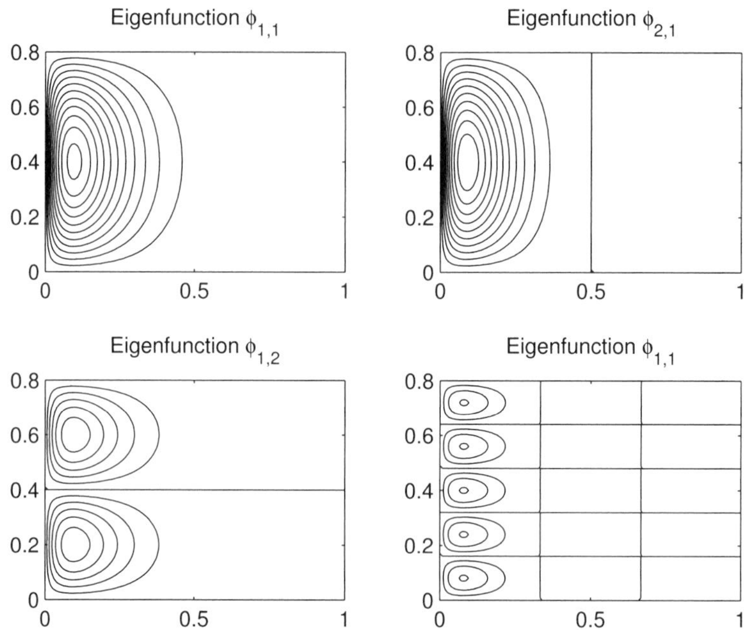

FIGURE 9.7: Contours of four eigenfunctions of L when $\alpha = 20$

9.5.2 Non-Dimensionalization

The variables in (9.71), whether dependent or independent, are dimensional. If we stay in this dimensional setting, we will end up having to deal with parameter values with magnitudes 10^8 (for λ) and 10^{-13} (for $f'(y)$), a range of twenty-one orders of magnitude. Although this range does not particularly create a major difficulty for MATLAB, it turns out we gain more insight into the impact of the three forces that balance each other in (9.71) on the behavior of solutions of the BVP, including why a boundary layer is generated, if we first non-dimensionalize the variables.

We non-dimensionalize x, y and ψ by defining new variable $\bar{x}$, $\bar{y}$ and $\bar{\psi}$ through the following relations:

$$\lambda\bar{x} = x, \quad \lambda\bar{y} = y, \quad \psi_0\bar{\psi}(\bar{x}, \bar{t}) = \psi(x, y),$$

where ψ_0 is a typical value of the stream function. Differentiating the last relation twice with respect to x, and noting that $\frac{\partial x}{\partial \bar{x}} = \lambda$, leads to

$$\psi_{xx} = \frac{\psi_0}{\lambda^2}\bar{\psi}_{\bar{x}\bar{x}}.$$

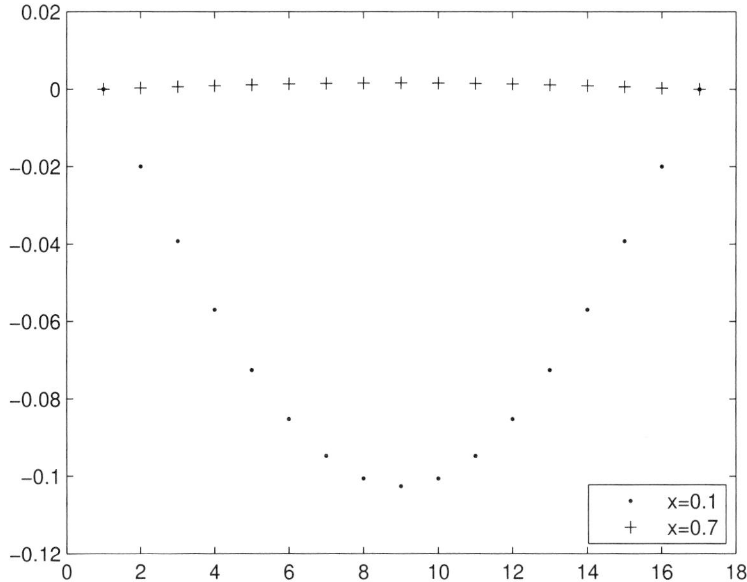

FIGURE 9.8: Graphs of $\phi_{1,1}(x, y) - \phi_{2,1}(x, y)$with $\alpha = 20$ and for $y \in (0, \tau)$ when $x = 0.1$ (graphed with "." dots), near the boundary layer, and at $x = 0.7$ (graphed with "+" dots).

Similarly

$$\psi_{yy} = \frac{\psi_0}{\lambda^2}\bar{\psi}_{\bar{y}\bar{y}}.$$

Hence the Laplace operator Δ in (9.71) transforms to

$$\frac{\psi_0}{\lambda^2}\bar{\Delta},$$

where

$$\bar{\Delta} = \frac{\partial^2}{\partial \bar{x}^2} + \frac{\partial^2}{\partial \bar{y}^2}.$$

In this way the dimensional boundary value problem (9.71) transforms to the non-dimensional one

$$\bar{\Delta}\bar{\psi} + \bar{\alpha}\bar{\psi}_{\bar{x}} = \bar{A}\sin\frac{\pi\bar{y}}{L}, \tag{9.79}$$

where L is the aspect ratio of the horizontal parts of the domain, i.e.,

$$L = \frac{b}{\lambda}, \tag{9.80}$$

and

$$\bar{\alpha} = \lambda\alpha, \quad \bar{A} = \frac{\lambda^2}{\psi_0} A. \tag{9.81}$$

Equation (9.79) is now supplemented by the boundary conditions

$$\bar{\psi}(\bar{x}, 0) = \bar{\psi}(\bar{x}, L) = \bar{\psi}(0, \bar{y}) = \bar{\psi}(1, \bar{y}) = 0, \tag{9.82}$$

In the next section we find the solution to (9.79)–(9.82).

9.5.3 Solution to the BVP

Before proceeding to obtain the solution to (9.79)–(9.82), we rename all of the variables by removing the bars from these variables. Hence we seek a solution to

$$\Delta\psi + \alpha\psi_x = A \sin\frac{\pi y}{L}, \tag{9.83}$$

subject to the boundary conditions

$$\psi(x, 0) = \psi(x, L) = \psi(0, y) = \psi(1, y) = 0, \tag{9.84}$$

where (see (9.72))

$$\alpha = \frac{f'(y)d\lambda}{R}, \quad A = \frac{\gamma\pi\lambda^2}{Rb\psi_0}. \tag{9.85}$$

We begin by noting that the PDE (9.83) is a Poisson equation. We could seek the solution to the boundary value problem (9.83)–(9.84), as we did in Section 9.4, in terms of the eigenfunctions of the operator

$$L(\phi) = \Delta\phi + \alpha\phi_x,$$

subject to the boundary conditions in (9.84)(see Problem 3, Section 9.5.1), but instead we follow the technique H. Stommel applied in his paper, namely, by seeking first the general solution of (9.83). To that end we look for solutions in the form

$$\psi(x, y) = \psi_h(x, y) + \psi_p(x, y), \tag{9.86}$$

where ψ_h is the solution of the *homogeneous* part of the (9.83), namely, the solution to

$$\Delta\psi + \alpha\psi_x = 0, \tag{9.87}$$

while ψ_p is any *particular* solution of (9.83). Neither ψ_h nor ψ_p may satisfy the boundary condition (9.84). Our strategy is to select the arbitrary parameters in (9.86) appropriately so the boundary conditions (9.84) hold.

9.5.3.1 Determining the Particular Solution ψ_p

Since ψ_p does not have to satisfy the boundary condition (9.84) we can essentially guess its form. Noting that $A\sin\frac{\pi y}{L}$ is the forcing term in (9.83), we consider a particular solution ψ_p in the form $\psi_p(x,y) = B\sin\frac{\pi y}{L}$, and determine the constant B so that ψ_p is a solution of (9.83). Substituting this template into (9.83) results in $B = -\frac{AL^2}{\pi^2}$. Hence

$$\psi_p(x,y) = -\frac{AL^2}{\pi^2}\sin\frac{\pi y}{L}. \tag{9.88}$$

9.5.3.2 Determining the Homogeneous Solution ψ_h

To determine the solution to (9.87) we apply the method of separation of variables to this equation, namely, assume a solution $\psi_h(x,y)$ of the form $F(x)G(y)$. Substituting the expression

$$\psi_h(x,y) = F(x)G(y)$$

into $\Delta\psi + \alpha\psi_x = 0$, we see that F and G must solve the equation

$$F''G + G''F + \alpha F'G = 0,$$

for all x and y in the domain $(0,1)\times(0,L)$. Dividing this expression by FG results in

$$(\frac{F''}{F} + \alpha\frac{F'}{F}) + \frac{G''}{G} = 0.$$

Because F only depends on x, G only depends on y and α is constant, it follows that the terms involving F must equal a constant, which we denote by μ^2, so that

$$\frac{F''}{F} + \alpha\frac{F'}{F} = \mu^2, \qquad \frac{G''}{G} = -\mu^2. \tag{9.89}$$

The above equations lead to the following set of second order differential equations

$$F'' + \alpha F' - \mu^2 F = 0, \qquad G'' + \mu^2 G = 0. \tag{9.90}$$

It is a simple computation to show that the following functions are the general solutions of the above equations:

$$G(y) = c_1\cos\mu y + c_2\sin\mu y, \quad F(x) = c_3 e^{m_1 x} + c_4 e^{m_2 x}, \tag{9.91}$$

where m_1 and m_2 are

$$m_1 = \frac{-\alpha + \sqrt{\alpha^2 + 4\mu^2}}{2}, \qquad m_2 = \frac{-\alpha - \sqrt{\alpha^2 + 4\mu^2}}{2}.$$

Referring back to (9.86), we now see that the general solution of the homogeneous part of the Poisson equation is (recall that $\psi_h = FG$)

$$\psi_h(x,y) = (c_3 e^{m_1 x} + c_4 e^{m_2 x})(c_1 \cos \mu y + c_2 \sin \mu y). \tag{9.92}$$

Since we have explicit formulas for ψ_h, (9.92), and for ψ_p, (9.88), we combine them to obtain the general solution of the full PDE (9.83):

$$\psi(x,y) = -\frac{AL^2}{\pi^2} \sin \frac{\pi y}{L} + (c_3 e^{m_1 x} + c_4 e^{m_2 x})(c_1 \cos \mu y + c_2 \sin \mu y). \tag{9.93}$$

9.5.3.3 Applying the Boundary Conditions

It remains to select c_1, c_2, c_3, c_4, and μ to enforce the boundary condition (9.84).

We begin by applying the first boundary condition, that $\psi(x, 0) = 0$ for all $0 < x < 1$: Evaluate (9.93) at $y = 0$ to get

$$0 = \psi(x,0) = c_1(c_3 e^{m_1 x} + c_4 e^{m_2 x}), \quad \text{for all } x \in (0,1),$$

which results in $c_1 = 0$. Thus ψ in (9.93) reduces to

$$\psi(x,y) = -\frac{AL^2}{\pi^2} \sin \frac{\pi y}{L} + (d_1 e^{m_1 x} + d_2 e^{m_2 x}) \sin \mu y, \tag{9.94}$$

where $d_1 = c_2 c_3$ and $d_2 = c_2 c_4$. Next, we apply the boundary condition $\psi(x, L) = 0$ to (9.94) to get

$$0 = \psi(x,L) = (d_1 e^{m_1 x} + d_2 e^{m_2 x}) \sin \mu L = 0, \quad \text{for all } x \in (0,1),$$

which implies that μ should be chosen in such a way that $\mu L = n\pi$, with $n = 1, 2, ...$ Hence, referring back to (9.94), we obtain infinitely many candidates for the solution to (9.83), which we index by n:

$$\psi_n(x,y) = -\frac{AL^2}{\pi^2} \sin \frac{\pi y}{L} + (d_1 e^{m_1 x} + d_2 e^{m_2 x}) \sin \frac{n\pi y}{L}, n = 1, 2, 3, ... \tag{9.95}$$

Next, we apply the boundary condition $\psi(0, y) = 0$ to (9.95):

$$-\frac{AL^2}{\pi^2} \sin \frac{\pi y}{L} + (d_1 + d_2) \sin \frac{n\pi y}{L} = 0 \quad \text{for all } y \in (0, L). \tag{9.96}$$

The above expression is of the form

$$\beta_1 \sin \frac{\pi y}{L} + \beta_2 \sin \frac{n\pi y}{L} = 0 \quad \text{for all } y \in (0, L), \tag{9.97}$$

for appropriate β_1 and β_2, both constants. The two functions $\sin \frac{\pi y}{L}$ and

$\sin \frac{n\pi y}{L}$ are linearly independent on the interval $(0, L)$ unless $n = 1$. Hence, unless we select n to be 1, the coefficients β_1 and β_2 must both vanish. We know, however, that $\beta_1 = -\frac{AL^2}{\pi^2}$ is nonzero. We have no choice other than to select only the first mode $n = 1$ from the infinitely many modes available in (9.95). Hence, $\psi = \psi_1$, or

$$\psi(x, y) = -\frac{AL^2}{\pi^2} \sin \frac{\pi y}{L} + (d_1 e^{m_1 x} + d_2 e^{m_2 x}) \sin \frac{\pi y}{L}, \tag{9.98}$$

is the solution to (9.83)–(9.84), satisfying three of the four boundary conditions in (9.84). Returning to (9.96), with n set equal to 1, we find that

$$\left(-\frac{AL^2}{\pi^2} + (d_1 + d_2) \right) \sin \frac{\pi y}{L} = 0, \quad \text{for all } y,$$

which implies that

$$d_1 + d_2 = \frac{AL^2}{\pi^2}. \tag{9.99}$$

Using this information in (9.98), we conclude that ψ takes the form

$$\psi(x, y) = \left(-\frac{AL^2}{\pi^2} + d_1 e^{m_1 x} + (\frac{AL^2}{\pi^2} - d_1) e^{m_2 x} \right) \sin \frac{\pi y}{L}. \tag{9.100}$$

Finally, we apply the boundary condition $\psi(1, y) = 0$ to (9.100) to get

$$-\frac{AL^2}{\pi^2} + d_1 e^{m_1} + (\frac{AL^2}{\pi^2} - d_1) e^{m_2} = 0$$

which results in $d_1 = \frac{AL^2 (e^{m_2} - 1)}{\pi^2 (e^{m_2} - e^{m_1})}$ and, from (9.99), that $d_2 = \frac{AL^2 (1 - e^{m_1})}{\pi^2 (e^{m_2} - e^{m_1})}$. Hence the final form of ψ is

$$\psi(x, y) = \frac{AL^2}{\pi^2} [-1 + \kappa e^{m_1 x} + (1 - \kappa) e^{m_2 x}] \sin \frac{\pi y}{L}, \tag{9.101}$$

where κ is

$$\kappa = \frac{e^{m_2} - 1}{e^{m_2} - e^{m_1}}.$$

We have proved the following theorem.

Theorem 9.5.1 (Stommel's Stream Function)
Consider the boundary-value problem (9.83)–(9.84). The function ψ defined in (9.101) is a solution of this problem.

The function ψ in (9.101) defines the stream function for Stommel's model of the Gulf Stream. To see that it embodies some of the salient features of this current, we now use MATLAB and plot its streamlines.

9.6 MATLAB Programs

The following program defines the stream function ψ from (9.101) to MATLAB. First we store the various physical constants and defined parameters in the file StommelConstants.m as follows (the parameter values used in this program have been taken verbatim from [1]):

```
psi0=10^9;                  % Setting the stream value scale,
                            % an arbitrary value.
lambda=10^9;                % Basin's length (in centimeters)
b=2*pi*10^8;                % Basin's width
d=20000;                    % Basin's depth
gamma=1;                    % Wind stress
fprime=2*10^(-13);          % Coriolis parameter
R=0.02;                     % Bottom friction parameter
%%%
%%% Formulas obtained in the text
%%%
A=(gamma*pi*lambda^2)/(R*b*psi0);
alpha=lambda*fprime*d/R;
mu=pi*lambda/b;
m1=(-alpha-sqrt(alpha^2+4*mu^2))/2;
m2=(-alpha+sqrt(alpha^2+4*mu^2))/2;
c1=(exp(m2)-1)/(exp(m2)-exp(m1));
c2=(1-exp(m1))/(exp(m2)-exp(m1));
%%%
```

The program StommelContours.m defined below, begins with StommelConstant.m and proceeds to generate a 100 by 100 grid of the domain to plot the contours of ψ:

```
n = 100; m = 50  % Grid points in horizontal and vertical directions
%%% Defining domain in MATLAB;
L=b/lambda;
[x,y]=meshgrid(0:1/(n+1):1,0:L/(m+1):L);
%%% Evaluation of the Stream function
z=A*L^2/(pi^2)*(-1+c1*exp(m1*x)+c2*exp(m2*x)).*sin(pi*y/L);
zz=z';
contour(x,y,zz)
```

MATLAB returns Figure 9.9. Executing the lines

```
[c, hh]= contour(x,y,zz);
clabel(c, hh)
```

adds the value of the contour levels to the contours in Figure 9.9.

Note that the streamline values are all negative, indicating a clockwise rotation. The most striking feature of the streamlines is of course their behavior near the western boundary, where, analogous to the Gulf Stream, they bunch up to form a boundary layer. Since the flux remains constant between any two streamlines, the fact that the area between any two neighboring streamlines narrows near the western boundary indicates that the velocity of the fluid flow must increase substantially relative to, say, the eastern boundary. It is worth emphasizing the significance of Stommel's analysis in [1], which demonstrated that the presence of this boundary layer is solely due to the intricate interplay among the Coriolis force, the wind stress, and bottom friction. In particular, because of the various scales and dimensions involved, the parameter α takes the value of 200 in its non-dimensional form (see 9.85). The size of this parameter ends up being the key factor that is responsible for the appearance of the boundary layer in Figure 9.9, as some of the exercises at the end of this section will demonstrate.

The solution we obtained in (9.101) is exact and in closed form. One can simply differentiate and substitute it into the original boundary value problem in (9.83) to verify that (9.101) is indeed a solution. Another advantage of having a formula for the solution is that it is then relatively simple to compute various properties of the flow associated with (9.101). For example, recalling that the velocity field $\mathbf{v} = \langle u, v\rangle$ is related to ψ by $u = \frac{\partial \psi}{\partial y}$ and $v = -\frac{\partial \psi}{\partial x}$, we find that

$$u = \frac{AL}{\pi}\left(-1 + \frac{e^{m_2}-1}{e^{m_2}-e^{m_1}}e^{m_1 x} + \frac{1-e^{m_1}}{e^{m_2}-e^{m_1}}e^{m_2 x}\right)\cos\frac{\pi y}{L},$$

$$v = -\frac{AL^2}{\pi^2}\left(-1 + \frac{e^{m_2}-1}{e^{m_2}-e^{m_1}}m_1 e^{m_1 x} + \frac{1-e^{m_1}}{e^{m_2}-e^{m_1}}m_2 e^{m_2 x}\right)\sin\frac{\pi y}{L}.$$

The `quiver` command of MATLAB when combined with the latter formulas leads to Figure 9.10. This figure is the output of the following program:

```
global psi0 lambda b d gamma fprime R A
global alpha mu m1 m2 c1 c2

StommelConstants

L=b/lambda;
%
```

```
[x,y]=meshgrid(0.01:0.1:0.99,0.01:L/10:L-0.01);
[u,v]=StommelVelocity(x,y);
%
quiver(x,y,u/norm(u),v/norm(v))
```

This program calls on `StommelVelocity.m` listed below:

```
function [u,v]=StommelVelocity(x,y);

global psi0 lambda b d gamma fprime R A
global alpha mu m1 m2 c1 c2

L=b/lambda;
term=(-1+c1*m1*exp(m1*x)+c2*m2*exp(m2*x));
u = A*L/pi*term.*cos(pi*y/L);
v=-A*L^2/(pi^2)*term.*sin(pi*y/L);
```

Now that we have the velocity field $\mathbf{v}$ in hand, we can compute the relative vorticity, $\omega = \nabla \times \mathbf{v}$, which is equivalent to $(-\Delta\psi)\,\mathbf{k}$. We suspect that the amplitude of the relative vorticity, $-\Delta\psi$, will also favor the western boundary. To verify this, we compute this quantity:

$$-\Delta\psi = -\frac{AL^2}{\pi^2}\left((\frac{\pi^2}{L^2} + (m_1^2 - \frac{\pi^2}{L^2})\frac{e^{m_2}-1}{e^{m_2}-e^{m_1}} e^{m_1 x} + \right.$$

$$\left. (m_2^2 - \frac{\pi^2}{L^2})\frac{1-e^{m_1}}{e^{m_2}-e^{m_1}} e^{m_2 x}) \right) \sin\frac{\pi y}{L}.$$

The graph of this function is shown in Figure 9.11.

As pointed out earlier, the advantage of having the exact solution for the stream function ψ is that we can infer a considerable amount of information about the physical problem by direct computations involving ψ. Several of the problems at the end this section are intended to drive this point home. It is obvious, however, that we cannot be as successful if we are able to compute the solution of an initial-boundary value problem only approximately. In the next section we consider computing the solution to (9.83) using a standard finite difference method. We then discuss the obstacles we may encounter when we attempt to determine the vorticity, say, when we only have an approximate knowledge of the stream function.

Problems 9.6

1. Verify the statements in (9.90) and (9.91).

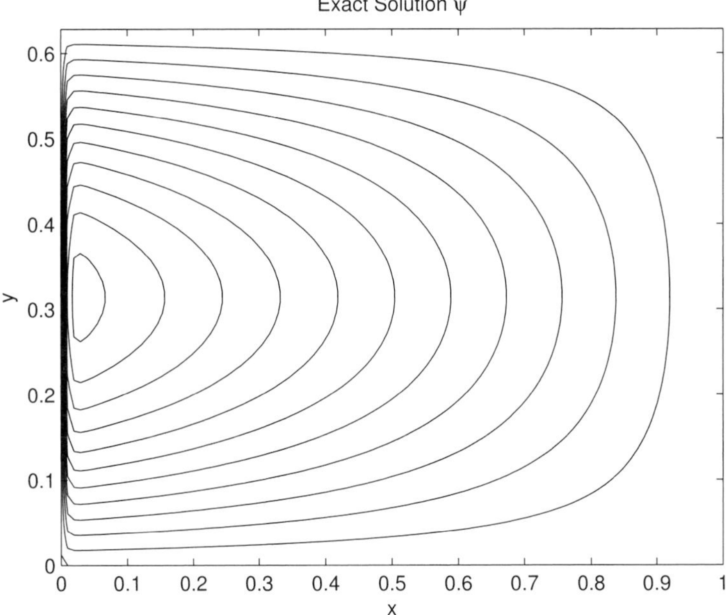

FIGURE 9.9: Streamlines of ψ.

2. Verify by direct substitution that (9.93) satisfies (9.83).

3. Verify by direct substitution that (9.101) satisfies (9.83).

4. Show that $\sin y$ and $\sin ny$ are linearly independent in the interval (a, b) unless $n = 1$.

5. In applying the separation of variables method to (9.83) we made the tacit assumption that μ^2, the constant of separation of variables, is positive. Consider instead the following two cases, construct the solution ψ of the BVP in each case and report on any barriers you encounter when attempting to solve for the boundary conditions:

   (a) $\mu = 0$.

   (b) $\mu^2 = -\delta^2$, with δ a real number, i.e., assuming that the constant of separation of variables is negative.

6. Experiment with the various parameter values in `StommelConstants.m`

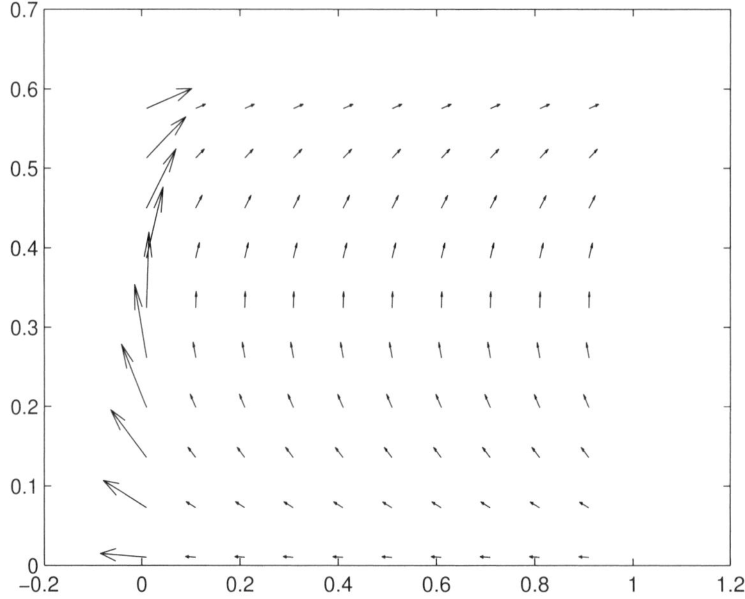

FIGURE 9.10: Velocity field associated with the stream function ψ.

to see the impact of α on the size of the boundary layer. In particular, plot the contours of ψ_α with $\alpha = 0, 10, 20, 50, 100, 200, 500$ and 1000.

7. With the constants as defined in `StommelConstants.m`, and the stream function is defined in (9.83), apply MATLAB's `ode45` to plot the trajectory of the particle that is located at the position $P = (0.9, \frac{L}{2})$, where L is the aspect ratio of the basin. This trajectory should follow the path of a streamline through P in clockwise direction. Experiment with the value of `tfinal` in `ode45` to get the trajectory to cover about

   (a) a quarter of the streamline,

   (b) half of the streamline,

   (c) a full streamline.

8. Repeat the computations of the above problem with the point $P = (0.8, \frac{L}{2})$. Report on the difference between the various times it takes these trajectories to cover the corresponding distances.

9. Write a MATLAB program that plots the position of set of N

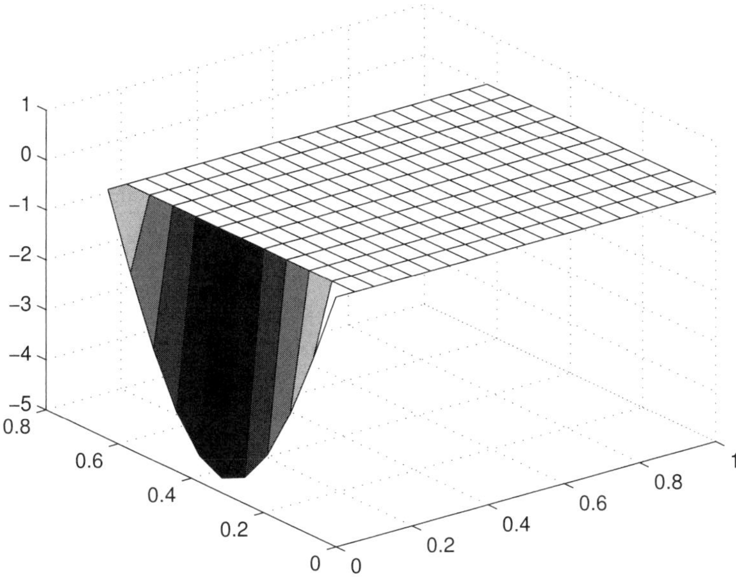

FIGURE 9.11: The relative vorticity $-\Delta\psi$ associated with the stream function ψ.

points whose initial positions are equally distributed on a circle of radius 0.01 and centered at $P = (0.9, \frac{L}{2})$. Experiment with the `tfinal` in `ode45` and display the snapshots of the initial circular parcel of fluid as time evolves. Report on the deformation of this parcel fluid in relation to the general characteristics of the flow, especially its vorticity, and especially in connection with the flow in the boundary layer. Experiment with the `options` in `ode45` to gain confidence in the numerical values you are displaying.

9.7 Stommel Model—A Numerical Approach

In the previous section we obtained the exact solution to the boundary value problem (9.83)–(9.84):

$$\Delta\psi + \alpha\psi_x = A\sin\frac{\pi y}{L},$$

subject to the boundary conditions

$$\psi(x, 0) = \psi(x, L) = \psi(0, y) = \psi(1, y) = 0.$$

where α and A are defined in (9.85).

Our success in getting hold of the solution analytically is owed to several factors: first, the original PDE is linear and second order, which lent itself to the method of separation of variables, which in turn allowed us to obtain the general solution of the PDE in terms of the normal modes or basis functions of the PDE. Second, the geometry of the domain, a cube (a rectangle, in actuality), is simple enough that it allowed us to ensure that the boundary conditions are satisfied exactly. Third, the forcing function in the Stommel model is very special, in fact it is proportional to a single normal mode, so we are able to obtain the analytic solution (9.101) in closed form. What we propose to do in this section is to explore the challenges we encounter when we violate some of the conditions under which we are able to construct the solution.

To start, we consider a natural generalization of the Poisson equation to

$$a_1 \Delta\psi + a_2\psi_x + a_3\psi_y + a_4\psi = f(x, y), \qquad u|_{\partial\Omega} = g(x, y), \tag{9.102}$$

where Ω is the rectangle $(0, a) \times (0, b)$. Since the boundary of Ω consists of four lines, the function g in (9.102) may be expressed as

$$g(x, 0) = g_{_S}(x), \quad g(x, b) = g_{_N}(x), \quad g(0, y) = g_{_W}(y),$$
$$g(a, y) = g_{_E}(y), \tag{9.103}$$

where the subscripts are intended to remind the reader of the north, south, east and west location of the four boundaries. The finite difference methodology we developed and implemented in earlier chapters is natural for this problem.

Let

$$x_0 = 0,\ x_1,\ x_2,\ ...,\ x_i,\ ...,\ x_{n-1},\ x_n,\ x_{n+1} = a,$$

be a discretization of the horizontal axis $(0, a)$; we consider only a uniform mesh and let

$$h = x_{i+1} - x_i$$

denote the step-size. Similarly let

$$y_0 = 0,\ y_1,\ y_2,\ ...,\ y_j,\ ...,\ y_{m-1},\ y_m,\ y_{m+1} = b,$$

be a discretization of the vertical axis $(0, b)$ with

$$k = y_{j+1} - y_j$$

the step-size in the vertical direction. Note that the discretized domain is a lattice of points (x_i, y_j), having nm interior points, those corresponding to the indices $i = 1, ..., n$ and $j = 1, ..., m$. The value of ψ at these points are the ones we aim to compute. The four sets of boundary points constitute points of the form (x_0, y_j), $j = 1, ..., m$, or (x_i, y_{m+1}), $i = 1, ..., n$, where the value of ψ is known from the boundary conditions specified in (9.103).

With (x_i, y_j) representing a typical point in the interior of the discretized domain, we replace the various derivatives in (9.102) by their finite difference approximations, keeping the truncation error of each term in mind to design a numerical scheme that has the desired truncation error for the entire PDE in (9.102). For instance, recalling that

$$\psi_x(\xi, \eta) = \lim_{h \to 0} \frac{\psi(\xi + h, \eta) - \psi(\xi - h, \eta)}{2h},$$

we replace $\psi_x(x_i, y_j)$, the value of ψ_x at a typical point (x, y_j) in the discretized domain, by

$$\psi_x(x_i, y_j) \approx \frac{\psi(x_{i+1}, y_j) - \psi(x_{i-1}, y_j)}{2h}.$$

We introduce the short-hand notation

$$\Psi_{i,j} = \psi(x_i, y_j),$$

and note that the above expression takes the form

$$\psi_x(x_i, y_j) \approx \frac{\Psi_{i+1,j} - \Psi_{i-1,j}}{2h}. \tag{9.104}$$

A simple calculation shows that the above approximation of ψ_x is second order, that is

$$\psi_x(x_i, y_j) - \frac{\Psi_{i+1,j} - \Psi_{i-1,j}}{2h} = -\frac{h^2}{6}\psi_{xxx}(x, y) + ... \tag{9.105}$$

Similarly, we replace $\psi_y(x_i, y_j)$ by the second order finite difference approximation

$$\psi_y(x_i, y_j) \approx \frac{\Psi_{i,j+1} - \Psi_{i,j-1}}{2k}, \tag{9.106}$$

from which we deduce the truncation error

$$\psi_y(x_i, y_j) - \frac{\Psi_{i,j+1} - \Psi_{i,j-1}}{2k} = -\frac{k^2}{6}\psi_{yyy}(x, y) + ... \tag{9.107}$$

The finite difference approximations to $\psi_{xx}(x_i, y_j)$ and $\psi_{yy}(x_i, y_j)$ are obtained similarly. Since

$$\psi_{xx}(\xi, \eta) = \lim_{h \to 0} \frac{\psi(\xi + h, \eta) - 2\psi(\xi, \eta) + \psi(\xi - h, \eta)}{h^2},$$

we approximate $\psi(x_i, y_j)$ by the centered finite difference formula

$$\psi_{xx}(x_i, y_j) \approx \frac{\Psi_{i+1,j} - 2\Psi_{i,j} + \Psi_{i-1,j}}{h^2}. \tag{9.108}$$

Applying the Taylor series formula to (9.108) leads to the truncation error in the finite difference approximation of $\psi_{xx}(x_i, y_j)$:

$$\psi_{xx}(x_i, y_j) - \frac{\Psi_{i+1,j} - 2\Psi_{i,j} + \Psi_{i-1,j}}{h^2} = -\frac{h^2}{12}\psi_{xxxx}(x_i, y_j) + ... \tag{9.109}$$

which demonstrates that the right side of (9.108) is a second order approximation of ψ_{xx}, compatible with the order of approximation of ψ_x by the right side of (9.104). Similarly, ψ_{yy} is approximated as follows:

$$\psi_{yy}(x_i, y_j) \approx \frac{\Psi_{i,j+1} - 2\Psi_{i,j} + \Psi_{i,j-1}}{k^2}, \tag{9.110}$$

with the truncation error

$$-\frac{k^2}{12}\psi_{yyyy}(x_i, y_j) + ... \tag{9.111}$$

Returning now to the original PDE in (9.102), we evaluate this expression at (x_i, y_j) and replace all derivatives by their finite difference approximation (9.104)–(9.110) to obtain the following finite difference approximation of (9.102):

$$a_1 \left(\frac{\Psi_{i+1,j} - 2\Psi_{i,j} + \Psi_{i-1,j}}{h^2} + \frac{\Psi_{i,j+1} - 2\Psi_{i,j} + \Psi_{i,j-1}}{k^2} \right) +$$
$$+a_2 \left(\frac{\Psi_{i+1,j} - \Psi_{i-1,j}}{2h} \right) + a_3 \left(\frac{\Psi_{i,j+1} - \Psi_{i,j-1}}{2k} \right) + a_4 \Psi_{i,j} = F_{i,j}, \tag{9.112}$$

for $i = 1, ..., n$ and $j = 1, ..., m$, where

$$F_{i,j} = f(x_i, y_j). \tag{9.113}$$

To simplify notation we introduce

$$\alpha_1 = -2(\frac{a_1}{h^2} + \frac{a_1}{k^2} - \frac{a_4}{2}), \quad \alpha_2 = \frac{a_1}{h^2} + \frac{a_2}{2h},$$
$$\alpha_3 = \frac{a_1}{h^2} - \frac{a_2}{2h}, \quad \alpha_4 = \frac{a_1}{k^2} + \frac{a_3}{2k}, \quad \alpha_5 = \frac{a_1}{k^2} - \frac{a_3}{2k}, \tag{9.114}$$

in terms of which (9.112) becomes

$$\alpha_1 \Psi_{i,j} + \alpha_2 \Psi_{i+1,j} + \alpha_3 \Psi_{i-1,j} + \alpha_4 \Psi_{i,j+1} + \alpha_5 \Psi_{i,j-1} = F_{i,j}. \tag{9.115}$$

Expression (9.115) is the fundamental relation that contains all of the information in the Stommel model at each node (i, j). In the next section we construct a matrix representation of this relation and prepare it for application of the various suites of tools in MATLAB. The approximate solution $\Psi_{i,j}$ that we obtain using the finite difference approach turns out to be quite comparable to the exact solution (9.101) we obtained in the previous section.

9.7.1 Constructing the System $A\Psi = B$

The equations (9.115) form a system of linear algebraic equations and can easily be converted to the form

$$A\Psi = B, \tag{9.116}$$

which is suitable for MATLAB, as we show next. The matrix A will be an nm by nm matrix of coefficients whose structure we will display below, the vector Ψ will be an nm by 1 matrix (a column vector) of the unknowns, consisting of the values of ψ corresponding to the interior points in the lattice:

$$\Psi_{1,1}, \Psi_{2,1}, ..., \Psi_{n,1}, \Psi_{1,2}, \Psi_{2,2}, ..., \Psi_{n,2}, ..., ..., \Psi_{1,m}, \Psi_{2,m}, ..., \Psi_{n,m}, \tag{9.117}$$

and B will be an nm by 1 matrix of the known quantities $F_{i,j}$ as well as the boundary data g's from (9.103).

To illustrate, and to get a feel for the structure of the matrix A, let us begin by letting $i = 1$ and $j = 1$ in (9.115). We get

$$\alpha_1 \Psi_{1,1} + \alpha_2 \Psi_{2,1} + \alpha_3 \Psi_{0,1} + \alpha_4 \Psi_{1,2} + \alpha_5 \Psi_{1,0} = F_{1,1}. \tag{9.118}$$

Each term in (9.118) with a zero index represents an evaluation on a boundary of Ω and is therefore known (see (9.103)). For example, $\Psi_{1,0} = g_S(x_1)$ and $\Psi_{0,1} = g_W(y_1)$. With this in mind, (9.118) reduces to

$$\alpha_1 \Psi_{1,1} + \alpha_2 \Psi_{2,1} + \alpha_4 \Psi_{1,2} = F_{1,1} - \alpha_3 g_W(y_1) - \alpha_5 g_S(x_1). \tag{9.119}$$

The above expression defines the first row of the matrix A in (9.116) as

$$\alpha_1, \alpha_2, 0, 0, ..., 0, \alpha_4, 0, 0, ..., 0, ..., ..., 0, 0, ..., 0, \tag{9.120}$$

and the first entry of B as

$$F_{1,1} - \alpha_3 g_W(y_1) - \alpha_5 g_S(x_1). \tag{9.121}$$

The second row of A, corresponding to $i = 2$ and $j = 1$, looks slightly different. With this choice of (i, j) the expression (9.115) results in

$$\alpha_1 \Psi_{2,1} + \alpha_2 \Psi_{3,1} + \alpha_3 \Psi_{1,1} + \alpha_4 \Psi_{2,2} = F_{2,1} - \alpha_5 g_S(x_2). \tag{9.122}$$

Hence the second row of A from (9.122) is

$$\alpha_3, \alpha_1, \alpha_2, 0, 0, ..., 0, \alpha_4, 0, 0, ..., 0, ..., ..., 0, 0, ..., 0 \tag{9.123}$$

with

$$F_{2,1} - \alpha_5 g_S(x_2) \tag{9.124}$$

as B's second entry. The next $n-3$ rows of A and B have a similar character: with $3 \leq i \leq n-1$ the i-th row of A will have four nonzero entries

$$\alpha_3, \alpha_1, \alpha_2, \alpha_4, \tag{9.125}$$

located at the $(i, i-1), (i, i)$, $(i, i+1)$, and the $(i, i+n)$ positions, respectively. The corresponding B_i values are

$$F_{i,1} - \alpha_5\, g_S(x_i). \tag{9.126}$$

The n-th row of A, corresponding to the point (x_n, y_1), resembles the first row of A in that it will contain only three nonzero entries located at $(n, n-1)$, (n, n) and $(n, 2n)$. The corresponding B_n is

$$F_{n,1} - \alpha_2\, g_E(y_1) - \alpha_5\, g_S(x_n). \tag{9.127}$$

So far we have demonstrated the first n rows of A and B. Following this line of reasoning the reader can arrive at the complete description of the remaining $(m-1)n$ rows of A and B: Here we write down their descriptions for $n = 4$ and $m = 3$. The matrix A is

$$\begin{bmatrix}
\alpha_1 & \alpha_2 & 0 & 0 & \alpha_4 & 0 & 0 & 0 & 0 & 0 & 0 & 0 \\
\alpha_3 & \alpha_1 & \alpha_2 & 0 & 0 & \alpha_4 & 0 & 0 & 0 & 0 & 0 & 0 \\
0 & \alpha_3 & \alpha_1 & \alpha_2 & 0 & 0 & \alpha_4 & 0 & 0 & 0 & 0 & 0 \\
0 & 0 & \alpha_3 & \alpha_1 & 0 & 0 & 0 & \alpha_4 & 0 & 0 & 0 & 0 \\
\alpha_5 & 0 & 0 & 0 & \alpha_1 & \alpha_2 & 0 & 0 & \alpha_4 & 0 & 0 & 0 \\
0 & \alpha_5 & 0 & 0 & \alpha_3 & \alpha_1 & \alpha_2 & 0 & 0 & \alpha_4 & 0 & 0 \\
0 & 0 & \alpha_5 & 0 & 0 & \alpha_3 & \alpha_1 & \alpha_2 & 0 & 0 & \alpha_4 & 0 \\
0 & 0 & 0 & \alpha_5 & 0 & 0 & \alpha_3 & \alpha_1 & 0 & 0 & 0 & \alpha_4 \\
0 & 0 & 0 & 0 & \alpha_5 & 0 & 0 & 0 & \alpha_1 & \alpha_2 & 0 & 0 \\
0 & 0 & 0 & 0 & 0 & \alpha_5 & 0 & 0 & \alpha_3 & \alpha_1 & \alpha_2 & 0 \\
0 & 0 & 0 & 0 & 0 & 0 & \alpha_5 & 0 & 0 & \alpha_3 & \alpha_1 & \alpha_2 \\
0 & 0 & 0 & 0 & 0 & 0 & 0 & \alpha_5 & 0 & 0 & \alpha_3 & \alpha_1
\end{bmatrix} \tag{9.128}$$

and the column vector B is

$$B = \mathbf{F} - \alpha_2 \mathbf{E} - \alpha_3 \mathbf{W} - \alpha_4 \mathbf{N} - \alpha_5 \mathbf{S}$$

where the column vectors $\mathbf{F}, \mathbf{E}, \mathbf{W}, \mathbf{N}, \mathbf{S}$, are, respectively

$$\begin{bmatrix} F_{1,1} \\ F_{2,1} \\ F_{3,1} \\ F_{4,1} \\ F_{1,2} \\ F_{2,2} \\ F_{3,2} \\ F_{4,2} \\ F_{1,3} \\ F_{2,3} \\ F_{3,3} \\ F_{4,3} \end{bmatrix}, \begin{bmatrix} 0 \\ 0 \\ 0 \\ g_E(y_1) \\ 0 \\ 0 \\ 0 \\ g_E(y_2) \\ 0 \\ 0 \\ 0 \\ g_E(y_3) \end{bmatrix}, \begin{bmatrix} g_W(y_1) \\ 0 \\ 0 \\ 0 \\ g_W(y_2) \\ 0 \\ 0 \\ 0 \\ g_W(y_3) \\ 0 \\ 0 \\ 0 \end{bmatrix}, \begin{bmatrix} 0 \\ 0 \\ 0 \\ 0 \\ 0 \\ 0 \\ 0 \\ 0 \\ g_N(x_1) \\ g_N(x_2) \\ g_N(x_3) \\ g_N(x_4) \end{bmatrix}, \begin{bmatrix} g_S(x_1) \\ g_S(x_2) \\ g_S(x_3) \\ g_S(x_4) \\ 0 \\ 0 \\ 0 \\ 0 \\ 0 \\ 0 \\ 0 \\ 0 \end{bmatrix}$$

There are several features in this example that are worth emphasizing because these features are independent of the specific example, where we chose $n = 4$ and $m = 3$, and also because these features become computationally more significant when m and n become large. First notice that the $nm \times nm$ matrix A is *banded* and *sparse*. It is banded because nonzero entries appear on diagonals and subdiagonals only, and sparse because so many of A's entries are zero. These features will play significant roles when we solve $A\Psi = B$ in the MATLAB. Second, note that A is not symmetric unless $\alpha_2 = \alpha_3$ and $\alpha_4 = \alpha_5$, which happens, as (9.114) shows, if and only if, $a_2 = a_3 = 0$.

An alternative way of viewing the matrix A in the above example is to think of it as consisting of a 3 by 3 (recall that $m = 3$) collection of block matrices $\mathcal{A}_{ij}$, each $\mathcal{A}_{ij}$being a 4×4 matrix (recall that $n = 4$):

$$A = \begin{bmatrix} \mathcal{A}_{11} & \mathcal{A}_{12} & \mathcal{A}_{13} \\ \mathcal{A}_{21} & \mathcal{A}_{22} & \mathcal{A}_{23} \\ \mathcal{A}_{31} & \mathcal{A}_{32} & \mathcal{A}_{33} \end{bmatrix}.$$

In our example $\mathcal{A}_{11}$ is the following 4×4 matrix

$$\mathcal{A}_{11} = \begin{bmatrix} \alpha_1 & \alpha_2 & 0 & 0 \\ \alpha_3 & \alpha_1 & \alpha_2 & 0 \\ 0 & \alpha_3 & \alpha_1 & \alpha_2 \\ 0 & 0 & \alpha_3 & \alpha_1 \end{bmatrix} \qquad (9.129)$$

The two block matrices $\mathcal{A}_{22}$ and $\mathcal{A}_{33}$ are the same as $\mathcal{A}_{11}$. Hence the three diagonal block matrices in A consist of identical 4×4 matrices, given by (9.129), which we now denote by $\mathcal{A}$:

$$A = \begin{bmatrix} \mathcal{A} & \ldots & \ldots \\ \ldots & \mathcal{A} & \ldots \\ \ldots & \ldots & \mathcal{A} \end{bmatrix}.$$

The 4×4 matrix $\mathcal{A}_{12}$ has even a simpler characterization; it is simply

$$\alpha_4 I_4$$

where I_4 is the 4×4 identity matrix. Similarly, $\mathcal{A}_{23} = \mathcal{A}_{12}$. In what follows we denote this matrix by $\mathcal{B}$. Furthermore, the matrices $\mathcal{A}_{21} = \mathcal{A}_{32} = \alpha_5 I_4$, which we denote by $\mathcal{C}$. The remaining two matrices $\mathcal{A}_{13}$ and $\mathcal{A}_{31}$ are zero matrices, which we denote by 0_4. Putting all of this information together, the 12×12 matrix A can now be rewritten as

$$A = \begin{bmatrix} \mathcal{A} & \mathcal{B} & 0_4 \\ \mathcal{C} & \mathcal{A} & \mathcal{B} \\ 0_4 & \mathcal{C} & \mathcal{A} \end{bmatrix} \tag{9.130}$$

This structure is solely dependent on n and m. In the general case when A is $nm \times nm$ it takes the form

$$A = \begin{bmatrix} \mathcal{A} & \mathcal{B} & 0_n & \cdots & \cdots & \cdots & \cdots & \cdots & \cdots \\ \mathcal{C} & \mathcal{A} & \mathcal{B} & 0_n & \cdots & \cdots & \cdots & \cdots & \cdots \\ 0_n & \cdots & \cdots & \cdots & \cdots & \cdots & \cdots & \cdots & \cdots \\ \cdots & \cdots & \cdots & \cdots & \cdots & \cdots & \cdots & \cdots & \cdots \\ \cdots & \cdots & 0_n & \mathcal{C} & \mathcal{A} & \mathcal{B} & 0_n & \cdots & \cdots \\ \cdots & \cdots & \cdots & \cdots & \cdots & \cdots & \cdots & \cdots & \cdots \\ \cdots & \cdots & \cdots & \cdots & \cdots & \cdots & \cdots & \cdots & 0_n \\ \cdots & \cdots & \cdots & \cdots & \cdots & 0_n & \mathcal{C} & \mathcal{A} & \mathcal{B} \\ \cdots & \cdots & \cdots & \cdots & \cdots & \cdots & 0_n & \mathcal{C} & \mathcal{A} \end{bmatrix}, \tag{9.131}$$

where each $\mathcal{A}$ is the tridiagonal $n \times n$ matrix

$$\mathcal{A} = \begin{bmatrix} \alpha_1 & \alpha_2 & 0 & 0 & \cdots & \cdots & 0 \\ \alpha_3 & \alpha_1 & \alpha_2 & 0 & \cdots & \cdots & 0 \\ 0 & \alpha_3 & \alpha_1 & \alpha_2 & 0 & \cdots & 0 \\ \cdots & \cdots & \cdots & \cdots & \cdots & \cdots & \cdots \\ \cdots & \cdots & \cdots & \cdots & \cdots & \cdots & \cdots \\ 0 & \cdots & \cdots & 0 & \alpha_3 & \alpha_1 & \alpha_2 \\ 0 & \cdots & \cdots & \cdots & 0 & \alpha_3 & \alpha_1 \end{bmatrix} \tag{9.132}$$

and

$$\mathcal{B} = \alpha_4 I_n, \qquad \mathcal{C} = \alpha_5 I_n. \tag{9.133}$$

9.8 MATLAB Program for the Stommel Model

MATLAB is suited well for solving systems of linear simultaneous equations of the form (9.116),

$$A\Psi = B$$

especially when A is a large matrix having the special structure in (9.128), in the special example we considered earlier, or in (9.131), in the general case, when A is banded and sparse. We first present a MATLAB program that generates the contours similar to the ones we obtained when using the exact solution (see Figure 9.9), without taking advantages of MATLAB's capability of handling sparseness of A, and then a second MATLAB program that relies on the function

```
sparse
```

which gives us a considerably more efficient result, by reducing the computational time substantially as well as by allowing us to consider relatively large matrices A.

We first state the MATLAB program that is a straightforward implementation of the $A\Psi = B$, one that does not take advantage of the sparseness of A:

```
clear all
clf
psi0=10^9;
lambda=10^9;
b=2*pi*10^8;
L=b/lambda;
d=20000;
gamma=1;
fprime=2*10^(-13);
R=0.02;
AA=(gamma*pi*lambda^2)/(R*b*psi0);
alpha=lambda*fprime*d/R;
%
n=50;m=30;
nm=n*m;
%
h=1/(n+1); k=L/(m+1);
a1=1;a2=alpha;a3=0;a4=0;
%
```

```
alpha1=-2*(a1/h^2+a1/k^2 -a4/2);
alpha2=a1/h^2+a2/(2*h);
alpha3=a1/h^2-a2/(2*h);
alpha4=a1/k^2+a3/(2*k);
alpha5=a1/k^2-a3/(2*k);
%
% define A
matrix1=alpha1*diag(ones(nm,1));
matrix2=alpha2*diag(ones(nm-1,1),1);
matrix3=alpha3*diag(ones(nm-1,1),-1);
matrix4=alpha4*diag(ones(nm-n,1),n);
matrix5=alpha5*diag(ones(nm-n,1),-n);
A=matrix1+matrix2+matrix3+matrix4+matrix5;

%
% fix for the boundary conditions
%
for j=1:m-1
    A(j*n,j*n+1)=0;
    A(j*n+1,j*n)=0;
end
% define right side
bb=[];
for j=1:m
   for i=1:n
      bb=[bb,AA*sin(pi*j*k/L)];
   end
end
%
% Solve A psi = bb
%
psi=(A\bb')';
%
%  Convert vector psi to matrix PSI
for j=1:m
    for i=1:n
   PSI(i,j)=psi((j-1)*n+i);
    end
end

newpsi=zeros(n+2,m+2);
%
% Introduce Boundary Conditions
```

```
%
for i=2:n+1
    for j=2:m+1
    newpsi(i,j)=PSI(i-1,j-1);
    end
end
[x,y]=meshgrid(0:h:1,0:k:L);
contour(x,y,newpsi')
title('Numerical Solution \psi with n = 50, m = 30')
xlabel('x'); ylabel('y')
colorbar
```

The output of this program is shown in Figure 9.12. Before addressing the quality of this output, we pause to comment on the structure of the above code. First note the use of the function

```
diag(v)
```

where `v` is a vector of size n, which allows us to create an $n \times n$ matrix of zeros with `v` occupying its diagonal. Since

```
ones(nm,1)
```

results in an nm-vector of ones, the expression

```
diag(ones(nm,1))
```

leads to an $nm \times nm$ matrix of zeros with ones on its diagonal, which is of course the same as I_{nm}. We could have alternatively obtained this matrix by invoking `eye(nm)`. However, the significance of `diag` is in its ability to generate matrices with special sub- and super-diagonal entries. So

```
diag(v,1)
```

again assuming `v` is vector of size n, results in an $(n+1) \times (n+1)$ matrix with `v` on the super-diagonal just above the main diagonal. The lines that define `matrix1` through `matrix5` each constructs a segment of the banded matrix A in (9.131). The matrix `A` is almost the matrix A we are seeking except for the extraneous α_2's and α_3's that must be replaced by 0's, those that separate the blocks denoted by $\mathcal{A}$ in (9.131). The purpose of the lines

```
for j=1:m-1
    A(j*n,j*n+1)=0;
    A(j*n+1,j*n)=0;
end
```

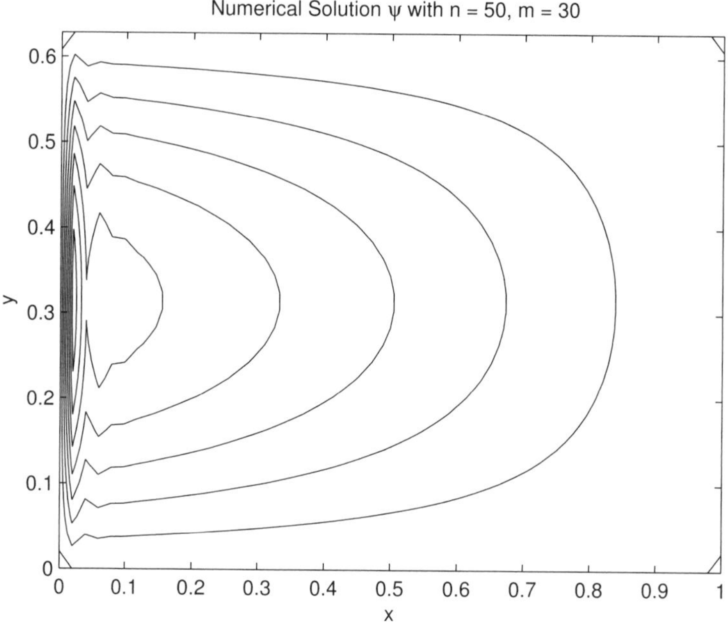

FIGURE 9.12: The finite-difference numerical solution of the Stommel BVP with $n = 50$ and $m = 30$. Note that the grid resolution is not high enough to capture the geometry of the streamlines properly (compare with Figure 9.9).

is to accomplish just that. The remainder of the code is simply intended for making `psi` available to `contour`.

The quality of the output in Figure 9.12 suffers from our choice of the size of n and m, in this case $n = 50$ and $m = 30$. While this choice leads to a matrix A, having the formidable size of 1500 by 1500, the choice of n is not large enough to adequately resolve the boundary layer near $x = 0$. Taking a larger value of n will help, but it has the adverse effect of leading to a larger and larger matrix A, and unless one has access to a computer with very large memory, the size of A will eventually result in MATLAB running out of memory on most desktops and laptops. On the other hand, we note that the great majority of the entries of A are zero, hence it stands to reason to seek a utility within MATLAB that would take advantage of the sparsity of A. The functions

```
sparse
```

or

```
spdiags
```

are exactly the right tools we need. Fortunately we only need to replace the five lines in the above code that define `matrix1` through `matrix5` with the following lines

```
matrix1=sparse(1:nm,1:nm,alpha1*ones(nm,1),nm,nm);
matrix2=sparse(1:nm-1,2:nm,alpha2*ones(nm-1,1),nm,nm);
matrix3=sparse(2:nm,1:nm-1,alpha3*ones(nm-1,1),nm,nm);
matrix4=sparse(1:nm-n,n+1:nm,alpha4*ones(nm-n,1),nm,nm);
matrix5=sparse(n+1:nm,1:nm-n,alpha5*ones(nm-n,1),nm,nm);
```

Figure 9.13 shows the output of the revised code with $n = 200$ and $m = 200$, which has resulted in a considerably better approximate solution. Figure 9.14 shows the error between the exact solution and the approximate solution when $n = 200$ and $m = 200$, demonstrating the impact of higher resolution in capturing the correct behavior near the boundary layer without compromising the accuracy in the rest of the domain.

Problems 9.8

1. Verify the truncation error formulas in (9.105) and (9.107).

2. Verify the truncation error formulas in (9.109) and (9.111).

3. Write down the matrix A and B when $a = 1$, $b = \frac{\pi}{5}$, $\alpha = 200$, $\beta = \gamma = 0$, $f(x, y) = \tau \sin 5y$, $m = 4$ and $n = 4$.

4. Write a MATLAB program to regenerate a figure similar to Figure 9.12. Experiment with the values of m and n and write a report on the influence of these two parameters on the quality of the approximate solution when compared with the exact solution.

5. Use `sparse` and write a MATLAB program to regenerate a figure similar to Figure 9.13. Experiment with the values of m and n and report on how large a value of m and n your computing resource allows you to implement. Also report on whether the quality of the approximate solution keeps getting better with increasing values of n and m or if you begin to observe the influence of round-off error after some threshold on n or m.

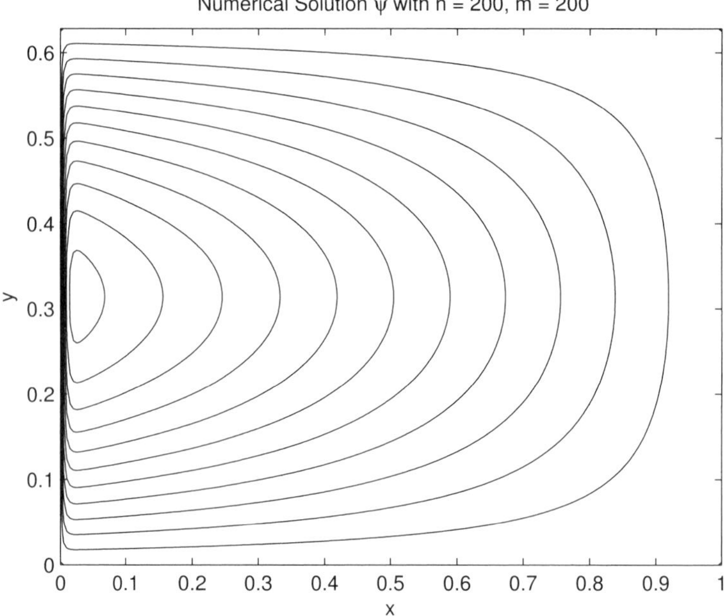

FIGURE 9.13: The finite difference numerical solution of the Stommel BVP with $n = 200$ and $m = 200$, implemented using MATLAB's `sparse`. The higher grid resolution has resulted in capturing the boundary layer more accurately. Compare with Figure 9.9.

9.9 Munk Model of Wind-Driven Circulation

In [2], which appeared nearly two years after Stommel's paper [1], Walter Munk offered an alternative approach to modeling the frictional force, whether these forces are due to the interaction between the fluid and its surrounding solid boundaries and the bathymetry, or the intermolecular forces that are present in any viscous flow. This approach ended up leading to the fourth order PDE

$$A\Delta^2\psi - \beta\frac{\partial\psi}{\partial x} = -\mathrm{curl}_z\tau, \tag{9.134}$$

where $\Delta^2\psi = \Delta(\Delta\psi)$ is the biharmonic operator. Because the PDE in (9.134) is a fourth order differential operator two sets of boundary conditions are needed. In [2], Dirichlet and Neumann boundary conditions

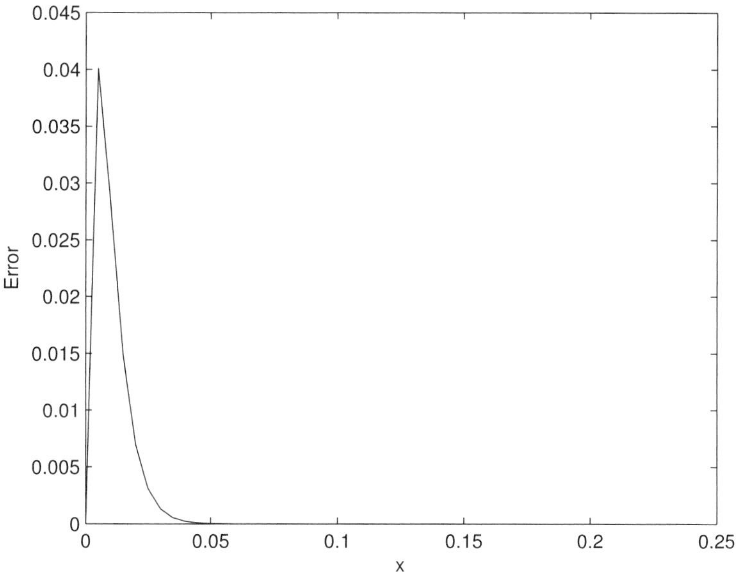

FIGURE 9.14: The error between the exact solution and the finite-difference numerical solution of the Stommel BVP when $n = 200$ and $m = 200$ at $y = \frac{b}{2}$.

were imposed on ψ:

$$\psi|_{\partial\Omega} = 0, \quad \nabla\psi \cdot \mathbf{n}|_{\partial\Omega} = 0, \tag{9.135}$$

where $\mathbf{n}$ is a unit normal to the boundary of the domain.

Instead of reconstructing the derivation of (9.134) from [2], which was our approach when introducing Stommel's model, here we choose to present the derivation of Pedlosky in [4] for two reasons: The latter paper's derivation introduces an approach that provides a unified derivation of the Stommel and Munk Models, and because the mathematical technique presented is broad enough that it is of interest in its own right. In particular, it gives us the point of view of creating a hierarchy of models for the same phenomenon, in this case wind driven circulation, which is an approach with quite a bit of appeal in applied mathematics in general; there is the hope that each model has more information in it than its predecessor.

We begin by recalling the equations of motion of GFD:

$$\left\{\begin{array}{rcl}\frac{\partial u}{\partial t}+u\frac{\partial u}{\partial x}+v\frac{\partial u}{\partial y}-fv & = & -\frac{1}{\rho_0}\frac{\partial p}{\partial x}-Ru+A\Delta u+F_1\\ \frac{\partial v}{\partial t}+u\frac{\partial v}{\partial x}+v\frac{\partial v}{\partial y}+fu & = & -\frac{1}{\rho_0}\frac{\partial p}{\partial y}-Rv+A\Delta v+F_2\\ 0 & = & \frac{1}{\rho_0}\frac{\partial p}{\partial z}-\rho_0 g\\ \frac{\partial u}{\partial x}+\frac{\partial v}{\partial y} & = & 0,\end{array}\right. \tag{9.136}$$

where ρ_0 is the density of the fluid, R is the coefficient of frictional drag, the same concept we saw in Stommel's model, and A represents molecular viscosity, or internal friction, due to fluid motion, $\mathbf{F}$ is external forcing, and f is the Coriolis parameter. As in our discussion on the Shallow Water Equation, our emphasis here is on horizontal motion, based on which we have essentially ignored the influence of w in the system. As we have seen on many occasions, the last equation allows us to seek a stream function (we note that the definition of ψ here is the negative of the corresponding definition given in [4])

$$\psi(x, y, t)$$

such that

$$u=\frac{\partial\psi}{\partial y}, \qquad v=-\frac{\partial\psi}{\partial x}.$$

Substituting these expressions in the first two equations in (9.136) leads to the following system:

$$\frac{\partial\psi_y}{\partial t}+\psi_y\frac{\partial\psi_y}{\partial x}-\psi_x\frac{\partial\psi_y}{\partial y}+f\psi_x=-\frac{1}{\rho_0}\frac{\partial p}{\partial x}-R\psi_y+A\Delta\psi_y+F_1, \tag{9.137}$$

and

$$-\frac{\partial\psi_x}{\partial t}-\psi_y\frac{\partial\psi_x}{\partial x}+\psi_x\frac{\partial\psi_x}{\partial y}+f\psi_y=-\frac{1}{\rho_0}\frac{\partial p}{\partial y}+R\psi_x-A\Delta\psi_x+F_2. \tag{9.138}$$

Next, to eliminate the pressure p from (9.137)–(9.138), we differentiate the first equation with respect to y, the second with respect to x and subtract the two resulting equations, and after some algebra and rearranging of terms, we arrive at

$$\frac{\partial q}{\partial t}+Rq+\beta\psi_x+(\psi_y q_x-\psi_x q_y)=A\Delta q+F_{1,y}-F_{2,x} \tag{9.139}$$

where

$$q=\Delta\psi, \tag{9.140}$$

is often called the *potential vorticity* of the flow.

The expression $F_{1,y}-F_{2,x}$ is the same external force we came across when simplifying the Stommel model (see (9.67) for comparison). Following the approach of [4], we acknowledge the surface influence of wind, and denote its shearing component by $\tau(x, y, t)$, and relate it to $\mathbf{F}$ as follows:

$$F_{1,y} - F_{2,x} = -\frac{\tau}{\rho_0 H}, \tag{9.141}$$

where H is the basin's depth. Finally, using the traditional definition of Jacobian, that $J(a, b)$ is

$$J(a, b) = a_x b_y - a_y b_x$$

we can rewrite (9.139) as

$$\frac{\partial q}{\partial t} + Rq + \beta\psi_x - J(\psi, q) = A\Delta q - \frac{\tau}{\rho_0 H}, \qquad q = \Delta\psi. \tag{9.142}$$

Note that we obtain Munk's equation (9.134) from (9.142) by taking only time independent processes, ignoring the nonlinearities in J, and by setting $R = 0$.

As we proceeded in the case of Stommel model, the first step in starting the process of studying (9.142) is to non-dimensionalize these equations. As suggested in [4], we select the following relations between the dimensional variables ψ, τ, x, y, and t and their non-dimensional counterparts Ψ, T, x', y' and t':

$$x = Lx',\ y = Ly',\ t = \frac{1}{\beta L}t', \tag{9.143}$$

and

$$\tau = \frac{\tau_0}{L}T(x', y', t'), \quad \psi = \frac{\tau_0}{\rho_0 \beta H}\Psi, \tag{9.144}$$

where L is a typical length scale, say one of the horizontal dimensions of the basin, and τ_0 is a typical traction force imparted by the wind on the surface of the basin. Note how t, time, is scaled by β and L, thus we have brought in the important scales from the rotation of the planet and the size of the basin, and how τ_0, β and H are present in the non-dimensionalization of ψ, the stream function. Hence a change of one unit of t' carries with it a passage of time t proportional to βL, and a change of one unit of Ψ is related to how the typical wind stress τ_0 relates to the basin's depth H and the location of the fluid parcel on the surface of the planet through β.

The procedure for determining the non-dimensional equation from (9.142) is the same as the one employed with the Stommel model. We apply the chain rule of differentiation to each term in (9.142) and transform all derivatives to those with respect to x', y' and t'. Note that by

(9.143) and (9.144)

$$\frac{\partial}{\partial x} = \frac{1}{L}\frac{\partial}{\partial x'}, \quad \frac{\partial}{\partial y} = \frac{1}{L}\frac{\partial}{\partial y'}$$

and

$$\frac{\partial}{\partial t} = \beta L \frac{\partial}{\partial t'}.$$

Hence, since $q = \Delta\psi$, we have

$$q = \frac{\tau_0}{\rho\beta H L^2}(\frac{\partial^2 \Psi}{\partial x'^2} + \frac{\partial^2 \Psi}{\partial y'^2}). \tag{9.145}$$

We define q' by

$$q' = \Delta' \Psi.$$

With this definition and (9.145), the relationship between the dimensional potential vorticity q and its non-dimensional counterpart q' is

$$q' = \frac{1}{\tau_0}\rho\beta H L^2 q. \tag{9.146}$$

Similarly, $\frac{\partial q}{\partial t}$ is related to its non-dimensional counterpart by

$$\frac{\tau_0}{\rho_0 H L}\frac{\partial q'}{\partial t'}.$$

Following in this manner, every term in (9.142) is transformed to its equivalent non-dimensional term. We end up with the following equation:

$$\frac{\partial q'}{\partial t'} + \delta q' + \Psi'_x - Ro\, J'(\Psi, q') = A'\Delta' q - T, \quad q' = \Delta'\Psi, \tag{9.147}$$

where Ro, called the *Rossby number*, is

$$Ro = \frac{\tau_0}{\rho_0 \beta^2 H L^3}$$

and

$$\delta = \frac{R}{\beta L}, \quad A' = \frac{A}{\beta L^3}.$$

The system of equations in (9.147) is nonlinear in (Ψ, q') because of the presence of the Jacobian term

$$Ro\, J'(\Psi, q').$$

If the Rossby number Ro were zero, the system of equations in (9.147) would lend itself to the methods we have developed in this book, especially the method of separation of variables and Fourier series. The Rossby number of course does not vanish, but in certain circumstances it is small enough that one could take advantage of how weakly the nonlinear term $J'(\Psi, q')$ enters into the problem. The paper [4] and several subsequent papers did exactly that, by considering situations where Ro is small, and obtaining solutions of (9.147) in terms of a Taylor series expansions in Ro. We complete the section by examining this method in detail.

In what follows we omit the primes on all of the variables, replace Ψ with ψ, and simply work with

$$\frac{\partial q}{\partial t} + \delta q + \psi_x - \epsilon\, J(\psi, q) = A\Delta q - T, \quad q = \Delta\psi, \tag{9.148}$$

where we have replaced Ro with ϵ to emphasize that this term is going to be small. As in our approach with the Stommel model, we consider the domain

$$B = \{(x, y) | \, 0 < x < \lambda, \, 0 < y < b\} \tag{9.149}$$

and introduce Dirichlet boundary conditions

$$\psi(x, 0) = \psi(0, y) = \psi(\lambda, y) = \psi(x, b) = 0,$$

and Neumann boundary conditions

$$\frac{\partial \psi}{\partial y}\Big|_{x=0} = \frac{\partial \psi}{\partial x}\Big|_{y=0} = \frac{\partial \psi}{\partial y}\Big|_{x=\lambda} = \frac{\partial \psi}{\partial x}\Big|_{y=b} = 0.$$

The second boundary conditions are needed if A in (9.148) does not vanish.

We seek solutions ψ of (9.148) of the form

$$\psi(x, y, t) = \sum_{n=0}^{\infty} \epsilon^n \psi_n(x, y, t), \tag{9.150}$$

The goal here is to look for functions ψ_n that will presumably play a role similar to the normal modes we encountered in an earlier chapter, satisfying relatively simple PDEs that hopefully we can solve or compute easily. What we will find out shortly is that the equation for ψ_n depends only on ψ_i for $i < n$, thus giving us an iterative algorithm for obtaining ψ_n.

We begin by substituting the template given by (9.150) into the nonlinear PDEs in (9.148). Note that

$$q = \sum_{n=0}^{\infty} \epsilon^n \Delta\psi_n,$$

$$q_t = \sum_{n=0}^{\infty} \epsilon^n \frac{\partial \Delta \psi_n}{\partial t}, \quad \psi_x = \sum_{n=0}^{\infty} \epsilon^n \frac{\partial \psi_n}{\partial x}, \quad \Delta q = \sum_{n=0}^{\infty} \epsilon^n \Delta^2 \psi_n.$$

The expression $J(\psi, q)$ needs a bit more careful calculation. Recall that $J(\psi, q)$ is

$$J(\psi, q) = \psi_x q_y - q_x \psi_y. \tag{9.151}$$

Substitute the series solution of each of the four terms in (9.151) into this expression:

$$J = \sum_{n=0}^{\infty} \epsilon^n \frac{\partial \psi_n}{\partial x} \sum_{n=0}^{\infty} \epsilon^n \frac{\partial \Delta \psi_n}{\partial y} - \sum_{n=0}^{\infty} \epsilon^n \frac{\partial \Delta \psi_n}{\partial x} \sum_{n=0}^{\infty} \epsilon^n \frac{\partial \psi_n}{\partial y}.$$

After expanding the above expression and rearranging and collecting terms that multiply ϵ^i, we end up with

$$J(\psi, q) = J(\psi_0, \Delta\psi_0) + \epsilon\left[J(\psi_1, \Delta\psi_0) + J(\psi_0, \Delta\psi_1)\right] +$$

$$\epsilon^2\left[J(\psi_0, \Delta\psi_2) + J(\psi_1, \Delta\psi_1) + J(\psi_2, \Delta\psi_0)\right] + ... \tag{9.152}$$

Since the jacobian in (9.148) is multiplied by a factor of ϵ, we note that the influence of the nonlinear term in (9.152) is a second order effect, i.e., the impact of J is not felt until we need to compute the equation for ψ_1. To see this better, substitute the series solution of each term in (9.148) to get

$$\sum_{n=0}^{\infty} \epsilon^n \left[\frac{\Delta\psi_n}{\partial t} + \delta\Delta\psi_n + \frac{\partial \psi_n}{\partial x} - A\Delta^2\psi_n\right] = -T + \epsilon J(\psi_0, \Delta\psi_0) +$$

$$\epsilon^2\left[J(\psi_1, \Delta\psi_0) + J(\psi_0, \Delta\psi_1)\right] +$$

$$\epsilon^2\left[J(\psi_0, \Delta\psi_2) + J(\psi_1, \Delta\psi_1) + J(\psi_2, \Delta\psi_0)\right] + ... \tag{9.153}$$

Equating powers of ϵ from each side of (9.153) results in the following set of iterative PDEs:

$$\frac{\Delta\psi_0}{\partial t} + \delta\Delta\psi_0 + \frac{\partial \psi_0}{\partial x} - A\Delta^2\psi_0 = -T. \tag{9.154}$$

$$\frac{\Delta\psi_1}{\partial t} + \delta\Delta\psi_1 + \frac{\partial \psi_1}{\partial x} - A\Delta^2\psi_1 = J(\psi_0, \Delta\psi_0). \tag{9.155}$$

$$\frac{\Delta\psi_2}{\partial t} + \delta\Delta\psi_2 + \frac{\partial \psi_2}{\partial x} - A\Delta^2\psi_2 = J(\psi_0, \Delta\psi_2) +$$

$$J(\psi_1, \Delta\psi_1) + J(\psi_2, \Delta\psi_0). \tag{9.156}$$

The equation in (9.154) solely depends on ψ_0, hence its solution, ψ_0, can be determined by solving this PDE subject to the specified initial and boundary conditions. Once ψ_0 is obtained, the equation in (9.155) can then be investigated, because its right side, $J(\psi_0, \Delta\psi_0)$, is now known. This recursive process will lead to solutions of all other equations for $n > 1$.

The method we have described is enormously powerful. It is one of several techniques used in applied mathematics to obtain reduced models of complex problems once a physical parameter, in this case the Rossby number, is identified and is observed to be smaller than some of the other physical parameters in the problem. What is remarkable in the development we have presented here is that the very first equation, the lowest order equation, (9.154), that results from this method, includes both models there were studied by Stommel and by Munk; that is

$$\psi_0(x, y, t)$$

contains all of the features that the earlier models captured. Presumably, the higher order corrections, namely,

$$\psi_0 + \epsilon\psi_1$$

and

$$\psi_0 + \epsilon\psi_1 + \epsilon^2\psi_2,$$

obtained by taking into account the solutions to the PDEs in (9.155), (9.156), etc., bring out, in a hierarchical manner, more features of the solution of the fully nonlinear problem in (9.148).

Problems 9.9

1. Complete the calculations that lead to (9.139).

2. Verify that t', T and Ψ defined in (9.143) and (9.144) are dimensionless.

3. Suppose L is 1000 kilometers and that a parcel of fluid is being observed at the 45 degrees parallel. Refer to (9.143) and determine the passage of time t in hours if t' has changed by one unit.

4. Complete the calculation that leads to the non-dimensional equation (9.147).

5. Complete the calculations that lead to the PDEs in (9.154) through (9.156).

9.10 Project A: Stommel Model with a Nonuniform Mesh

The finite difference method that was introduced in this chapter to address the Stommel problem (9.83)–(9.84) was designed for a uniform mesh in the domain. The goal of this project is to develop a nonuniform mesh, especially designed for the horizontal axis where we expect the formation of the boundary layer near the western boundary. Ideally, we would like to have more sample points of the solution near $x = 0$ where we expect sharp transitions in the stream function ψ, and not so many sample points away from the western boundary where ψ is nearly constant.

We consider the domain

$$B = \{(x, y) | \, 0 < x < 1, 0 < y < L\},$$

where L was defined in Section 9.7. Consider the uniform mesh $Z = \{0, z_1, z_2, ..., z_i, ..., z_n, 1\}$ where $z_{i+1} - z_i = h$ and $h = \frac{1}{n+1}$. The function

$$x = z^2, \tag{9.157}$$

maps the uniform mesh in Z to a nonuniform mesh $X = \{0, x_1, x_2, ..., x_i, ..., x_n, 1\}$ of the interval $(0, 1)$.

1. Is X denser near the origin than it is near $x = 1$?
2. Consider a function $\Psi(x)$ defined on the interval $(0, 1)$.
   (a) Write down the Forward Euler Method formula for $\Psi'(x_i)$.
   (b) Write down the Centered Euler Method formula for $\Psi'(x_i)$.
3. Recall that when dealing with a uniform mesh, we estimated $\Psi''(x_i)$ by
$$\frac{\Psi_{i+1} - 2\Psi_i + \Psi_{i-1}}{h^2}.$$
   Write down the equivalent formula when x_i is the nonuniform mesh introduced by the function in (9.157).
4. Discretize the domain in B by (x_i, y_j), where $\{y_i\}$ is a uniform mesh. Apply this mesh to the Stommel model PDE
$$\Delta\psi + \alpha\psi_x = A \sin \frac{\pi y}{L}, \tag{9.158}$$

with the same parameter values defined in (9.85). Write the resulting algebraic system in the form

$$AX = B$$

as we in did Section 9.7. Explain the details of the structure of the matrix A. Is it banded? Is it sparse?

5. Write a MATLAB program to find the approximate solution to the Stommel BVP using this nonuniform mesh. Report on the efficiency of this method relative to the uniform mesh solution. Is the solution from the nonuniform mesh with $n = 100$ more accurate than the corresponding solution when the same size uniform mesh is used?

6. Replace the mesh function $x = z^2$ with $x = z^4$. How does this improve, if any, the performance of the MATLAB code that plots the contours of ψ.

9.11 Project B: Munk Model and the Finite Difference Method

The goal of this project is to develop a MATLAB program that is capable of finding the approximate solution to the BVP

$$\Delta^2\psi + \alpha\psi_x = A\sin\frac{\pi y}{L}, \tag{9.159}$$

subject to the boundary conditions

$$\psi|_{\partial B} = 0, \tag{9.160}$$

and

$$\frac{\partial\psi}{\partial\mathbf{n}}|_{\partial B} = 0. \tag{9.161}$$

The domain is the usual rectangle

$$B = \{(x, y)|\, 0 < x < 1, 0 < y < L\}$$

and the aspect ratio L is defined in Section 9.7.

When applying the finite difference method to (9.159) we need to pay attention to two significant issues relative to our approach for the Stommel model, one how to discretize $\Delta^2\psi$ and the second how to discretize the boundary conditions.

1. Consider the uniform n by n mesh x_i, y_j in B. Write down a discretization of
$$\Delta^2\psi|_{(x_i,y_j)}.$$
Hint: Recall that $\Delta^2\psi = \Delta(\Delta\psi)$. Apply twice the centered difference formula we obtained previously for $\Delta\psi$.

2. Explain how many points in the neighborhood of an interior point (x_i, y_j) are needed to in order to estimate the biharmonic operation of ψ at that point.

3. Discretize the Neumann boundary condition (9.161) for points (x_i, y_j) that are one step removed from the boundary . Recall that the boundary condition (9.160) already dictates the value ψ assumes on the boundary. For example, consider the boundary condition
$$\frac{\partial\psi}{\partial x}|_{(a,b)} = 0.$$
Applying the centered difference formula to this relation yields
$$\frac{\psi(a+h,b) - \psi(a-h,b)}{2h} \approx 0.$$
We therefore conclude that this particular Neumann boundary condition is approximated by the relation
$$\psi(a+h,b) = \psi(a-h,b).$$
Apply this result to the case when $a = 0$ and note that we are able to define the value of ψ at the "ghost" point $(-h, b)$, a point outside of the domain B. These ghost points are needed in the evaluation of the discretized biharmonic operator.

4. Combine the information from the ghost points and the stencil for the biharmonic to convert the BVP in (9.159), (9.160)–(9.161) to the matrix form
$$AX = B.$$
Describe the structure of A. If it is banded, how many subdiagonals are occupied?

5. Consult the original paper of Walter Munk, [2] to determine the appropriate values for α and the amplitude A in (9.159) after the stream function has been non-dimensionalized.

6. Write a MATLAB program to solve the matrix relation $AX = B$ and plot the contours of the stream function. Compare the contours you obtain with those in Fig 2. on page 83 of [2]. In particular, does the MATLAB contours have the "dip" (i.e, the nonconvex part) in the midlatitude contours seen in [2]?

9.12 Project C: Galerkin Method and the B. Saltzman and E. Lorenz Equations

In the paper [5] B. Saltzman presented and studied the following system of PDEs as a model for the behavior of convection in the atmosphere:

$$\begin{cases} \dfrac{\partial}{\partial t}(\Delta\psi) + J(\psi, \Delta\psi) - \nu\Delta^2\psi - g\alpha\dfrac{\partial\theta}{\partial x} & = 0, \\ \dfrac{\partial\theta}{\partial t} + J(\psi, \theta) - \dfrac{\delta T}{H}\dfrac{\partial\psi}{\partial x} - \kappa\Delta\theta & = 0, \end{cases} \tag{9.162}$$

where θ is the temperature and ψ is the stream function, related to the velocity $\mathbf{v} = \langle u, w\rangle$ by

$$u = -\frac{\partial\psi}{\partial z}, \quad w = \frac{\partial\psi}{\partial x}. \tag{9.163}$$

The fluid flow is assumed to be two-dimensional and incompressible. The domain is the rectangular region

$$D = \{(x, z)|\, 0 \le x \le \frac{H}{a},\ 0 \le z \le H\}.$$

The operator Δ is the usual two-dimensional laplacian, $\frac{\partial^2}{\partial x^2} + \frac{\partial^2}{\partial z^2}$, and J$(a, b)$ stands for the Jacobian of a and b, i.e., $J(a, b) = a_x b_z - a_z b_x$. The parameters ν and κ are the viscosity and thermal diffusivity of the fluid. The parameter δT denotes the temperature difference between the two boundaries $z = 0$ and $z = H$, which is assumed large enough to induce the so-called Rayleigh-Bénard instability in the flow and initiate convection in the region. Finally, g stands for the acceleration of gravity and α is the coefficient of volume expansion, which is part of the definition of the equation of state, relating density to temperature (see Equation (6), page 330 of [5]).

System (9.162) is augmented by the boundary conditions

$$\psi(x, z, t) = 0, \quad (x, z) \in \partial D, \tag{9.164}$$

and

$$\theta(x, 0, t) = \theta(x, H, t) = 0, \ \frac{\partial\theta}{\partial x}(x, 0, t) = \frac{\partial\theta}{\partial x}(x, H, t) = 0. \tag{9.165}$$

The study in [5] involves deriving approximate solutions to (9.162)–(9.165) using Fourier modes in the spatial domain and applying the method of lines in time, obtaining solutions for ψ and θ of the form

$$\begin{cases} \psi(x,z,t) &= \sum_{m=1}^{\infty}\sum_{n=1}^{\infty} a_{mn}(t)\sin(\frac{m\pi a}{H}x)(\sin\frac{n\pi}{H}z), \\ \theta(x,z,t) &= \sum_{m=0}^{\infty}\sum_{n=1}^{\infty} b_{mn}(t)\cos(\frac{m\pi a}{H}x)\sin(\frac{n\pi}{H}z). \end{cases} \tag{9.166}$$

The goal of this project is to follow that approach in [5] and seek solutions in the Fourier modes listed in (9.166) but to apply the Galerkin method instead to obtain the set system of ODES for the coefficients $A_{m,n}$ and $B_{m,n}$.

1. We seek solutions of the form

$$\psi(x,z,t) = \sum_{m}^{N}\sum_{n}^{N} A_{m,n}(t)\,\phi 1_{m,n}(x,z) \tag{9.167}$$

and

$$\theta(x,z,t) = \sum_{m}^{N}\sum_{n}^{N} B_{m,n}(t)\,\phi 2_{m,n}(x,z) \tag{9.168}$$

where $A_{m,n}(t)$ and $B_{m,n}(t)$ are the time dependent coefficients, yet to be determined, and $\phi 1_{m,n}(x,z)$ and $\phi 2_{m,n}(x,z)$ are the basis functions we use to represent the spatial dimensions. Following [5] we choose the same Fourier basis system in (9.166), which complies with the boundary conditions:

$$\phi 1_{m,n}(x,z) = \sin\left(\frac{m\pi a}{H}x\right)\sin\left(\frac{n\pi}{H}z\right) \tag{9.169}$$

and

$$\phi 2_{m,n}(x,z) = \cos\left(\frac{m\pi a}{H}x\right)\sin\left(\frac{n\pi}{H}z\right) \tag{9.170}$$

Let $N = 2$. Note that this results in using only the first 10 basis functions. Substitute (9.167)–(9.168) into (9.162). Take the inner product of each of the resulting 10 equations by the appropriate basis function (i.e., by ϕ_1 for the balance of linear momentum equation, and by ϕ_2 for the balance of energy equation), and integrate over the domain. All dependencies on x and z are now integrated out and we are left with 10 ODEs in the variables $A_{m,n}$ and $B_{m,n}$. Verify that two of these equations, the ones for $A_{1,1}$ and $B_{2,2}$, are

$$\frac{a\pi^2}{4}A'_{1,1}(t) = -\frac{\nu\pi^4}{4aH^2}A_{1,1}(t) - \frac{a\nu\pi^4}{2H^2}A_{1,1}(t) -$$

$$\frac{a^3\nu\pi^4}{4H^2}A_{1,1}(t) - \frac{9\pi^4}{16H^2}A_{1,2}(t)\,A_{2,1}(t) + \frac{9a^2\pi^4}{16H^2}A_{1,2}(t) +$$
$$\alpha g H\pi B_{1,1}(t) - \frac{\pi^2}{4a}A_{1,1}(t)$$

and

$$\frac{H^2}{4a}B'_{2,2}(t) = \frac{\Delta T\pi}{2}A_{2,2}(t) - \frac{\pi^2}{4}A_{2,1}(t)\,B_{0,1}(t) -$$
$$\frac{\kappa\pi^2}{a}B_{2,2}(t) - a\kappa\pi^2 B_{2,2}(t).$$

2. Instead of considering all ten variables $A_{i,j}$, $B_{i,j}$, consider only the following three

$$A_{1,1},\ B_{1,1},\ B_{0,2}$$

and set the remaining 7 variables to zero. Show that we obtain the following system of equations:

$$\begin{cases} A'_{1,1} &= \sigma(B_{1,1} - A_{1,1}) \\ B'_{1,1} &= rA_{1,1} - B_{1,1} - A_{1,1}B_{0,2} \\ B'_{0,2} &= A_{1,1}B_{1,1} - \dfrac{4}{1+a^2}B_{0,2} \end{cases} \tag{9.171}$$

which is precisely the celebrated system of equations we have encountered earlier, derived by E. Lorenz, and today is known as the system of Lorenz Equations. These equations display the "butterfly" effect and the ensuing chaos. Write down what σ and r are in terms of the physical parameters in this problem. Are these parameters dimensionless?

3. Let $a = 2$. Write a MATLAB program to solve the system of ODEs in (9.171) for a variety of values for σ and r. In particular start with initial data

$$A_{1,1} = 0.1,\ B_{1,1} = 0.1,\ B_{0,2} = 0.1,$$

set $\sigma = 10$ and vary r from 1 to 30 and report on the change of behavior of the solution to (9.171).

4. Write a MATLAB program to display the particle trajectories of the system of ODEs in (9.163), where the coefficients functions $A_{1,1}(t)$, $B_{1,1}(t)$ and $B_{0,2}(t)$ are obtained from executing the MATLAB program in the previous problem. Plot several representative particle trajectories. Report on whether your see chaotic behavior in the particle paths.

5. Write a MATLAB program that displays an animation of the evolution of $(A_{1,1}(t), B_{1,1}(t), B_{0,2}(t))$ and $(x(t), z(t))$, for initial data of your own choosing. The purpose of this exercise is to observe if there is a correlation between the chaotic behavior in the coefficients and the behavior of a typical particle in the (x, z) domain.

6. Write a MATLAB program to solve the initial value problem for the above system of ten equations. Let the three parameters of the Lorenz system have the same values. Solve the initial value problem, selecting random but small initial values of for the seven variables that were ignored in the process of arriving at the Lorenz system. Plot the graph of the three variables selected for the Lorenz system. Report on the cases where the butterfly effect seems to persist for the larger system, and especially comment on the cases where this effect disappears for the larger system.

7. There are two directions this project can move in at this stage. One can a) explore what happens to the particle paths when N, the number of modes, is allowed to assume values larger than 2, and b) one can explore choices of basis functions other than Fourier modes. In either direction, the interesting question to explore is to what extent the chaotic behavior observed in the Lorenz attractor is a function of the number of modes in the truncation of the approximate solution in (9.166) or the basis used in (9.169)–(9.170).

9.13 References

1. Stommel, H., "The western intensification of wind-driven ocean currents," *Transactions of the American Geophysical Union*, 29, pp. 202–206, 1948.

2. Munk, W., "On the wind-driven ocean circulation," *Journal of Meteorology*, Vol 7, No. 2, pp. 79 – 93, 1950.

3. Weinberger, H., *A First Course in Partial Differential Equations: with Complex Variables and Transform Methods*, Dover, 1965.

4. Pedlosky, J., "A study of the time dependent ocean circulation," *Journal of the Atmospheric sciences*, Vol 22, pp. 267–272, 1965.

5. Saltzman, B, "Finite amplitude free convection as an initial value problem – I," *J. Atmos. Sci.*, Vol 19, 1962, pp. 329–341.

Chapter 10

Some Special Topics

Most of what we have covered in the previous chapters has dealt with how applied mathematics has contributed to our understanding of some of the basic and fundamental problems in physical oceanography. We have applied methods from linear algebra and matrix theory, from ordinary and partial differential equations, and obtained approximate solutions to circulation problems ranging from the reduced models of the Gulf Stream to flows in shallow water regions. We have also used several MATLAB tools based on sophisticated algorithms, algorithms whose design rely heavily on rigorous mathematical theories from matrix theory and differential equations, to obtain solutions of partial differential equations. In addition we employed many visualization utilities within MATLAB whose development are also based entirely on advanced mathematics.

The problems we have addressed are examples of "reduced" models, often called "toy" models colloquially. They are intended to draw out some of the salient properties of the systems of equations considered, to allow a researcher to compare these properties with some of our intuitive expectations of the oceanographic phenomena in which we are interested. Examples of such properties are the commonly accepted speed of the fluid flow in the Gulf Stream, or its relative size when viewed as a boundary layer, or its vorticity, all of which are predicted rather well from a model as simple as the one proposed by Stommel. If, however, we are interested in more detailed information about the Gulf Stream, such as the rate at which fluid is transported along or across this stream, a reduced model such as the Stommel model just does not have enough physical fidelity to allow us to investigate such questions. It would be unreasonable to expect, for example, that transport and mixing can be independent of the variation in the bathymetry, or assume that the flow is essentially time-independent, or that it is not affected by the delicate nonlinearities that we ignored in setting up the models we investigated. The next step in this development requires including features and terms which we ignored in our first attempt, and instead investigate the full set of PDEs in complex geometries and solve them with high enough temporal and spatial resolutions to allow comparisons of numerical data with observational ones. This program has been the focus of many investigators for the past three decades and is the subject of several advanced books listed in the references. See [1], [2] and [3], for examples. The area of high-fidelity and high-performance computing of oceanographic and atmospheric modeling continues to be an area of intense current research, and while enormous strides have been made in obtaining physically accurate solutions, significant mathematical and computational obstacles remain.

We now describe, albeit briefly, a different direction of development, namely how questions initiated in physical oceanography have contributed to the growth of new analytical and computational directions in mathematics. We review here some of these advances, especially those in the past two decades, and point to several seminal papers that provide substantive introductions to these topics.

10.1 Finite-Time Dynamical Systems

Much of our attention in this book has been focused on computing solutions of initial and boundary values involving PDEs where the solution often represents the velocity field $\mathbf{v}$ of a flow. This vector field gives rise to a dynamical system

$$\frac{d\mathbf{x}}{dt} = \mathbf{v}(\mathbf{x}, t), \quad \mathbf{x}(t_0) = \mathbf{x}_0, \tag{10.1}$$

so that $\mathbf{x}(t)$ denotes the particle trajectory of the particle whose position at time t_0 is $\mathbf{x}_0$. Hence, once the equations of motions that govern geophysical fluid flows are solved, one ends up with a system of ordinary differential equations for particle trajectories. This latter system is often nonlinear, even when the original PDE system is linear. Today we are fairly good at computing approximate solutions of initial-value problems of the type shown in (10.1); in fact we have done just that for many examples in this book, by appealing to MATLAB's `ode45`.

The interest expressed by investigators in the past few decades goes considerably beyond the information one gets by simply integrating a system like (10.1). That is because knowing the location of a single trajectory, or even an ensemble of trajectories, is not always enough to shed light on the general behavior of solutions of (10.1), of the kind that would help us characterize or distinguish one vector field $\mathbf{v}$ from another. We are often interested in detailed information about the behavior of solutions of (10.1) near equilibrium points, and about identifying regions that remain invariant, much like the regions in the normal modes of the Flow-in-the-Bay problem we studied earlier. Much effort in the theory of dynamical systems has gone into developing tools such as computing Liapunov exponents, invariant stable and unstable manifolds, distinguished hyperbolic trajectories, and Lagrangian coherent structures, to allow us to classify one dynamical system from another. The development of many of these quantities was directly motivated by the desire to find mathematical tools that would correspond to physical phenomena we observe in nature, such as eddies, gyres, fronts and jets. See [4] for an introduction to what a Lyapunov exponent is and several different methods for computing it, and [5], [6], [7] for an excellent collections of accessible books on dynamical systems, chaos and invariance properties. The latter texts are currently available as Google® books on the internet.

To illustrate some of these points, consider the velocity field we obtained by solving the PDE that arises in the Stommel model (see equation (9.101) in Chapter 9) whose stream function ψ solves the PDE

$$\Delta\psi + \alpha\psi_x = A \sin\frac{\pi y}{L},$$

and we ended up with

$$\psi(x, y) = \frac{AL^2}{\pi^2}\left[-1 + \kappa e^{m_1 x} + (1-\kappa)e^{m_2 x}\right]\sin\frac{\pi y}{L}. \tag{10.2}$$

The stream function ψ is related to $\mathbf{v}$ by

$$u = \frac{\partial\psi}{\partial y}, \quad v = -\frac{\partial\psi}{\partial x}.$$

It should be clear that the stream function in (10.2) will give rise to a highly nonlinear system of ODEs, whose solution will require applying a tool such as `ode45`, a scenario we have encountered repeatedly earlier.

There are several aspects of the above example worth noting: First, the velocity field $\mathbf{v}$ of (10.1) is available to us through analytical formulas, so introducing these equations as input to `ode45` is simple. Second, the velocity field given by (10.2) is stationary, independent of t. It turns out that analyzing solutions of a two-dimensional autonomous (time-independent) system of ODEs is considerably simpler than time-dependent systems of equations, where chaotic behavior is commonly encountered, as we have seen in a few examples in the text. Because of these attributes of the Stommel model, it is not too difficult to find the equilibrium points, say, of the flow, and study their stability properties.

Unfortunately the situation we have seen in the context of the Stommel model, that the velocity field is known in closed form and written in terms of a few relatively simple analytical functions, almost never happens for real geophysical flows. Moreover, geophysical fluid flows are almost always time-dependent. And finally, and this is perhaps the most significant difference between realistic applications and the ones we have developed, most geophysical velocity fields are known only for discrete values of time and for discrete values in the domain, the main reason being that either we obtain these velocity fields as the result of solving the governing GFD equations numerically, or identifying them by collecting measurements. A striking attribute of either a numerically obtained velocity field, or one obtained by field measurement, is that the vector field is known for only a finite amount of time and defined over a discrete domain.

Much mathematical research in the past two decades has gone into understanding *finite-time* and discerte dynamical systems. The main area of mathematical development is concerned with notions such as equilibrium points, invariant curves or manifolds, *finite-time Liapunov Exponents (FTLE)*, *Lagrangian Coherent Structures* (LCS) and *Distinguished Hyperbolic Trajectories* (DHT) for such systems. These notions have resulted in a suite of computational tools that have paid off enormously for a class of transport problems in physical oceanography, a small sample of which we address now.

We present a survey of the efforts in this area by referring to a series of papers by G. Haller and co-authors, in particular to [8] and [12], where the theory, and its practical implication, are introduced for finite-time transport, as well as what an LCS is for approximate velocity fields. In a separate development, led by J. E. Marsden, see [14], the authors view Lagrangian Coherent Structures for time-dependent, or aperiodic, vector fields or flows, as the analog of the stable and unstable manifolds of equilibrium points of time-independent flows. In particular, they propose the important point of view of an LCS being a curve, a ridge, across which the behavior of finite-time Liapunov exponents change sharply. One of the rewarding aspects of studying [14] is learning about the following illustrative model, where the velocity field is associated with the time-dependent stream function

$$\psi(x, y, t) = A \sin(\pi f(x, t)) \sin(\pi y)$$

where

$$f(x, t) = a(t)x^2 + b(t)x, \quad a(t) = \epsilon \sin \omega t, \quad b(t) = 1 - 2\epsilon \sin \omega t,$$

which we explored in a set of exercises in a previous chapter. In [14] the authors point out all of the attributes of their new tools in the context of this example.

Finally, we mention the development by S. Wiggins and co-authors, especially in the area of Distinguished Hyperbolic Trajectories for time-dependent flows. The monograph [15] is a readily accessible source where the authors, by introducing a series of simple examples, show what a distinguished hyperbolic trajectory is, why it is important in its own right, but particularly that it is an important notion in oceanic flows. The tutorial in [15] also is a resource for studying *lobe dynamics*, a tool that in the past decade or so has proved quite useful in identifying coherent structures in time-dependent vector fields. In particular, in [16], the authors show how to use lobe dynamics to detect DHTs in a realistic flow in the Mediterranean Sea, and discover the behavior of front-eddy interaction that is simply not observable by looking at the time series, or snapshots, of the underlying velocity fields.

We have only touched on a few, albeit important and seminal, contributions in the interplay between mathematics of finite-time vector fields and problems that arise in physical oceanography. It is worth re-emphasizing that these developments are leading to practical computational tools that will enhance our understanding of complex fluid flows, not just in the field of oceanography, but in any field where there is the potential of obtaining partial information about a flow from field measurements.

10.2 Data Assimilation and Filtering

Another area where computational mathematics and physical oceanography are having a strong interplay is in data assimilation, where investigators are interested in introducing attributes of field measurements into mathematical models in order to improve them and to arrive at more reliable and predictive models. And conversely, use attributes of models to design measurement devices that are better suited for the phenomenon under study, as well as identifying optimal location to deploy these devices.

Data assimilation is a relatively mature field in oceanography with a rich history, dating back fifty years or so when temperature, salinity, and velocity data collected at specific locations were introduced to models. The techniques have been primarily based on statistical and optimization methods, not so much to calibrate the models, but more toward selecting available free parameters in the models in order to guide them toward the truth represented by nature.

One way to think about data assimilation is to imagine having a class of models, where a typical member of the class captures a substantial amount of the physics of the problem, but not all of the physics. The lack of knowledge or the uncertainty could truly be because some of the physics is just too hard to model (the air-sea interface is a good example), or the model has been made artificially simple for computational reasons. The latter happens, for example, when we model a physical setting by a two-dimensional vector field, when a three-dimensional model would be more realistic. This was the case with the Stommel or Munk model, where we convinced ourselves that the variation in the radial direction is considerably smaller than the variations in the horizontal directions. We may then look for introducing a parameter into the reduced system and select that parameter, by taking into account field measurements, in such a way that our model output remains close to the observed data.

Until recently, most data assimilation efforts have been of the Eulerian type, that is measurements are made at certain positions, and velocity fields are then updated using the data. As mentioned, the mathematical techniques in the updating stage are of a statistical nature, and the Kalman filter, dating to the 1960s, was one of the original and very successful techniques in incorporating Eulerian data into models. In the past few decades, however, new technologies have emerged that allow us to collect data when the instrument actually moves with the flow. Gliders and drifters are examples of such data collecting devices. The book [17] has an excellent introduction to these devices. The type of data one collects from gliders and drifters is of a Lagrangian nature, and it would stand to reason that a different set of mathematical ideas are needed in order to accommodate them into models. The mathematical work of C. K. R. T. Jones and co-authors, among others, has been motivated by exactly this question. In [18], for example, new mathematical techniques are introduced to assimilate Lagrangian data, collected at discrete times and positions, into a model that is primarily an Eulerian model.

Now that we have access to Lagrangian data, an interesting mathematical question arises: is one set of data better than another in terms of helping with guiding a model toward the true physical representation we are seeking? In other words, are there mathematical strategies we can employ that, considering that data collection is not inexpensive, would allow us to collect a minimal amount of data that would lead to an optimal representation? This is an important area of current research. An example of success in this direction is reported in [19] where a strategy is introduced, taking advantage of knowledge of how trajectories separate from each other, reminiscent of the dynamical systems ideas listed in the previous section, to identify the optimal positions where data should be collected in order to achieve a stable assimilation methodology.

In yet another direction, A. Majda and his collaborators have begun a new area of investigation where turbulence is the dominant feature of the flow. Because turbulent flows are very complex, their study does not lend itself well to the type of deterministic representations we have concentrated on in this book. Instead, statistical methods take center stage and the main objective can be paraphrased by asking how one should *filter* a partially observed data to obtain the best possible estimate of the natural system of interest. In the book [23], the authors propose a set of strategies that have the potential of providing real-time filtering of turbulent signals.

The viability of these strategies are demonstrated in the context of two examples, the Lorenz 96 model, which we introduced in (4.49) as part of Project B in Chapter 4, and the following two-layer quasi-geostrophic model,

$$\frac{\partial q_1}{\partial t} + J(\psi_1, q_1) + U\frac{\partial q_1}{\partial x} + (\beta + \alpha U)\frac{\partial \psi_1}{\partial x} = F_1, \tag{10.3}$$

$$\frac{\partial q_2}{\partial t} + J(\psi_2, q_2) - U\frac{\partial q_2}{\partial x} + (\beta - \alpha U)\frac{\partial \psi_2}{\partial x} = F_2, \tag{10.4}$$

where ψ_i is the stream function in each layer (and is related to the velocity $\mathbf{v}_i$ by $u_i = -\frac{\partial \psi_i}{\partial y}$, $v_i = \frac{\partial \psi_i}{\partial x}$), β is the rate of change of the Coriolis parameter with latitude, U is a mean longitudinal shear, and q_i is the potential vorticity in the i-the layer, i.e.,

$$q_i = \beta y + \Delta\psi_i + \frac{\alpha}{2}(\psi_{i-3} - \psi_i), \quad i = 1, 2.$$

The operator J is the Jacobian, and the constant α and the forcing terms F_1 and F_2 are chosen suitably for different regimes of interest; typically F_i is chosen proportional to $\Delta^4 q_i$, called hyperviscosity, in order to damp out the high wavenumber oscillations. As for boundaries, the bottom of the domain is assumed flat, the top has a rigid lid, and the layers have equal depth H. As for boundary conditions, q_i is assumed doubly periodic in x and y. Although this model is physically unrealistic, it has emerged as an appropriate model for studying observed turbulent data in the ocean and the atmosphere.

The techniques we have developed in this book, the Galerkin approach and the method of lines, work exceptionally well with the two-layer model, but even in the simple case of a discretization that involves 100 points in the x and y directions, one may end up with well over 10,000 unknowns. One strategy would be to improve the numerical schemes with the goal of obtaining high resolution solutions to the PDEs in (10.3)–(10.4), or its more physically relevant generalizations. This approach is being pursued by several research groups, especially those in national laboratories where there is a requirement for having operational codes for specific regions and basins around the world.

The approach promoted in [23], however, is quite different and is motivated by the desire to get as much information out of available data as possible, by devising statistical tools that are suitable for very large and strongly chaotic dynamical systems, meaning systems with large positive Liapunov exponents. The bulk of the development in [23] is concerned with presenting reduced models, either in the form of the Lorenz 96, or as systems of stochastic differential equations, where parameter values are selected with enough care to mimic the complexities observed in the ocean and atmosphere. The radical filtering strategies are then tested on these reduced models, and their success, as well as their shortcomings, are presented.

10.3 Normal Modes and Data

In the Flow-in-the-Bay boundary value problem, as well as in the context of applying the Galerkin method, we have already come across the notion of normal modes as a set of functions, generally orthonormal in the domain of interest, which satisfy a requisite set of boundary conditions. These functions are often the eigenfunctions of the Laplacian, and satisfy Dirichlet, Neumann, or a mixed boundary condition. Normal modes are the building block in terms of which we construct solutions of initial-boundary value problems with arbitrary initial data.

We have demonstrated that normal modes can be determined, for instance, by applying separation of variables, if the domain is geometrically simple, such as a rectangle or a cube, or if it has special symmetries. Determining normal modes in complex geometries, such as regions with coastlines of lakes and estuaries, is considerably more difficult and requires numerical approximation.

In a series of papers by A. D. Kirwan, see [20] and [21], the mathematical groundwork was established that showed the relationship between normal modes and velocity fields. Specifically, it is observed that any incompressible vector field $\mathbf{v}$ can be written as

$$\mathbf{v} = \nabla \times \left[(\mathbf{n}\psi) + \nabla \times (\mathbf{n}\phi)\right],$$

where ϕ is a potential function and ψ is a stream function. Next it is noted that we can construct a set of normal modes $\{\phi_n\}$ and $\{\psi_n\}$, associated with ϕ and ψ, as eigenfunctions of the Laplacian with Dirichlet and Neumann boundary conditions. And finally, *any* velocity field $\mathbf{v}$ can be reconstructed as a linear combination of the (ϕ_n, ψ_n), i.e.,

$$\mathbf{v} = \sum_{n=1}^{\infty} A_n(t)\nabla\phi_n + B_n(t)\nabla^{\perp}\psi_n,$$

reminiscent of the Galerkin method. The coefficients (A_n, B_n) need to be determined somehow.

In [20] and [21], the authors showed how the knowledge of normal modes associated with the Black Sea could be combined with information about the location of a handful of Lagrangian trajectories to establish the underlying velocity vector field globally, temporally and spatially. In a separate development, in [22], the authors extended this approach to show how the normal mode analysis can be blended with HF Radar data to reconstruct a surface velocity field from incomplete data collected over a significantly large portion of the Monterey Bay. In [24] the same strategy was applied to the Chesapeake Bay, where eigenfunctions of the Laplace operator were computed with Dirichlet and Neumann boundary conditions.

The primary goal in the above efforts is to develop a representation for the velocity field in a region from data . The eigenfunction approach is particularly effective when the available data is sparse. Recently, a different approach is being pursued when satellite imagery is available. The efforts in [25] and in [26] are designed to combine mathematical models with information stored in images to extract velocity fields. The key mathematical idea is that information stored in images does not vary substantially from frame to frame, so one could use techniques from the field of Computer Vision, and especially from the theory of *optical flow*, to efficiently construct the rates at which physical quantities change in images; for example, in [26] the optical flow approach is being applied to constructing two-dimensional velocity vector fields from hyperspectral imagery. In cases where images are readily available, such as images of river flows or sea-surface temperature profiles of coastline regions, the optical flow approach to constructing velocity fields could have an enormous pay-off, a development well worth watching in the coming years.

10.4 Concluding Remarks

The material presented in the references in this chapter, as well as those in the early chapters, are sources of topics and problems for future studies. It is hoped that the mathematical techniques presented in this book, especially when combined with the capabilities of a software like MATLAB, provide the first steps in encouraging the reader to consult some of these books and papers, to select a direction to continue learning about the fascinating subject of physical oceanography.

10.5 References

1. Vallis, G., *Atmospheric and Ocean Dynamics*, Cambridge University Press, 2006.

2. Bennett, A., *Lagrangian Fluid Dynamics*, Cambridge University Press, 2006.

3. Miller, R., *Numerical Modeling of Ocean Circulation*, Cambridge University Press, 2007.

4. Ramasubramanian, K., Sriram, M. S. "A comparative study of computation of Lyapunov spectra with different algorithms," *Physica D*, Vol 139, 2000, pp. 72–86.

5. Wiggins, S., *Introduction to Applied Nonlinear Dynamical Systems and Chaos*, 2nd edition, Springer, 2003.

6. Wiggins, S., *Chaotic Transport in Dynamical Systems*, Springer-Verlag, 1992.

7. Wiggins, S., *Normally Hyperbolic Invariant Manifolds in Dynamical Systems*, Springer-Verlag, 1994.

8. Haller, G., Poje, A., "Finite-time transport in aperiodic flows," *Physica D*, Vol 119, 1998, pp. 352–380.

9. Haller, G., "Finding finite-time invariant manifolds in two-dimensional velocity fields," *Chaos*, Vol 10, 2000, pp. 99–108.

10. Haller, G., Yuan, G.,"Lagrangian coherent structures and Mixing in two-dimensional turbulence," *Physica D*, Vol 147, 2000, pp. 352–370.

11. Haller, G., "Distinguished material surfaces and coherent structures in 3d fluid flows," *Physica D*, Vol 149, 2001, pp. 248–277.

12. Haller, G., "Lagrangian coherent structures from approximate velocity data," *Phys. Fluids A.*, Vol 14, 2002, pp. 1851–1861.

13. Haller, G., "Exact theory of unsteady separation for two-dimensional flows," *J. Fluid Mechanics*, Vol 512, 2004, pp. 257–311.

14. Shadden, S., Lekien, F., Marsden, J., "Definition and properties of Lagrangian coherent structures from finite-time Lyapunov exponents in two-dimensional aperiodic flows," *Physica D*, Vol 212, 2005, pp. 271–304.

15. Mancho, A., Small, D., Wiggins, S., "A tutorial on dynamical systems concepts applied to Lagrangian transport in oceanic flows defined as finite time data sets: Theoretical and computational Issues," *Physics Reports*, Vol 437, Nos 3-4, 2006, pp. 55–124.

16. Branicki, M., Mancho, A., Wiggins, S., "A Lagrangian description of transport associated with a fronteddy interaction: Application to data from the north-western Mediterranean Sea," *Physica D*, Vol 240, 2011, pp. 282–304.

17. Griffa, A., Kirwan, A., Mariano, A., Ozgokmen, T., Rossby, T., eds, *Lagrangian Analysis and Predictability of Coastal and Ocean Dynamics*, Cambridge University Press, 2007.

18. Kuznetsov, L., Ide, K., Jones, C., "A method for assimilation of Lagrangian data," *Monthly Weather Review*, Vol 131, 2003, pp. 2247–2260.

19. Poje, A., Toner, M., Kirwan, A., Jones, C., "Drifter launch strategies based on Lagrangian templates," *J. Phys. Oceanography*, Vol 32, 2002, 1855–1869.

20. Eremeev, V., Ivanov, L., Kirwan, A., Reconstruction of oceanic flow characteristics from quasi-Lagrangian data - approach and mathematical methods, *J. Geophysical Research*, Vol 97, 1992, pp. 9733–9742.

21. Eremeev, V., Ivanov, L., Kirwan, A., Melnichenko, O., Kochergin, S., Stanichnaya, R., Reconstruction of oceanic flow characteristics from quasi-Lagrangian data - characteristics of the large-scale circulation of the Black Sea, *J. Geophysical Research*, Vol 97, 1992, pp. 9743–9753.

22. Lipphardt, B., Kirwan, A., Grosch, C., Lewis, J., Paduan, J., Blending HF radar and model velocities in Monterey Bay through

normal mode analysis, *J. Geophysical Research*, Vol 105, 2000, pp. 3425–3450.

23. Majda, A., Harlim, J., *Filtering Complex Turbulent Systems*, Cambridge University Press, 2012.

24. Mcilhany, K.,Gillary, G., Malek-Madani, R., "Normal mode analysis of the Chesapeake Bay using FEMLAB", Proceedings of the COMSOL Multiphysics User's Conference 2005 Boston.

25. Auraux, D., Fehrenbach, J., "Identification of velocity fields for geophysical fluids from a sequence of images,", *Experiments in Fluids*, Vol 50, 2010, pp. 313–328 (2010).

26. Luttman, A., Bollt, E., Basnayake, R., Kremer, S., Tufillaro, N., "A stream function framework for estimating fluid flow from digital imagery,", *Experiments in Fluids*, submitted.

Appendix

Answers to Selected Problems

Chapter 1: An Introduction to MATLAB®

Section 1.3

1c) The period of $\sin 2x$ is π, the period of $\sin 3x$ is $\frac{2\pi}{3}$, hence the period of g is 2π, the smallest common period.

5) Define f by

```
f=inline('exp(-x).*quadv(@(y) sin(y^2),0,x)','x')
```

Then $f(0) = 0$, $f(1) = 0.1141$, and $f(2) = 0.1089$. The graph of this function can be obtained from

```
x=0:0.01:10;
for i=1:length(x)
    y(i)=f(x(i));
end
plot(x,y)
title('Graph of exp(-x^2)\int_0^x sin(y^2) dy')
```

See Figure A.1.

Section 1.4

1c) Combine `plot3` with `title`:

```
t=0:0.01:2*pi;
plot3(sin(t.^2),cos(t.^2),t)
title(['Graph of the curve <sin(t^2), cos(t^2), t>')
```

See Figure A.2.
2d) Combine `plot3, ones, zeros` with parametrization $\mathbf{r}(t) = \langle a + \cos t, b\sin t, c\rangle$:

```
t=0:0.01:2*pi;
plot3(cos(t),3+sin(t),zeros(length(t)))
hold on
```

```
plot3(-2+cos(t),1+sin(t),ones(length(t)))
plot3(3+cos(t),4+sin(t),4*ones(length(t)))
```

Section 1.6

4) The vector `fib` below contains the first 100 Fibonacci numbers:

```
fib1=1;
fib2=1;
fib(1)=fib1;
fib(2)=fib2;
for i=1:98
    fib3=fib1+fib2;
    fib(i+2)=fib3;
    fib1=fib2;
    fib2=fib3;
end
```

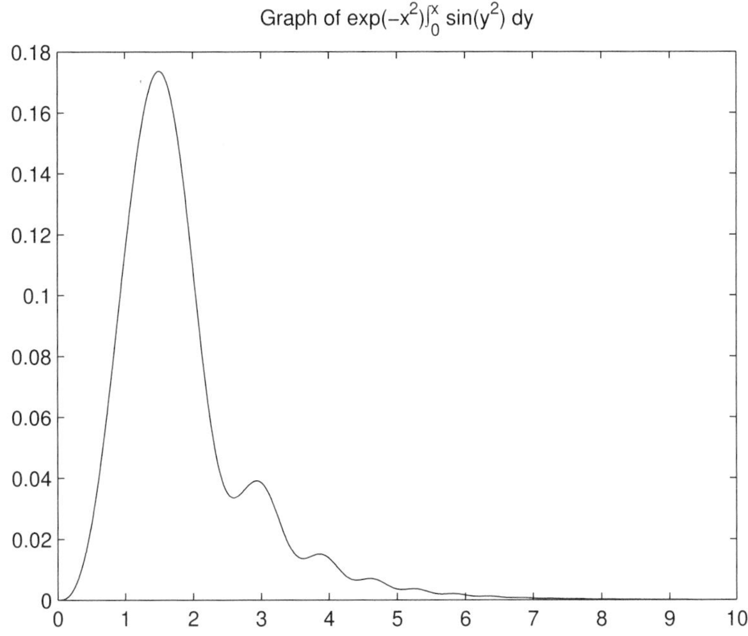

FIGURE A.1: Graph of the curve $e^{-x^2}\int_0^x \sin y^2\,dy$.

Section 1.8

4c) The graphs of S_{32} and C_{32} of $f(x) = x^2$ are shown in Figure A.3.

```
f=inline('x.^2','x');
b(1)=1/2*quadv(@(x) f(x), 0,2);
for i=1:32
     b(i+1)=2/2*quadv(@(x) f(x).*cos(i*pi*x/2),0,2);
     a(i)=2/2*quadv(@(x) f(x).*sin(i*pi*x/2),0,2);
end
x=0:0.01:2;
PartialSumSine=0;
PartialSumCosine=b(1);
for i=1:32
    PartialSumSine=PartialSumSine+a(i)*sin(i*pi*x/2);
    PartialSumCosine=PartialSumCosine+b(i+1)*cos(i*pi*x/2);
end
plot(x,f(x),'r',x,PartialSumSine,'b',x,PartialSumCosine,'g')
```

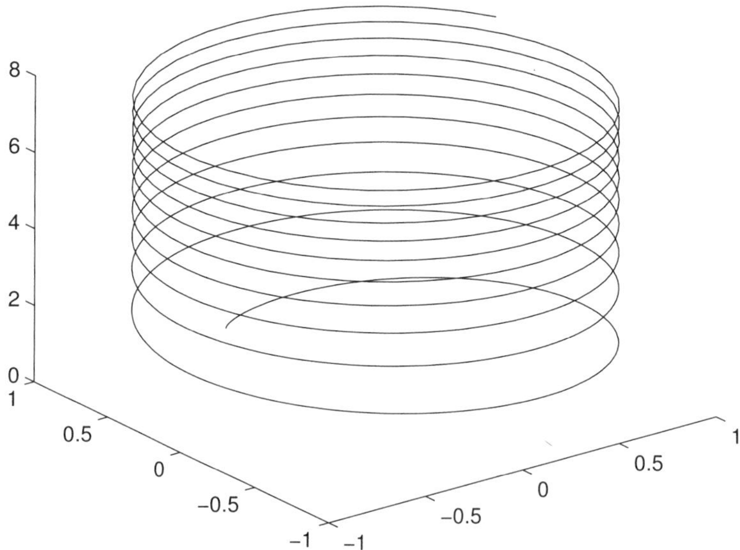

FIGURE A.2: Graph of the curve $\mathbf{r}(t) = \langle \sin(t^2), \cos(t^2), t \rangle$.

Chapter 2: Matrix Algebra

Section 2.2

5c) Start with the definitions of **a** and **b**:

$$\mathbf{a} \times \mathbf{b} = (a_1\mathbf{e}_1 + a_2\mathbf{e}_2 + a_3\mathbf{e}_3) \times (b_1\mathbf{e}_1 + b_2\mathbf{e}_2 + b_3\mathbf{e}_3).$$

Note that $\mathbf{e}_i \times \mathbf{e}_i = \mathbf{0}$ for each i. So, simply using the distributive rule of vector multiplication, the above expression reduces to

$$(a_1b_2)\,\mathbf{e}_1 \times \mathbf{e}_2 + (a_1b_3)\,\mathbf{e}_1 \times \mathbf{e}_3 + (a_2b_1)\,\mathbf{e}_2 \times \mathbf{e}_1 + (a_2b_3)\,\mathbf{e}_2 \times \mathbf{e}_3 +$$

$$(a_3b_1)\,\mathbf{e}_3 \times \mathbf{e}_1 + (a_3b_2)\,\mathbf{e}_3 \times \mathbf{e}_2.$$

the rest of the proof follows because each cross product of the form $\mathbf{e}_i \times \mathbf{e}_j$ can be written in terms of $\mathbf{e}_k$.

Section 2.3

6c) In general AB is not symmetric if A and B are. For example, $A =$

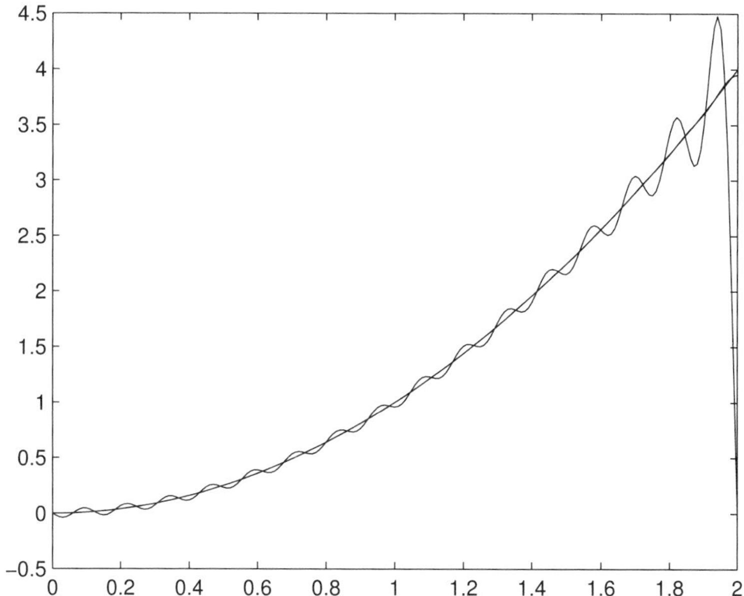

FIGURE A.3: Graph of $f(x) = x^2$ and its 32-partial sums, S_{32} and C_{32}. Note that C_{32} is a considerably better approximation to f, which is because the even-extension of f to the outside of the interval $(0, 2)$ is continuous while its odd extension is not.

$\begin{bmatrix} 1 & 2 \\ 2 & 1 \end{bmatrix}$ and $B = \begin{bmatrix} 1 & -2 \\ -2 & 0 \end{bmatrix}$. Then $AB = \begin{bmatrix} -3 & -2 \\ 0 & -4 \end{bmatrix}$, which is not symmetric.

10) Let $B = \frac{1}{2}(A + A^{\mathrm{T}})$ and $C = \frac{1}{2}(A - A^{\mathrm{T}})$.

Section 2.4

1b) This set is equivalent to the xy-plane in R^3.

1f) This set is equivalent to the plane $z = x + y$ in R^3.

6) Yes.

Section 2.5

2(iii) $A^{-1} = \frac{1}{a^2+b^2} \begin{bmatrix} a & b \\ b & -a \end{bmatrix}$.

3(iii) $A^{-1} = \begin{bmatrix} 9 & -36 & 30 \\ -36 & 192 & -180 \\ 30 & -180 & 180 \end{bmatrix}$

4c None.

Section 2.6

1d) $\begin{bmatrix} 0 & \frac{1}{b} & 0 \\ 0 & 0 & \frac{1}{c} \\ \frac{1}{a} & -\frac{1}{b} & 0 \end{bmatrix}$

Section 2.7

2b) The span is the xy plane.

6a) Suppose c_1, c_2 and c_3 satisfy $c_1 + c_2x + c_3x^2 = 0$ for all x. Let $x = 0$ in this expression, which results in $c_1 = 0$. The expression now reduces to $c_2x + c_3x^2 = 0$ for all x. Divide by x and then set $x = 0$.

7) $2n + 1$.

Section 2.9

2) $a! = 1$ and $a! = -1$. When $a = 1$, or $a = -1$, then if $b = 0$ then the system has infinitely many solutions.

Section 2.10

1e) The eigenvalues are $\lambda_1 = \frac{1}{2}\left(b - \sqrt{4a + b^2}\right)$, and $\lambda_2 = \frac{1}{2}\left(\sqrt{4a + b^2} + b\right)$.

2b) The eigenvalues are $\lambda_1 = 2$, $\lambda_2 = -1$ and $\lambda_3 = 1$.

0.1in **3)** The eigenvalues of A are

$$\lambda_1 = \frac{a+d}{2} + \frac{1}{2}\sqrt{a^2 - 2ad + 4bc + d^2}, \quad \lambda_2 = \frac{a+d}{2} - \frac{1}{2}\sqrt{a^2 - 2ad + 4bc + d^2}.$$

Therefore $\lambda_1 + \lambda_2 = a + d$ and $\lambda_1\lambda_2 = \frac{1}{4}(a+d)^2 - \frac{1}{4}(a^2 - 2ad + d^2) - bc$ which reduces to $ad - bc$ or det A.

Chapter 3: Differential and Integral Calculus

Section 3.1

2) By definition

$$f'(d) = \lim_{h \to 0} \frac{f(d+h) - f(d)}{h} = \lim_{h \to 0} \frac{[(d+h)^2 + b(d+h) + c] - [d^2 + bd + c]}{h} =$$

$$\lim_{h\to 0}(2d + ah + b) = 2d + b.$$

6) Let $a > 0$. Then, assuming h is small enough so that $a + h > 0$, we have

$$f'(a) = \lim_{h\to 0} \frac{(a+h)^2 - a^2}{h} = 2a = 2|a|.$$

Next, assume $a < 0$, and h small enough so that $a + h < 0$. Then

$$f'(a) = \lim_{h\to 0} \frac{-(a+h)^2 + a^2}{h} = -2a = 2|a|.$$

Finally, $f'(0) = \lim_{h\to 0} \frac{|h|h}{h} = \lim_{h\to 0} |h| = 0$. Hence $f'(a) = 2|a|$ for all a.

Section 3.2

1) $S_7(x) = x - x^2 + \frac{x^3}{3} - \frac{x^5}{30} + \frac{x^6}{90} - \frac{x^7}{630}$

2c) $\frac{f(a+2h)+f(a+h)-2f(a)}{3h} - f'(a) = \frac{5}{6}hf''(a) + \frac{1}{2}h^2 f''(a) + \dots$, so this Δf is order one.

Section 3.3

4) First show that the contour $x^2 + 3y^2 - 2x = 1$ can be parametrized by $\mathbf{r}(t)$ where

$$\mathbf{r}(t) = \langle 1 + \sqrt{2}\cos t, \sqrt{\frac{2}{3}}\sin t\rangle, \quad t \in [0, 2\pi].$$

Minimum length of gradient occurs at $t = 0$ and $t = \pi$.

5c) $\nabla(f(x,y,z)g(x,y,z)) = \langle \frac{\partial(fg)}{\partial x}, \frac{\partial(fg)}{\partial y}, \frac{\partial(fg)}{\partial z}\rangle = \langle \frac{\partial f}{\partial x}g + \frac{\partial g}{\partial x}f, \frac{\partial f}{\partial y}g + \frac{\partial g}{\partial y}f, \frac{\partial f}{\partial z}g + \frac{\partial g}{\partial z}f\rangle = g\langle \frac{\partial f}{\partial x}, \frac{\partial f}{\partial y}, \frac{\partial f}{\partial z}\rangle + f\langle \frac{\partial g}{\partial x}, \frac{\partial g}{\partial y}, \frac{\partial g}{\partial z}\rangle = g\nabla f + f\nabla g$.

6) $\mathbf{0} = c_1\mathbf{e}_r + c_2\mathbf{e}_\theta \implies c_1\cos\theta - c_2\sin\theta = 0$, $c_1\sin\theta + c_2\cos\theta = 0$. This system is equivalent to $A\mathbf{c} = \mathbf{0}$ where $A = \begin{bmatrix} \cos\theta & \sin\theta \\ -\sin\theta & \cos\theta \end{bmatrix}$ and $\mathbf{c} = \begin{bmatrix} c_1 \\ c_1 \end{bmatrix}$. Since A is nonsingular we have $c_1 = c_2 = 0$, or that $\mathbf{e}_r$ and $\mathbf{e}_\theta$ are linear independent.

7) $f_x = \frac{\partial F}{\partial r}\frac{\partial r}{\partial x} + \frac{\partial F}{\partial \theta}\frac{\partial \theta}{\partial x} = \frac{\partial F}{\partial r}\cos\theta - \frac{\partial F}{\partial \theta}\frac{\sin\theta}{r}$.

Section 3.4

1d) $\text{div}\,(\nabla f) = \text{div}\,(\langle f_x, f_y, f_z\rangle) = (f_x)_x + (f_y)_y + (f_z)_z = \Delta f$.

2) $u = \frac{y}{(2x^2-y^2)^{3/2}}$, $v = \frac{2x}{(2x^2-y^2)^{3/2}}$.

5a(i) Take the dot product of $\mathbf{v} = u\mathbf{i} + v\mathbf{j} = u_r\mathbf{e}_r + u_\theta\mathbf{e}_\theta$ with $\mathbf{e}_r$ to get $u_r = u\mathbf{i}\cdot\mathbf{e}_r + v\mathbf{j}\cdot\mathbf{e}_\theta$. Note that $\mathbf{i}\cdot\mathbf{e}_r = \cos\theta$ and $\mathbf{j}\cdot\mathbf{e}_r = \sin\theta$.

Section 3.5

1) $\nabla \times (\psi\mathbf{k}) = \nabla\langle 0, 0, \psi\rangle = \langle \frac{\partial\psi}{\partial y}, -\frac{\partial\psi}{\partial x}, 0\rangle$.

3b) $\nabla \times (\nabla\phi(x,y,z)) = \nabla \times \langle \phi_x, \phi_y, \phi_z\rangle = \langle \phi_{zy} - \phi_{yz}, \phi_{xz} - \phi_{zx}, \phi_{yx} - \phi_{xy}\rangle = \mathbf{0}$ because order of differentiation does not matter if ϕ is twice continuously differentiable.

Chapter 4: Ordinary Differential Equations

Section 4.1

1) Consider the expression

$$c_1\phi_1(x) + c_2\phi_2(x) + ... + c_n\phi_n(x) \equiv 0, \qquad (*)$$

which holds for all x. For the set $\{\phi_1, \phi_2, ..., \phi_n\}$ to be linearly independent we need to show that $c_1 = c_2 = ... = c_n = 0$. To that end we differentiate the expression in (*) $n-1$ times to get the system of linear equations

$$\left\{\begin{array}{rcl} c_1\phi_1(x) + c_2\phi_2(x) + ... + c_n\phi_n(x) & = & 0, \\ c_1\phi_1'(x) + c_2\phi_2'(x) + ... + c_n\phi_n'(x) & = & 0, \\ c_1\phi_1''(x) + c_2\phi_2''(x) + ... + c_n\phi_n''(x) & = & 0, \\ ... & ... & .... \\ c_1\phi_1^{(n-1)}(x) + c_2\phi_2^{(n-1)}(x) + ... + c_n\phi_n^{(n-1)}(x) & = & 0. \end{array}\right.$$

The above system is equivalent to $A\mathbf{c} = \mathbf{0}$ where A is the $n \times n$ matrix in the problem statement and $\mathbf{c}$ is the vector of the coefficients c_i. Since A is nonsingular the unique solution to $A\mathbf{c} = \mathbf{0}$ is $\mathbf{c} = A^{-1}\mathbf{0}$, which is the zero vector. Hence the set of functions $\{\phi_i\}$ is linearly independent.

3b(ii) Consider the expression

$$c_1\psi_1(x) + c_2\psi_2(x) + c_3\psi_3(x) \equiv 0 \qquad (**)$$

We need to show that $c_i = 0$ for all i. Applying the definition of each ψ_i, we see that (**) reduces to

$$(c_1 - c_3)\phi_1(x) + (-c_1 + c_2)\phi_2(x) + (-c_2 + c_3)\phi_3(x) \equiv 0.$$

Since the functions ϕ_i are linearly independent, the coefficients in the above expression must vanish, i.e,

$$\begin{cases} c_1 - c_3 &= 0, \\ -c_1 + c_2 &= 0, \\ -c_2 + c_3 &= 0. \end{cases}$$

This system is equivalent to $A\mathbf{c} = \mathbf{0}$ where $A = \begin{bmatrix} 1 & 0 & -1 \\ -1 & 1 & 0 \\ 0 & -1 & 1 \end{bmatrix}$. This matrix is nonsingular, hence $\mathbf{c}$ must vanish.

Section 4.2

1b) The Wronskian of y_1 and y_2 is

$$\det(\begin{bmatrix} e^{\lambda x} & xe^{\lambda x} \\ \lambda e^{\lambda x} & e^{\lambda x} + \lambda x e^{\lambda x} \end{bmatrix}) = e^{2ax},$$

which is nonzero.

2d) $y(x) = c_1 \sin 2x + c_2 \cos 2x + \sin x$

4a) $y(x) = -xe^{-x} + \frac{3}{5}e^{-x} + \frac{4}{5}\sin(2x) - \frac{3}{5}\cos(2x)$

6a) $y(x) = \frac{e^{-3x}\left(-2e^{2x} - e^{3x} - 4e^{2x+6} + e^{3x+4} + 2e^4 + 4e^6\right)}{3(e^4-1)}$

Section 4.4

1b) See Figure A.4. This figure was obtained by executing the following lines in `MATLAB`:

```
[t,x]=ode45(@Problem2,[0,3],[1 -1]);
plot(x(:,1),x(:,2))
```

where `Problem2.m` is

```
function xprime=Problem2Section4Ch4(t,x)
%
xprime=[x(2)-x(1); -2*x(1)-0.3*x(2)];
```

4) Let $A = 1$, $B = 0.1$, and $C = -0.2$. To get Figure A.5 execute the lines

```
global A B C
%
A=1; B=0.1; C=-0.2;
[t,x]=ode45(@abc,[0,50],[0.1 0.2 0.1]);
plot3(x(:,1),x(:,2),x(:,3))
```

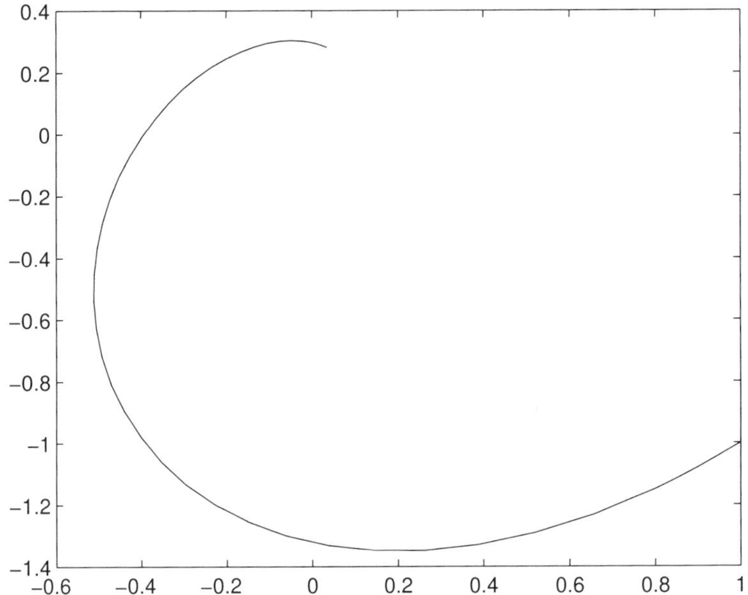

FIGURE A.4: Solution to $x_1' = x_2 - x_1, -2x_1 - 0.3x_2$, $x_1(0) = 1$, $x_2(0) = -1$

where `abc.m` is

```
function abcprime=abc(t,x);
%
global A B C
abcprime=[A*sin(x(3))+C*cos(x(2));B*sin(x(1))+A*cos(x(3));...
    C*sin(x(2))+B*cos(x(1))];
```

Section 4.5

1b) Equilibria are $P_1 = (0,0)$ and $P_2 = (1,0)$. Both eigenvalues of the linearized matrices about each of these equilibrium points have zero real parts so their state of stability cannot be determined from the theorem in this section. **2a)** First write the equation $x'' + ax' + bx = 0$ in the system form by defining $z_1 = x$, $z_2 = x'$ so that

$$z_1' = z_2, \quad z_2' = -az_2 - bz_1.$$

The equilibrium point $(0,0)$ is asymptotically stable if $a < 0$, unstable if $a > 0$ and stable if $a = 0$.

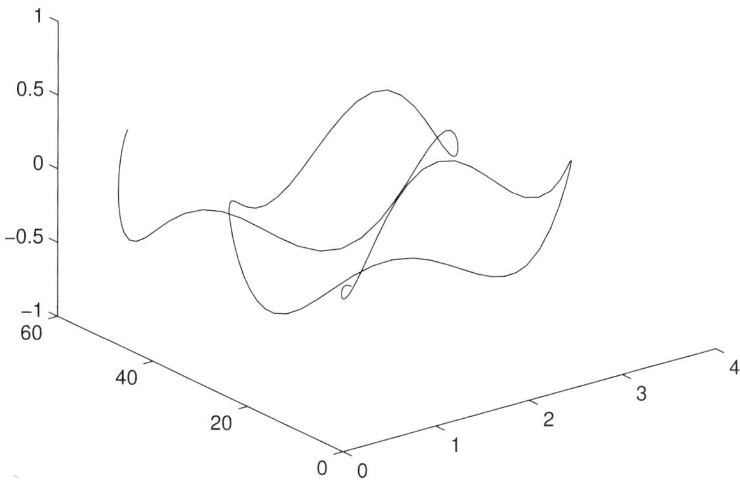

FIGURE A.5: Solution to the **ABC** flow problem.

Chapter 5: Numerical Methods for ODEs

Section 5.1

1) a) The exact solution is $y(t) = 3e^{0.2t}$. b) $y_n = (1+0.2h)y_{n-1}$. $y_1 = 3.06$, $y_2 = 3.1212$. $x_2 = 0.2$, $y(0.2) = 3.122432322577165$, absolute error $= |y_2 - y(0.2)| = 0.001232322577165$, relative error $= \frac{\text{absolute error}}{|y(0.2)|} = 3.946675059229967e-004$.

4) $y_n = y_{n-1} + h\sin(t_{n-1}y_{n-1})$.

5) $x_n = x_{n-1} + h * y_{n-1}$, $y_n = y_{n-1} - 0.1 * y_{n-1} + \sin x_{n-1}$.

Section 5

2) $y_i = -2(-3)^(i-1)$, $i = 1, 2, ...$

3) $\gamma_1 = -0.05 - 0.998749i$, $\gamma_2 = -0.05 + 0.998749i$, $y_i = c_1\gamma_1^i + c_2\gamma_2^i$ where $c_1 = 0.5 + 1.02628i$ and $c_2 = 0.5 - 1.02628i$. From this we get $y_3 = -1.88$, $y_4 = 1.388$, and $y_5 = 1.7412$, etc.

Section 5.6

1) In the complex plane, with coordinates denoted by (a, b), the equation for a circle of radius r centered at the point (c, d) is $(a-c)^2+(b-d)^2 = r^2$.

Divide $(1-ah)^2+b^2h^2=1$ by h^2 to get $(a-\frac{1}{h})^2+b^2=\frac{1}{h^2}$, which is the equation of a circle of $\frac{1}{h}$ centered at $(\frac{1}{h},0)$. Hence the inequality $(1-ah)^2+b^2h^2\geq 1$ corresponds to the region outside of this circle.

Section 5.7

3) Recall that

$$f(a+h)=f(a)+hf'(a)+\frac{h^2}{2}f''(a)+\frac{h^3}{6}f'''(a)+\frac{h^4}{24}f^{(\mathrm{iv})}(a)+\cdots$$

and

$$f(a-h)=f(a)-hf'(a)+\frac{h^2}{2}f''(a)-\frac{h^3}{6}f'''(a)-\frac{h^4}{24}f^{(\mathrm{iv})}(a)+\cdots$$

Hence $\frac{f(a+h)-f(a-h)}{2h}-f'(a)=\frac{h^2}{3}f''(a)+\cdots$.

Section 5.8

2a) The exact solution is $y(x)=\cos 2x+\tan(2)\sin 2x$.

Chapter 6: Equations of Fluid Dynamics

Section 6.1

2a) $\mathbf{V}=\langle(1-t)e^{-t}X_2,(1-2t)e^{-2t}X_1\rangle$. Now

$$X=\frac{e^{2t}\left(e^t x-ty\right)}{e^{3t}-t^2},\quad Y=\frac{e^t\left(e^{2t}y-tx\right)}{e^{3t}-t^2}.$$

Substituting the latter into the expression for $\mathbf{V}$ yields $\mathbf{v}$:

$$\mathbf{v}=\langle\frac{(t-1)\left(tx-e^{2t}y\right)}{e^{3t}-t^2},\frac{(2t-1)\left(ty-e^t x\right)}{e^{3t}-t^2}\rangle$$

1f) $\mathbf{v}=\langle-\frac{(y-xg(t))f'(t)}{f(t)g(t)-1},-\frac{(x-yf(t))g'(t)}{f(t)g(t)-1}\rangle$

4a) $\mathbf{v}=\langle -x,-y,0\rangle$.

4c) $a_1=\sin(x)\cos(y)\cos^2(z)-\cos(x)\sin^2(y)\sin(z)$, $a_2=\cos(x)\sin(y)\cos^2(z)-\sin(x)\cos(x)\sin(y)\sin(z)$, $a_3=\sin(x)\cos(x)\cos(y)\cos(z)-\sin(x)\sin^2(y)\cos(z)$.

Section 6.2

3) Let λ_1 and λ_2 be the eigenvalues of F. Recall that $\lambda_1+\lambda_2=F_{11}+F_{22}$ and $\lambda_1\lambda_2=\det F$. Differentiate the latter two expression with respect to t to get

$$\frac{d\lambda_1}{dt}+\frac{d\lambda_2}{dt}=\frac{\partial F_{11}}{\partial t}+\frac{\partial F_{22}}{\partial t},$$

and

$$\lambda_2 \frac{d\lambda_1}{dt} + \lambda_1 \frac{d\lambda_2}{dt} = (\text{div } \mathbf{v})\det F.$$

Solve for λ_1 and λ_2 and replace expressions in F with the appropriate ones for $\mathbf{v}$.

Section 6.4

2) $\nabla \times \mathbf{v}_A = -2\mathbf{k}$, $\nabla \times \mathbf{v}_B = -\frac{1}{\sqrt{x^2+y^2}}\mathbf{k}$. The vorticity of $\mathbf{v}_C$ is undefined at the origin and otherwise is zero. A better description of the latter is that the vorticity of $\mathbf{v}_C$ behaves like the delta function.

7) Recall (or prove) the identity div $\nabla p = \Delta p$.

8b) 0

8e) $\frac{2}{x^2+y^2+z^2}$.

13) Recall (or prove) the identity div $(\psi\mathbf{v}) = \nabla\psi \cdot \mathbf{v} + \psi$div $\mathbf{v}$.

Section 6.5

3d) $a_1 = B(A\cos x \cos z - C \sin x \sin y)$

4a) $\mathbf{v} = \langle y, -x\rangle$, $\mathbf{a} = \langle -x, -y\rangle$. $p(x,y) = -\frac{1}{2}(x^2+y^2)$.

Section 6.6

3a) $\mathbf{v} = \frac{1}{\sqrt{x^2+y^2}}\langle y, -x\rangle$, $\nabla\mathbf{v} = \begin{bmatrix} -\frac{xy}{(x^2+y^2)^{3/2}} & \frac{x^2}{(x^2+y^2)^{3/2}} \\ -\frac{y^2}{(x^2+y^2)^{3/2}} & \frac{xy}{(x^2+y^2)^{3/2}} \end{bmatrix}$. Therefore,

$$A = \begin{bmatrix} -\frac{xy}{(x^2+y^2)^{3/2}} & \frac{x^2-y^2}{2(x^2+y^2)^{3/2}} \\ \frac{x^2-y^2}{2(x^2+y^2)^{3/2}} & \frac{xy}{(x^2+y^2)^{3/2}} \end{bmatrix}, \quad D = \begin{bmatrix} 0 & \frac{1}{2\sqrt{x^2+y^2}} \\ -\frac{1}{2\sqrt{x^2+y^2}} & 0 \end{bmatrix}.$$

Section 6.7

4) On the plane $x = 1$, say, the normal is $\nabla 1, 0, 0\rangle$, and $\mathbf{t} = \rho g$. Therefore, $\int\int_S \mathbf{t} \cdot d\mathbf{A} = \int_{-1}^{1}\int_{-1}^{1} \rho g dy dz = 4\rho g$.

Section 6.8

3) $p(x,y) = -\frac{-2x^2+2y^2+1}{2(x^2+y^2)^2}$

4) Take the divergence of both sides of $\rho\frac{\partial\mathbf{v}}{\partial t} = -\nabla p + \mu\Delta\mathbf{v}$ and use the fact that div $\mathbf{v} = 0$ and that div $\nabla p = \Delta p$.

Chapter 7: Equations of Geophysical Fluid Dynamics

Section 7.2

4) Each $\mathbf{a}$ is a unit vector, so $\mathbf{a}\cdot\mathbf{a} = 1$. Differentiate this expression with the respect to t to $2\mathbf{a}\cdot\frac{d\mathbf{a}}{dt} = 0$.

Section 7.3

3) In rectangular coordinates $\nabla p = yz\mathbf{i} + xz\mathbf{j} + xy\mathbf{k}$. In spherical coordinates $p(\rho,\theta,\phi) = \rho^3 \sin\phi\cos^2\phi\sin\theta\cos\theta$. Using the formula in (3.27) in Chapter 3, where

$$\nabla f = \frac{\partial F}{\partial \rho}\mathbf{e}_\rho + \frac{1}{\rho}\frac{\partial F}{\partial \theta}\mathbf{e}_\theta + \frac{1}{\rho\sin\theta}\frac{\partial F}{\partial \phi}\mathbf{e}_\phi,$$

we have

$$\nabla p = (3\rho^2\sin\phi\cos^2\phi\sin\theta\cos\theta)\mathbf{e}_\rho + (\rho^2\sin\phi\cos^2\phi\cos 2\theta)\mathbf{e}_\theta +$$

$$(\frac{1}{2}\rho^2\cos\phi(3\cos 2\phi - 1)\cos\theta)\mathbf{e}_\phi.$$

Chapter 8: Shallow Water Equations

Section 8.6

1) $(\phi_m,\phi_n) = \int_0^L \sin\frac{n\pi x}{L}\sin m\pi x L dx = \frac{Ln\sin m\pi\cos n\pi - Lm\cos m\pi\sin n\pi}{\pi m^2 - \pi n^2}$ which vanishes if $m! = n$.

3) Note that $(f(x))^2 = \sum_{n=1}^{N}\sum_{m=1}^{N} a_n a_m(\phi_n,\phi_m)$. But $(\phi_n,\phi_m) = \frac{L}{2}\delta_{mn}$, the Kronecker delta function.

Section 8.9

1) Let $u(x,t) = F(x-ct)$. By applying the chain rule, we have $u_t = -cF'(x-ct)$ and $u_{tt} = c^2F''(x-ct)$. Similarly, $u_{xx} = F''(x-ct)$. Hence $u_{tt} = c^2 u_{xx}$.

Chapter 9: Wind-Driven Ocean Circulation

Section 9.2

3) Suppose $\frac{F''}{F} = -\lambda^2$. Then $F'' + \lambda^2 F = 0$ and $F(x) = c_1\cos\lambda x + c_2\sin\lambda x$. Hence $\psi(x,y) = (c_1\cos\lambda x + c_2\sin\lambda x)(c_3e^{\lambda y} + c_4e^{-\lambda y})$. The First boundary condition, that $\psi(x,0) = 0$ for all $x < 0$, implies that $c_3 + c_4 = 0$, so $\psi(x,y) = (A\cos\lambda x + B\sin\lambda x)\sinh\lambda y$. The second boundary, that $\psi(x,h) = 0$, implies that either $\sinh\lambda h = 0$ or $F(x) \equiv 0$. Either case results in $\psi \equiv 0$, which is not an eigenfunction.

5) $\psi_1 = \sinh x \sin y$. Then $\mathbf{v} = \langle \frac{\partial \psi}{\partial y}, -\frac{\partial \psi}{\partial x} \rangle = \langle \sinh x \cos y, -\cosh x \sin y \rangle$.

7a) $C_n = -\frac{2h^3(2\cos n\pi - 2)\operatorname{csch}\left(\frac{\pi a n}{h}\right)}{\pi^3 a n^3}$

4) The Δ operator does not have any positive eigenvalues because the general solution to $\Delta\psi = \lambda^2\psi$ is the product exponential functions, as opposed to trigonometric functions, and the boundary condition $\psi|_{\partial\Omega} = 0$ causes all coefficients c_i to vanish.

2) Note that $(\nabla \cdot e^{\alpha x} \nabla \psi) = e^{\alpha x} \Delta \psi + \alpha e^{\alpha x} \psi_x$.

4) Suppose $n > 1$ and $c_1 \sin y + c_2 \sin ny \equiv 0$. Let $\bar{y} = \frac{\pi}{n}$. Then $\sin \bar{y}! = 0$ but $\sin n\bar{y} = 0$. Hence $c_1 = 0$, which in turn implies $c_2 = 0$. Hence $\sin y$ and $\sin ny$ are linearly independent if $n > 1$.

Index